AF545240

# Thin Plate Design for In-Plane Loading

The Constrado Monographs deal with the application of steel in construction. They each treat a specific subject, and the texts are written with authority and expedition. Subjects are treated in depth and are taken to the point of practical application.

Advisory Editor
M. R. HORNE, MA, ScD, FICE, FIStructE
Professor of Civil Engineering, University of Manchester

CONSTRADO MONOGRAPHS

# Thin Plate Design for in-Plane Loading

D. G. WILLIAMS, BE, MS, DIC, PhD, MIE Aust.
General Manager
Australian National Railways, Adelaide, Australia

B. AALAMI, BSc, DIC, PhD, CEng, MICE, MRINA
Professor of Structural Mechanics
University of Technology, Tehran, Iran

A HALSTED PRESS BOOK

JOHN WILEY & SONS
New York

First Published in Great Britain by
Granada Publishing Limited – Technical Books Division 1979
Frogmore, St Albans, Herts AL2 2NF
and
3 Upper James Street, London W1R 4BP

Published in the U.S.A. by
Halsted Press, a Division of
John Wiley & Sons, Inc., New York



*Library of Congress Cataloging in Publication Data*
Williams, Donald G
Thin plate design for in-plane loading.

(Constrado monographs)
"A Halsted Press book."
Bibliography: p.
Includes index.
1. Plates (Engineering) 2. Structural design.
I. Aalami, B., joint author. II. Title.
TA660.P6W47 624'.1776 79-17855
ISBN 0-470-26834-4

Printed in Great Britain

# Contents

# Foreword

This volume is the second in the Constrado Monograph Series dealing with the elastic design of thin plates. The objective of these volumes is to provide engineers with a comprehensive range of design data covering plates subject to transverse and in-plane loading. The first volume by Aalami and Williams (1975) covered plates subject to transverse loading. In this volume results are presented for plates subject to in-plane loading.

The plate-panel imperfections which give rise to non-linear behaviour under in-plane loading are an inevitable by product of the fabrication and fixing process. These imperfections, by virtue of the resulting stresses, will become of increasing design significance as the need for lighter more economic structures dictates the use of thinner and stronger plate materials. Unfortunately the complexity of the required analysis is a barrier to this trend, since the range of readily available solutions is limited and the cost of one-off numerical solutions is prohibitive.

In this volume the authors present for the first time a comprehensive range of design data for imperfect plates subject to in-plane load. These data are readily applicable and their use is illustrated by typical practical examples. A summary of the theory is included, together with a discussion of design considerations, so as to provide the user with an appreciation of the design significance of his analysis.

This book represents a considerable computational effort on the part of the authors and it is to be hoped that this will be justified by way of more soundly based and economic design of thin-plate structures.

*March 1978* M. R. HORNE

# Preface

The economic and/or structural feasibility of many modern structures necessitates light-weight, thin-walled stiffened plate configurations. This trend will continue as higher-strength materials are utilised and hence it will become increasingly important in design to take account of the interaction between the in-plane plate loading and the initial distortions which are an inevitable by-product of the fabrication and fixing process. This interaction can significantly reduce the in-plane load-bearing capacity of plates when out-of-plane deformations are of the order of one tenth the plate thickness, at loads well below the critical buckling load.

The mathematical treatment of the interaction between the initial distortions, and the in-plane loading, leads to a set of non-linear algebraic equations, commonly referred to as large-deflection equations, for which no general solution is known. However, the ever-increasing computational capacity of computers has facilitated the derivation of specific solutions using numerical techniques and these are finding increasing practical application. This growing demand has come about largely as a result of recent box-girder bridge failures which were caused in part by a lack of appropriate design data.

These solutions also have application in ship, aircraft, container and dock-gate design, as well as sundry other stiffened-plate structures. Codes of practice in the bridge and ship fields are currently under review in many countries, with the objective of providing analytically based, experimentally verified criteria covering large-deflection effects. These criteria will provide a rational basis for utilising thinner plate and thus justify lighter-weight structures with resultant economies.

Unfortunately the development of large-deflection design criteria and its application has been limited by the computational effort required to provide specific solutions. The box girder bridge design rules produced under the auspices of the Merrison Committee in Britain make a significant contribution to thin-plate design based on a relatively limited number of solutions to more common cases. This book aims to provide a wide range of particular solutions for use in conjunction with the Merrison Rules and similar Codes as they are developed. The solutions presented are simple to utilise and their application is illustrated by way of practical examples. The book also includes an outline of the basic equations and a discussion of design considerations and criteria in order to give the user an appreciation of the phenomena, the mechanism and its reality in practical terms.

The authors would like to record their gratitude to Dr J. C. Chapman, formerly Reader in Structural Engineering at Imperial College, under whom they both received their initial training and developed an interest in this topic. Thanks are also due to the British Steel Corporation, the British Ship Research Association, Professor L. K. Stevens of Melbourne University, the Building Research Division of the Australian Commonwealth Scientific and Industrial Research Organization, the Control Data Corporation (Melbourne) and the Australian National Railways Commission which provided the funds and computing facilities that made the work possible.

Acknowledgement is due to the Controller of Her Majesty's Stationery Office for permission to use specified diagrams, text and formulae from the Merrison Committee Report (1973) 'Inquiry into the basis of design and method of erection of steel girder bridges'.

The authors are also especially grateful to the many people who assisted in the preparation of the manuscript, in particular Mr S. Rissini, Miss M. Krawec, Miss H. Maddern, Mrs H. Liddell and Mrs K. Strong.

D. G. Williams
B. Aalami

# Notation

| | |
|---|---|
| $A$ | Cross sectional area of weld material |
| $A_e$ | Effective cross sectional area of plate–stiffener combination |
| $A_s$ | Cross sectional area of stiffener |
| $a$ | Plate side in $x$-direction |
| $b$ | Plate side in $y$-direction; mean spacing between stiffeners |
| $D$ | Plate flexural stiffness = $Et^3/[12(1-\nu^2)]$ |
| $E$ | Young's modulus of elasticity |
| $F_c$ | Weld shrinkage force |
| $f$ | Airy's stress function |
| $G$ | Shear modulus = $E/[2(1+\nu)]$; gauge length for measuring imperfections |
| $I$ | Beam moment of inertia |
| $k$ | Critical load factor |
| $L_m$ | Clear distance between weld ends on one side |
| $L_w$ | Full length of intermittent weld |
| $L_x, L_y$ | Plate panel buckling lengths in $x$- and $y$-directions |
| $M_x, M_y, M_{xy}$ | Moments per unit length of plate |
| $m$ | Magnification factor = $1/(1-P/P_{cr})$ |
| $N_x, N_y, N_{xy}$ | Membrane forces per unit length of plate |
| $N$ | Load cycles to cause fatigue failure; number of longitudinal stiffeners in a stiffened panel |
| $n$ | Number of weld passes |
| $P$ | Applied in-plane load |
| $P_{cr}$ | Critical buckling load |
| $Q_x, Q_y$ | Shearing forces in $z$-direction |
| $q$ | Intensity of transverse loading |
| $S$ | $s/s_f$, Maximum stress level in fatigue calculations |

| | |
|---|---|
| $s$ | Shear displacement |
| $t$ | Plate thickness |
| $U$ | $u/u_f$ |
| $u$ | Mid-plane displacement in $x$-direction |
| $V$ | $v/v_f$ |
| $v$ | Mid-plane displacement in $y$-direction |
| $W$ | $w/t$ |
| $w$ | Out-of-plane deflection of plate; mean distance between welds |
| $w_g$ | Amplitude of initial geometric out-of-plane deformation of plate |
| $w_i$ | Initial out-of-plane deformation of plate at any point |
| $w_o$ | Total amplitude of initial out-of-plane deformation of plate, including influence of residual stresses (if any) |
| $Z_e$ | Stiffened plate section modulus |
| $x, y, z$ | Rectangular co-ordinates |
| $\alpha$ | $\sigma_{esy}/\sigma_y$ |
| $\gamma$ | Shearing strain |
| $\epsilon_x, \epsilon_y$ | Direct strains in $x$- and $y$-directions |
| $\lambda_\sigma$ | Plate stress effectiveness |
| $\lambda_s$ | Plate stiffness effectiveness |
| $\nu$ | Poisson's ratio |
| $\Delta$ | Stiffener imperfection |
| $\sigma_b$ | $\pi^2 E/[12(1-\nu^2)(b/t)^2]$ |
| $\sigma_{cr}$ | Critical buckling stress |
| $\sigma_{ec}$ | Applied stress to cause collapse |
| $\sigma_{em}$ | Equivalent membrane stress |
| $\sigma_{eR}$ | Reference equivalent stress |
| $\sigma_{es}$ | Equivalent surface stress |
| $\sigma_{esm}$ | Applied stress to cause membrane yield |
| $\sigma_{esy}$ | Applied stress to cause surface yield |
| $\sigma_m$ | Axial membrane stress |
| $\sigma_R$ | Reference axial stress; residual stress |
| $\sigma_{Reff}$ | Effective residual stress |
| $\sigma_{Rs}$ | Residual stress due to stiffener attachment welds only |
| $\sigma_x, \sigma_y$ | Stresses in $x$- and $y$-directions |
| $\sigma_y$ | Yield stress |
| $\tau$ | Shear stress |
| $\tau_m$ | Membrane shear stress |
| $\tau_y'$ | $\tau_{av}(\sigma_y/\sigma_{esy})$ |

**Superscripts**

| | |
|---|---|
| $(^-)$ | Non-dimensional quantities |
| $(^*)$ | Quantities referring to Poisson's ratio |

**Subscripts**

| | |
|---|---|
| 1, 2, 3, . . . | Refer to locations on plate |
| av | Average |
| b | Bending action |
| cr | Critical buckling |
| e | Equivalent stress |
| eff | Effective |
| f | Refers to flat-plate behaviour |
| m | Membrane action |
| max | Maximum |
| R | Reference stress; residual stress |
| s | Surface stress |
| $x, y, z$ | Co-ordinate directions |
| $y$ | Membrane or surface yield |
| (,) | Represents partial differentiation in turn with respect to each subscript variable following |

# Units and Conversion Factors

The design curves and tables presented in this volume are all dimensionless and may be used directly in any system of units. However, the numerical examples given are worked out in SI units followed by their Imperial equivalents in parenthesis. The following conversion table lists the major conversion factors used.

| To convert | to | multiply by |
|---|---|---|
| inches (in) | millimetres (mm) | 25·40 |
| millimetres (mm) | inches (in) | 0·039 37 |
| pound force (lbf) | newtons (N) | 4·45 |
| newtons (N) | pound force (lbf) | 0·224 7 |
| kilogram force (kgf) | newtons (N) | 9·806 |
| newtons (N) | kilogram force (kgf) | 0·102 0 |
| pounds per square inch (psi) | newtons per square millimetre ($N/mm^2$) | 0·006 895 |
| newtons per square millimetre ($N/mm^2$) | pounds per square inch (psi) | 145·0 |
| kilogram force per square centimetre ($kgf/cm^2$) | newtons per square millimetre ($N/mm^2$) | 0·098 06 |
| newtons per square millimetre ($N/mm^2$) | kilogram force per square centimetre ($kgf/cm^2$) | 10·20 |

CHAPTER ONE

# Large-deflection Behaviour

Economic forces have to some extent pre-empted the gradual evolution of design criteria for thin plate subject to in-plane loading. The result is that the average design engineer is required to assimilate and apply complex plate theory in the course of routine design, if he is to satisfy himself as to the adequacy of a structure.

Work is current in various countries on the development of simplified design rules which will be included in relevant codes of practice. Based on the work of the Merrison Committee, the British are probably the most advanced in the field at this stage, but meanwhile a need was seen for a book providing design data together with discussion and examples.

This chapter serves as a general introduction to large-deflection behaviour of imperfect plates subject to in-plane load, the mechanics of the problem, the equations which describe the behaviour and the method used to obtain the solutions to these equations given in this book.

## 1.1 Applications

Plate is used in modern structures to transmit lateral and/or in-plane load to adjacent supports. Applications include ships, aircraft, bridges and containers. Materials include steel, aluminium, fibreglass, plywood and glass. The term thin plate is applied when the plate develops its strength by a combination of flexural (bending) and membrane (in-plane) actions. Because the out-of-plane deformations which occur in such applications are large relative to the thickness of the plate, it is commonly called large-deflection behaviour.

The plate panel may be part of an array of panels separated by uniaxial stiffening members, e.g., ship's plating, or it may be a discrete panel such as a glass window. Loading may be purely lateral, deriving from a cladding or containing function, or purely in-plane, either directly applied or deriving from the panel's contribution to the overall stiffness of a larger structural component, or a combination of both.

This volume is concerned with panels which are subject to purely in-plane load. This load may be normal to the side of the plate (tension or compression) or tangential to the side of the plate (shear). Solutions are given for normal and tangential load applied separately and in combination. Practical applications are illustrated by numerical examples.

## 1.2 Large-deflection behaviour of plate panels

The in-plane edge stresses in plate panels either directly applied or due to overall deformation of major structural components are assumed to be uniform through the thickness of the plate. If the individual plate panels were perfectly flat this load would be carried or transmitted in a condition of plane stress, i.e., no bending would occur and a purely membrane stress distribution would apply throughout the panel. This type of behaviour is linear and relatively easy to analyse. In practice no plate panels are perfectly flat. Initial imperfections are inevitable as a result of the fabrication and fixing process, and these imperfections interact with the applied in-plane edge load, resulting in a redistribution of membrane loading as compared to the plane stress condition, and the development of secondary bending stresses. As a rough guide, these effects can become significant when out-of-plane deformations exceed one tenth of the plate thickness, depending on the span ($b$) to thickness ($t$) ratio, the panel boundary conditions and the material properties. Since initial imperfections often exceed $0{\cdot}1t$ it is apparent that there are many applications where large-deformation analysis cannot be avoided.

Large-deformation behaviour is non-linear and its analysis is complex. It will, however, become of increasing significance as economic and structural considerations dictate lighter weight structures. At the moment it is necessary to analyse individual panels but it is hoped that relatively simple design rules will be developed in the future.

## 1.3 Behaviour of plated structures

There are many practical applications where the overall flexural stiffness of the structure is dependent on the in-plane stiffness of local plate panels. Since the stiffness of local panels varies non-linearly with applied load and this load is determined by the overall analysis, it is apparent that an iterative procedure is involved.

Fortunately, in most cases the overall force distribution is insensitive to local stiffness variations for all practical purposes. Local in-plane panel forces due to overall deformation can therefore be derived from a global analysis which assumes plate panels are fully effective. However, it may be necessary in some cases to carry out a second global analysis incorporating the reduced or effective stiffness of local panels in order to get an accurate estimate of overall deformations.

## 1.4 The critical buckling load

In columns the critical buckling load is that load at which a perfectly straight column, given the slightest lateral impetus, would continue to deform to the point of collapse without any increase in axial load. No column is perfectly straight and the imperfections interact with the axial load from the onset of loading, the critical load representing an upper bound to the collapse load.

Provided the initial imperfections are not excessive it has been shown that this interaction with the axial load is basically a function of the critical load from the onset of loading. This is described in the well-known magnification expression:

$$w = \Delta/(1 - P/P_{cr}) \tag{1.1}$$

where $w$ = lateral deflection of column due to axial load
$\Delta$ = initial lateral imperfection of column
$P$ = axial load applied to column
$P_{cr}$ = critical buckling load for column.

Moments and stresses can then be directly computed.

Plates also have a critical buckling load, although its implications are less catastrophic than in the case of columns. This is because the biaxial geometry facilitates the shedding of load from the centre to the stiffer sides of the plate, giving a capacity to sustain load greater than the critical load provided excessive yielding has not occurred at sub-critical load. The column and plate critical load phenomena are illustrated in Fig. 1.1.

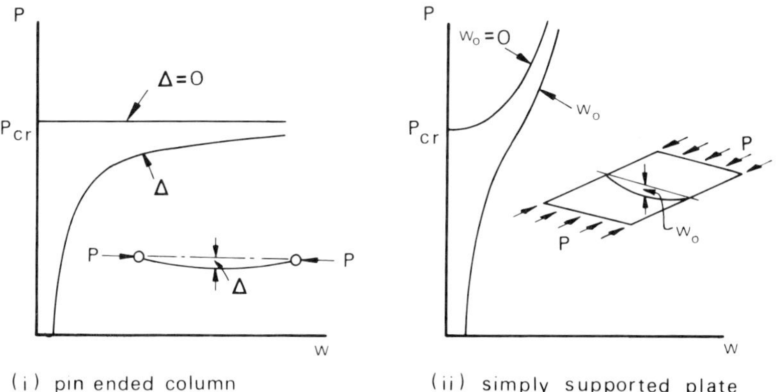


Fig. 1.1 Column and plate critical load behaviour

As might be expected, plates also exhibit the magnification phenomenon although the expression is slightly more complex, as follows:

$$[(w/t)^2 - (w_o/t)^2]^{1/2} = A\psi + B\psi^3 \qquad (1.2)$$

where $\psi = (P/P_{cr} - 1 + w_o/w)^{1/2}$
$w$ = lateral deflection of plate due to in-plane load
$w_o$ = initial lateral imperfection of plate
$P$ = in-plane loading applied to plate
$P_{cr}$ = critical buckling load for plate
$t$ = plate thickness
$A, B$ = constants for particular loading configuration.

More terms could be included on the right-hand side of the equation but experience has shown that two are quite adequate for design purposes. Plate forces, in-plane deformations and stresses can then be computed using the value for $w$ given by eq. (1.2) in related relatively simple equations incorporating coefficients for the particular load configuration.

The derivation and application of this simplified technique is described by Williams and Walker (1975, 1977). This brief discussion is included at this stage to impress upon the reader the fact that the critical buckling load has a pervasive influence on imperfect plate behaviour from the onset of loading.

## 1.5 Historical background

### 1.5.1 Equations

The differential equation most familiar to all structural engineers is that describing the small deflection of a beam element:

$$w_{,xx} = -M/EI \qquad (1.3)$$

where $w$ = beam deflection
$x$ = longitudinal beam axis
$M$ = bending moment applied to beam element
$EI$ = flexural rigidity of the beam in the plane of bending.

(Note that a comma followed by subscripts represents partial differentiation in turn with respect to each subscript variable.)

For uniformly distributed transverse load $q$, this can be rewritten:

$$w_{,xxxx} = q/EI \qquad (1.4)$$

Small deflections of isotropic plates are described by the following partial differential equation:

$$w_{,xxxx} + 2w_{,xxyy} + w_{,yyyy} = q_{xy}/D \qquad (1.5)$$

where $w$ = plate out-of-plane deflection
$x, y$ = plate co-ordinate axes
$q_{xy}$ = transverse pressure
$D$ = plate flexural stiffness.

This equation was derived by Lagrange (1811) and Navier (1820). Note that the first and third terms are analogous to simple beam bending in the directions of the two axes. The second term takes account of twisting action which develops because of varying deflection between adjacent strips and this was omitted in very early work on plates by Bernoulli (1789).

Saint Venant (1883) extended the above equation to include in-plane forces applied at the edges, acting in the middle plane of the plate, giving:

$$w_{,xxxx} + 2w_{,xxyy} + w_{,yyyy} = (q_{xy} + N_x w_{,xx} + 2N_{xy} w_{,xy} + N_y w_{,yy})/D \qquad (1.6)$$

where $N_x, N_y$ = tensile or compressive force per unit length
$N_{xy}$ = shear force per unit length.

The accuracy of the above equation is limited since it ignores the membrane forces developed within the plate. These forces can become significant when deflections reach the order of one tenth of the plate thickness, although as noted earlier in this chapter, this parameter varies depending on plate geometry and type of loading. Deflections are termed large when this stage is reached and the so-called large-deflection equations are as follows:

$$w_{,xxxx} + 2w_{,xxyy} + w_{,yyyy} = (q_{xy} + tf_{,yy}w_{,xx} - 2tf_{,xy}w_{,xy} + tf_{,xx}w_{,yy})/D \qquad (1.7)$$

$$f_{,xxxx} + 2f_{,xxyy} + f_{,yyyy} = E[(w_{,xy})^2 - w_{,xx}w_{,yy}] \qquad (1.8)$$

where $f$ = Airy's stress function
$N_x = tf_{,yy}$
$N_y = tf_{,xx}$
$N_{xy} = -tf_{,xy}$ (1.9)
$E$ = Young's modulus.

It can be seen that the in-plane forces are now coupled with the out-of-plane deflections. The equations were first derived in this form by Von Karman (1910) following on initial work on large deflections by Kirchhoff (1877) and Föppl (1907) on the use of stress functions.

The above equations describe the large-deflection behaviour of an initially flat plate; they take no account of initial out-of-plane deformations. Marguerre (1938) extended the Von Karman equations to include initial deflections, as follows:

$$w_{,xxxx} + 2w_{,xxyy} + w_{,yyyy} - t[f_{,yy}(w + w_0)_{,xx} - 2f_{,xy}(w + w_0)_{,xy} + f_{,xx}(w + w_0)_{,yy}]/D = q_{xy}/D \qquad (1.10)$$

$$f_{,xxxx} + 2f_{,xxyy} + f_{,yyyy} = E[(w_{,xy})^2 - w_{,xx}w_{,yy} - (w_{0,xy})^2 + w_{0,xx}w_{0,yy}] \qquad (1.11)$$

where $w$ = deflection due to applied load
$w_0$ = initial deflection or imperfection.

A detailed definition of plate forces and displacements is given in Section 1.6.

### 1.5.2 Solutions

Equations (1.10) and (1.11) are coupled and non-linear, that is to say the load-deformation relationships are non-linear and the principle of superposition does not apply. No closed form solution of the large-deflection equations is known and early investigations obtained approximate solutions using the Ritz energy method (Way 1938), infinite double Fourier series (Levy 1942) and finite differences (Kaiser 1936, Way 1938, Aalami and Chapman 1969). These solutions were for transversely loaded plates with simple boundary conditions. Subsequent developments in the field of transversely loaded plates are well documented elsewhere.

Bulson (1970) gives a comprehensive range of solutions and references for the critical buckling of flat plates subject to a wide variety of in-plane loading conditions. Rockey (1967) gives a review of work on shear buckling. In the field of initially deformed plates subject to in-plane load Hu et al. (1946), Falconer and Chapman (1953), Coan (1959) and Yamaki (1959) obtained solutions for plates subject to uniaxial compression.

The ever-increasing computational capacity of the digital computer and the refinement of numerical techniques such as finite differences and finite elements has greatly facilitated the solution of more general cases of loading and geometry. Finite differences have been applied to initially deformed plates subject to compression by Rushton (1969) and Williams (1971), to plates subject to shear by Williams (1973) and to combined compression and shear by Keays and Williams (1973). Finite elements have been applied to obtain elastic solutions to similar problems by many researchers (Mallet and Marcal 1968, Murray and Wilson 1969). Considerable progress has also been made in obtaining elasto-plastic solutions (Massonnet 1968, Dwight and Moxham 1969, Lin et al. 1972. Cristfield (1973) gives solutions and a comprehensive review of references related to elasto-plastic analysis.

A recent development in the field of elastic analysis has been the derivation by Williams and Walker (1975, 1977) of concise analytically based explicit expressions for various loadings applied to initially deformed plates.

## 1.6 Basic plate relationships and equations

The following outline is identical to that given in the authors' earlier volume on thin-plate design for transverse loading (Aalami and Williams 1975) with terms added to show how initial imperfections enter the equations.

### 1.6.1 Sectional actions

Consider a homogeneous rectangular isotropic plate with sides $a$ and $b$ and thickness $t$ as shown in Fig. 1.2. At any point at distances $x$ and $y$ from the co-ordinate axes, the acting components of forces and moments may be resolved

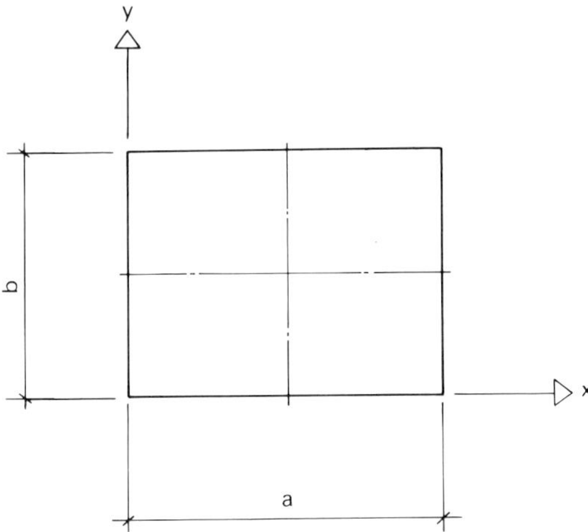


Fig. 1.2 Co-ordinate axes

into bending moments $M_x$ and $M_y$, twisting moments $M_{xy}$, normal shearing forces $Q_x$ and $Q_y$, direct membrane forces $N_x$ and $N_y$ and finally membrane shearing forces $N_{xy}$. These components are collectively called sectional actions and are normally grouped into those producing out-of-plane deformations ($M_x$, $M_y$, $M_{xy}$, $Q_x$, $Q_y$) and those giving rise to in-plane deformations of the mid-plane of the plate ($N_x, N_y, N_{xy}$).

Components of the first group are referred to as flexural actions, and together with the related transverse pressure are shown in Fig. 1.3(i) acting on an element of plate with their positive directions. The latter group, called membrane actions, are shown in Fig. 1.3(ii), also with their positive directions. Of course, these components all act on the same element, but for clarity are demonstrated separately. Note that in the figures double arrows are used as vectorial representation of moments in the direction of the arrow and in the sense given by the right-hand screw rule. For example, in Fig. 1.3(i), $M_x$ and $M_y$ are moments causing tension at the bottom face of the element.

### 1.6.2 Stresses on an element of plate

Each of the sectional actions introduced corresponds to a distribution of reacting stress through the depth of the plate. The relation between each action and the corresponding distribution of stress on the element face is such that the summation of the stresses on that face equals the action concerned. The distribution of stresses on an element of plate due to the positive direction of the sectional actions is shown in Fig. 1.4. For any given value of loading the complete state of stress of the plate element is the algebraic sum of the stresses shown. For example, the total direct stress at the lowest fibre of the element and on the face perpendicular to the $x$-axis and distance $dx$ from it, can be seen from Fig. 1.4 to be equal to $\sigma_{bx} + \sigma_{mx}$.

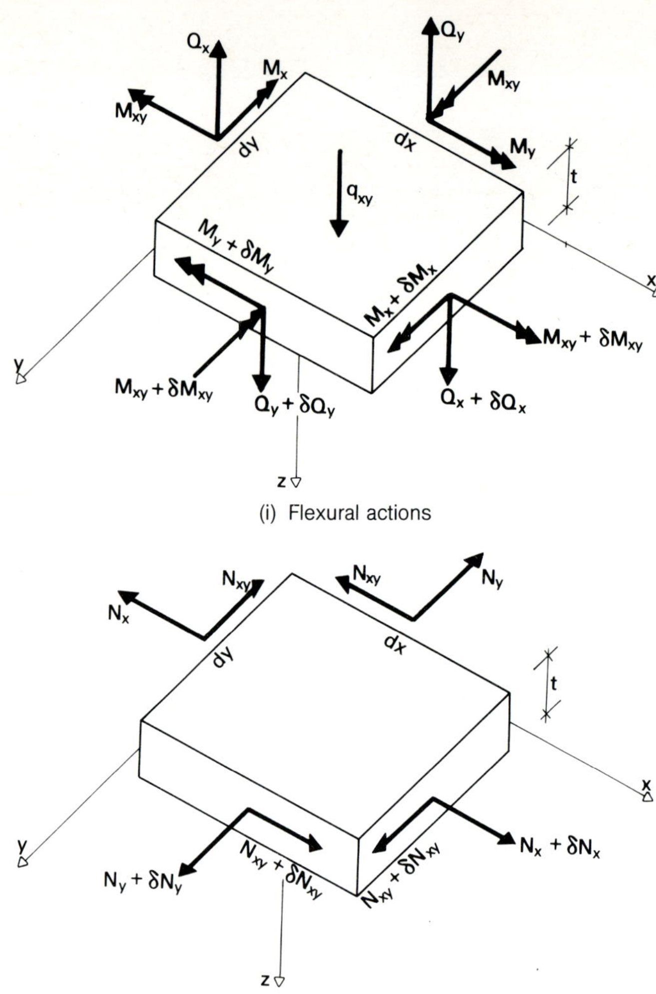


Fig. 1.3 Sectional actions

Among the stress components shown in Fig. 1.4, stresses $\tau_{xz}$ and $\tau_{yz}$ due to $Q_x$ and $Q_y$ respectively are in most practical cases insignificant compared to the other components of stress, and are normally disregarded in design.

### (1) Relationship between sectional actions and stresses

Maximum bending stresses at lowermost fibre due to $M_x$ and $M_y$

$$\sigma_{bx} = 6M_x/t^2 \qquad (1.12)$$

$$\sigma_{by} = 6M_y/t^2$$

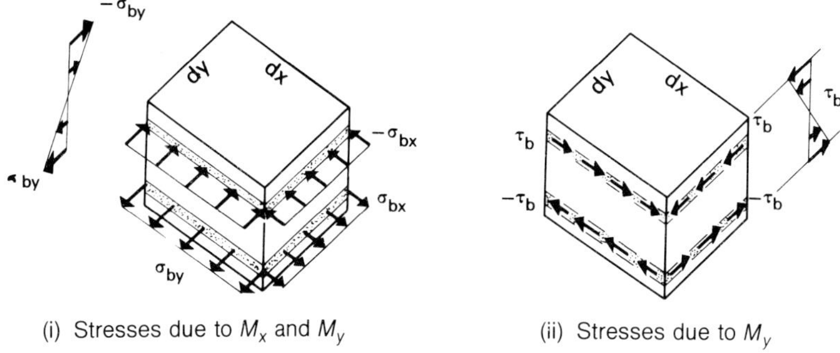


(i) Stresses due to $M_x$ and $M_y$ (ii) Stresses due to $M_y$

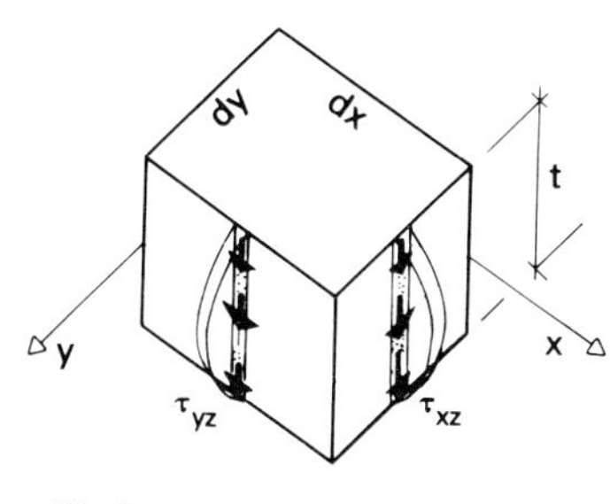


(iii) Stresses due to $Q_x$ and $Q_y$

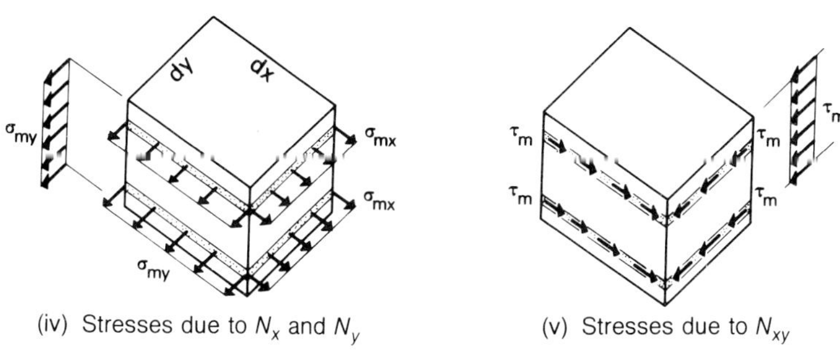


(iv) Stresses due to $N_x$ and $N_y$ (v) Stresses due to $N_{xy}$

Fig. 1.4 Stresses on an element of plate

Maximum membrane shearing stress due to $M_{xy}$

$$\tau_b = -6M_{xy}/t^2 \tag{1.13}$$

Maximum shearing stresses due to $Q_x$ and $Q_y$

$$\left.\begin{aligned} \tau_{xz} &= 1{\cdot}5Q_x/t \\ \tau_{yz} &= 1{\cdot}5Q_y/t \end{aligned}\right\} \tag{1.14}$$

Average direct membrane stresses and membrane shearing stresses due to $N_x$,

$N_y$ and $N_{xy}$

$$\left.\begin{aligned} \sigma_{mx} &= N_x/t \\ \sigma_{my} &= N_y/t \\ \tau_m &= N_{xy}/t \end{aligned}\right\} \quad (1.15)$$

henceforth, stresses $\tau_{xz}$ and $\tau_{yz}$ will be disregarded, because they are not critical to the cases considered in this book.

(2) Total stresses

Maximum total stresses ($\sigma_x, \sigma_y, \tau$) in the $x$- and $y$-directions occur either at the lowermost or the uppermost fibre of the element. From Fig. 1.5 these are as follows:
at the lowermost fibre

$$\left.\begin{aligned} \sigma_x &= \sigma_{bx} + \sigma_{mx} \\ \sigma_y &= \sigma_{by} + \sigma_{my} \\ \tau &= -\tau_b + \tau_m \end{aligned}\right\} \quad (1.16)$$

at the uppermost fibre

$$\left.\begin{aligned} \sigma_x &= -\sigma_{bx} + \sigma_{mx} \\ \sigma_y &= -\sigma_{by} + \sigma_{my} \\ \tau &= \tau_b + \tau_m \end{aligned}\right\} \quad (1.17)$$

### 1.6.3 Basic plate relationships

For a homogeneous isotropic plate, the basic relationships for stresses and deformations of a plate element as shown in Fig. 1.5 may be summarised as follows.

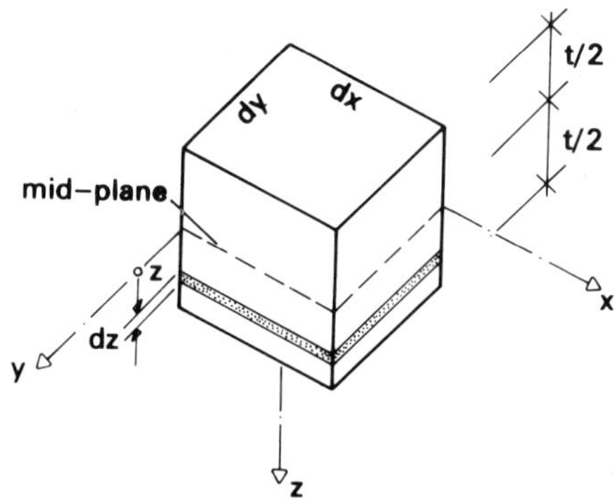


Fig. 1.5 An element of plate

### (1) Small deflections—flexural behaviour

The behaviour of plates in which the mid-plane undergoes only out-of-plane deformation ($w$ in the $z$-direction) is governed by the so-called small-deflection relationships and equations. For these plates the relationships of stresses and deformations with the transverse loading are linear.

The stresses and deformations of an element of plate due to the flexural actions alone ($M_x, M_y, M_{xy}, Q_x, Q_y$) are related to the transverse out-of-plane deflection of plate $w$ by the following relationships:

$$\left.\begin{array}{ll} \text{Rotation of mid-plane in } x\text{-direction} & w_{,x} \\ \text{Rotation of mid-plane in } y\text{-direction} & w_{,y} \end{array}\right. \tag{1.18}$$

Displacement of a point on a layer distance $z$ from the mid-plane (Fig. 1.5).

$$\left.\begin{aligned} u &= -zw_{,x} \\ v &= -zw_{,y} \end{aligned}\right\} \tag{1.19}$$

Hence strains at the same layer

$$\left.\begin{aligned} \epsilon_x &= u_{,x} = -zw_{,xx} \\ \epsilon_y &= v_{,y} = -zw_{,yy} \\ \gamma_{xy} &= u_{,y} + v_{,x} = -2zw_{,xy} \end{aligned}\right\} \tag{1.20}$$

Bending stresses at the same layer

$$\left.\begin{aligned} \sigma_x &= \frac{E}{(1-\nu^2)}(\epsilon_x + \nu\epsilon_y) = \frac{-zE}{(1-\nu^2)}(w_{,xx} + \nu w_{,yy}) \\ \sigma_y &= \frac{E}{(1-\nu^2)}(\epsilon_y + \nu\epsilon_x) = \frac{-zE}{(1-\nu^2)}(w_{,yy} + \nu w_{,xx}) \\ \tau_b &= G\gamma_{xy} = \frac{-zE}{(1+\nu)} w_{,xy} \end{aligned}\right\} \tag{1.21}$$

From integration of the stresses on the faces of the element and the equilibrium considerations (Timoshenko and Woinowsky-Krieger 1959) the flexural actions are

$$\left.\begin{aligned} M_x &= -D(w_{,xx} + \nu w_{,yy}) \\ M_y &= -D(w_{,yy} + \nu w_{,xx}) \\ M_{xy} &= D(1-\nu)w_{,xy} \\ Q_x &= -D(w_{,xxx} + w_{,xyy}) \\ Q_y &= -D(w_{,yyy} + w_{,xxy}) \end{aligned}\right\} \tag{1.22}$$

(2) Membrane behaviour

For purely membrane behaviour, the plate does not deform out of its plane. The deformations are confined to the $x$- and $y$-directions only, and are constant through the thickness of the plate for all the layers. There are several ways of expressing the membrane forces and stresses due to the membrane actions $N_x$, $N_y$ and $N_{xy}$. Herein Airy's stress function, which is defined below, is employed for the expression of the membrane actions, stresses and deformations.

Airy's stress function $f$ is defined as

$$\left.\begin{aligned} N_x &= t\sigma_{mx} = tf_{,yy} \\ N_y &= t\sigma_{my} = tf_{,xx} \\ N_{xy} &= t\tau_m = -tf_{,xy} \end{aligned}\right\} \tag{1.23}$$

Mid-plane strains in terms of forces and stress function

$$\left.\begin{aligned} \epsilon_x &= u_{,x} = \frac{1}{Et}(N_x - \nu N_y) = \frac{1}{E}(f_{,yy} - \nu f_{,xx}) \\ \epsilon_y &= v_{,y} = \frac{1}{Et}(N_y - \nu N_x) = \frac{1}{E}(f_{,xx} - \nu f_{,yy}) \\ \gamma_{xy} &= u_{,y} + v_{,x} = \frac{2(1+\nu)}{Et}N_{xy} = -\frac{2(1+\nu)}{E}f_{,xy} \end{aligned}\right\} \tag{1.24}$$

(3) Large-deflection behaviour—interaction between the flexural and membrane actions

In large-deflection behaviour the interaction between flexural and membrane actions is taken into account. In this case deflections and stresses vary in a non-linear manner with the magnitude of the transverse pressure.

For large deflections with out-of-plane deformations up to several times the plate thickness, the mid-plane strains should be modified in the following form (Von Karman 1910):

$$\left.\begin{aligned} \epsilon_x &= u_{,x} + 0{\cdot}5(w_{,x})^2 = \frac{1}{E}(f_{,yy} - \nu f_{,xx}) \\ \epsilon_y &= v_{,y} + 0{\cdot}5(w_{,y})^2 = \frac{1}{E}(f_{,xx} - \nu f_{,yy}) \\ \gamma_{xy} &= u_{,y} + v_{,x} + w_{,x}w_{,y} = -\frac{2(1+\nu)}{E}f_{,xy} \end{aligned}\right\} \tag{1.25}$$

The expressions given for the sectional actions in terms of out-of-plane deformation $w$ and the stress function $f$ are equally valid in the large-deflection range. In other words, bending moments and normal shearing forces are expressed by the relationship (1.22) and the membrane actions are given by (1.23). In the relationships discussed herein, the effects of shearing stresses $\tau_{xz}$ and $\tau_{yz}$ due to $Q_x$ and $Q_y$ are neglected due to their insignificant influence in homogeneous isotropic plates considered in this book.

(4) Large-deflection behaviour—introduction of initial imperfection

For large-deflections of initially-deformed plates the mid-plane strains are modified as follows (Marguerre 1938):

$$\left.\begin{aligned} \epsilon_x &= u_{,x} + 0{\cdot}5(w_{,x})^2 + w_{,x}w_{0,x} \\ \epsilon_y &= v_{,y} + 0{\cdot}5(w_{,y})^2 + w_{,y}w_{0,y} \\ \gamma_{xy} &= u_{,y} + v_{,x} + w_{,x}w_{,y} + w_{,x}w_{0,y} + w_{0,x}w_{,y} \end{aligned}\right\} \quad (1.26)$$

### 1.6.4 Governing plate equations

For a flat isotropic plate, the derivation of the governing equations is described in detail in Timoshenko and Woinowsky-Krieger (1959).

For small deflections,

$$w_{,xxxx} + 2w_{,xxyy} + w_{,yyyy} = q_{xy}/D \quad (1.27)$$

For membrane action alone,

$$f_{,xxxx} + 2f_{,xxyy} + f_{,yyyy} = 0 \quad (1.28)$$

For large-deflections the behaviour is expressed by the two coupled partial differential equations (1.7) and (1.8), and when initial imperfections are included these become (1.10) and (1.11).

### 1.6.5 Boundary conditions

(1) Flexural boundary conditions

For rotationally free edges

At $x = 0,a$, $M_x = 0$, which gives

$$w_{,xx} + \nu w_{,yy} = 0 \quad (1.29)$$

At $y = 0,b$, $M_y = 0$, which gives

$$w_{,yy} + \nu w_{,xx} = 0 \quad (1.30)$$

For rotationally fixed edges, rotation across the boundary is equated to zero.

At $x = 0,a$

$$w_{,x} = 0 \quad (1.31)$$

At $y = 0,b$

$$w_{,y} = 0 \tag{1.32}$$

For edges on rigid supports, both at $x = 0,a$ and $y = 0,b$,

$$w = 0 \tag{1.33}$$

For edges with no normal support (no support in $z$-direction).

At $x = 0,a$

$$w_{,xxx} + (2 - \nu)w_{,xyy}\ 0 \tag{1.34}$$

At $y = 0,b$

$$w_{,yyy} + (2 - \nu)w_{,xxy} = 0 \tag{1.35}$$

(2) Membrane boundary conditions

For defining the edge displacement of the boundaries perpendicular to the line of support and in the plane of the plate any one of the following three alternatives may be applicable.

(a) For stress-free edges

At $x = 0,a$, $\sigma_{mx} = 0$, which gives

$$f_{,yy} = 0 \tag{1.36}$$

At $y = 0,b$, $\sigma_{my} = 0$, which gives

$$f_{,xx} = 0 \tag{1.37}$$

(b) For zero displacement of the edge across the boundary (edge remains straight and in position)

At $x = 0, a$

$$u = \int u_{,x}\, dx = \frac{1}{E}\int [(f_{,yy} - \nu f_{,xx}) - 0{\cdot}5(w_{,x})^2]\, dx = 0 \tag{1.38}$$

At $y = 0,b$

$$v = \int v_{,y}\, dy = \frac{1}{E}\int [(f_{,xx} - \nu f_{,yy}) - 0{\cdot}5(w_{,y})^2]\, dy = 0 \tag{1.39}$$

(c) For the edge remaining straight and displacing perpendicular to the line of support in the plane of the plate, with the integral of the membrane direct stress along the edge equal to zero

At $x = 0,a$

$$u = \frac{1}{E}\int [(f_{,yy} - \nu f_{,xx}) - 0{\cdot}5(w_{,x})^2]\, dx = \text{constant} \tag{1.40}$$

$$\int_0^b t\sigma_x dy = t \int_0^b f_{,yy}\, dy = 0 \tag{1.41}$$

At $y = 0,b$

$$v = \frac{1}{E} \int [(f_{,xx} - \nu f_{,yy}) - 0{\cdot}5(w_{,y})^2]\, dy = \text{constant} \tag{1.42}$$

$$\int_0^a t\sigma_y\, dx = t \int_0^a f_{,xx}\, dx = 0. \tag{1.43}$$

For the membrane shear boundary conditions the following two cases are considered.

For the stress-free edges, both at $x = 0,a$ and $y = 0,b$, the membrane shearing stress $\tau_m$ is equated to zero, hence

$$-f_{,xy} = 0 \tag{1.44}$$

For the extreme case of the plate being connected to a rigid boundary such that no displacement of the boundary along the line of support may take place

At $x = 0,a, v_{,y} = 0$, which gives

$$f_{,xx} - \nu f_{,yy} = 0{\cdot}5E(w_{,y})^2 \tag{1.45}$$

Similarly at $y = 0,b, u_{,x} = 0$

$$f_{,yy} - \nu f_{,xx} = 0{\cdot}5E(w_{,x})^2 \tag{1.46}$$

## 1.7 Numerical analysis

### 1.7.1 Deformations and stresses

The numerical analysis used to obtain the solutions given in this book is based on a finite-difference representation of the Von Karman equations, written in terms of the local moments, mid-plane forces and displacements. Initial imperfections are included, in the form of the critical buckling mode for each load case.

The finite-difference equations are solved, for a given value of load and amplitude of initial deflection, using the dynamic relaxation method. In essence this is an iterative method, the rate of convergence of which depends on the mesh size and the set values of the dynamic parameters. This explicit method of solution requires only a simple substitution routine in the computer, thus avoiding the solution of a fully populated matrix problem required by implicit formulation, and therefore results in great savings in computer storage.

The accuracy of a solution depends on the order of the finite-difference approximation and the mesh size. First-order finite differences are used for all derivatives in the equilibrium equations, except on the boundaries, where higher-order differences were used to increase accuracy.

Figure 1.6 shows a comparison with corresponding results obtained by Coan (1959) and Yamaki (1959) using a series solution for a square plate with stress-free sides subject to uniaxial compression. It can be seen that agreement is good even for loads well in excess of the critical buckling load.

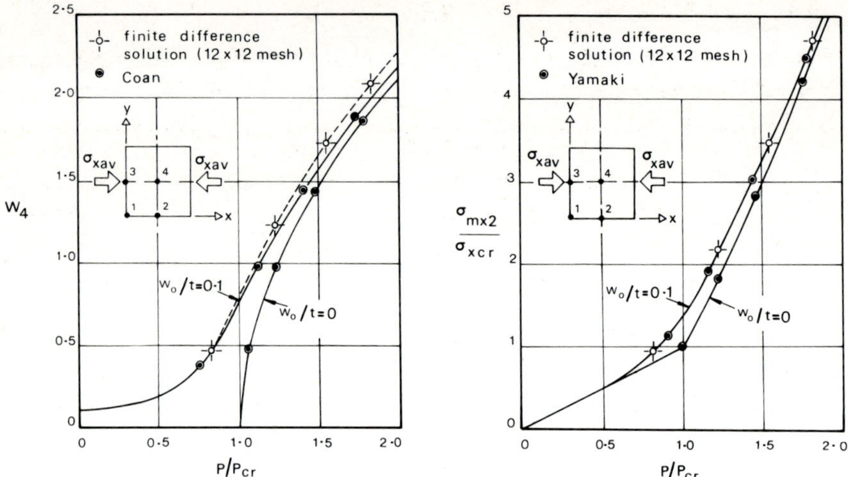


Fig. 1.6 Comparison of solutions for deflection and stress for square plate subjected to uniaxial compressive displacement, unloaded sides stress free

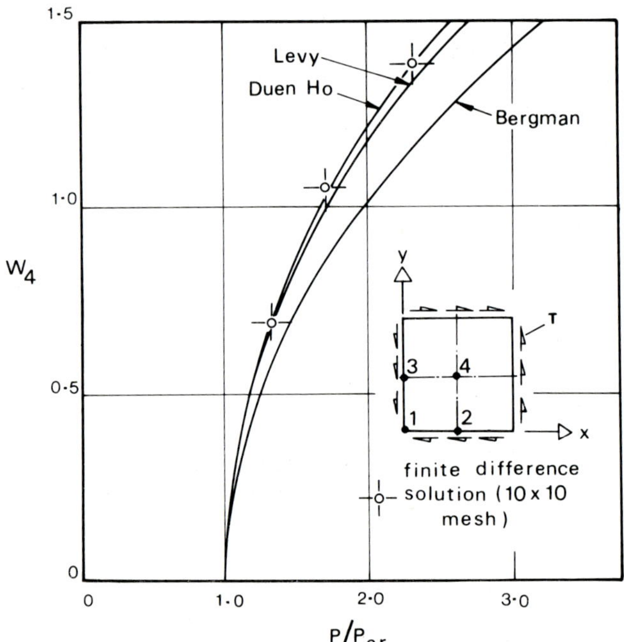


Fig. 1.7 Comparison of solutions for deflection for square plate subjected to shear stress

Figure 1.7 shows a comparison with corresponding results obtained by Bergman (1948), Duen Ho (1964) and Levy et al. (1945) for the case of a square plate subject to shear stress (sides unrestrained). The finite-difference results agree well with the Duen Ho and Levy solutions.

Table 1.1

**Dynamic relaxation solutions for square simply supported plate subjected to uniaxial compression, unloaded sides stress free ($u/u_{cr}$ = 0·885, $w_0/t$ = 0·25)**

| Mesh intervals | Maximum deflection $w/t$ | Corner membrane stress $\sigma_{mx1}/\sigma_{xav}$ | Maximum membrane stress $\sigma_{mx2}/\sigma_{av}$ | Corner twisting moment $M_{xy1}(b^2/Et^4)$ |
|---|---|---|---|---|
| 6 x 6 | 0·520 | 1·14 | 1·62 | 0·407 |
| 8 x 8 | 0·492 | 1·11 | 1·66 | 0·363 |
| 10 x 10 | 0·482 | 1·09 | 1·69 | 0·348 |
| 12 x 12 | 0·476 | 1·06 | 1·70 | 0·340 |
| 14 x 14 | 0·472 | 1·05 | 1·71 | 0·336 |
| 16 x 16 | 0·470 | 1·04 | 1·71 | 0·334 |
| 18 x 18 | 0·470 | 1·04 | 1·72 | 0·332 |

Table 1.1 gives a measure of convergence as a function of mesh size. It can be seen that a 10 x 10 mesh for a square plate gives results to within about 3% of the eventual asymptotes. The mesh divisions used in each of the load cases for which results are given in this book are noted on the figures defining the load cases in Chapter 4.

### 1.7.2 Critical buckling load and mode profile

The finite-difference program was also used to obtain values of critical buckling load for cases not given in the literature. The particular load configuration was applied in increments and a lateral disturbance applied at each stage. This disturbance was then removed and the critical buckling load was defined as the load at which the plate continued to deform out-of-plane rather than resume a flat condition which is the case at sub-critical loads. The latter procedure was also used to obtain profiles of critical buckling mode for all cases except uniform compression which could be described to the required accuracy by trigonometric functions.

CHAPTER TWO

# Design Considerations

Until the advent of the Merrison Rules the only cognisance paid to large-deformation effects in plate panels subject to in-plane load was by way of reduced plate effectiveness to account for the loss of stiffness. The reduction factors incorporated in the British and American Bridge Codes are typical. These rules are however very limited, because they apply only to the simplest case of square panels subject to uniaxial compression and take no account of variation of effectiveness with load level and magnitude of imperfection.

German practice is more comprehensive. In addition to loss-of-effectiveness rules similar to those in other Codes, the designer is required to check that the factor against critical buckling exceeds specified values of 1·71 for compression and 1·35 for shear and combined stress conditions. Allowance can be made for yielding by way of reduction of the critical load. These rules combined with limits on span-to-thickness ratio ($b/t$) go some way to minimising the significance of secondary effects; they also of course inhibit design of lightweight structures.

The main problem in designing for imperfections is the complexity of the analysis and the variety of load cases which occur in practice. Computer programs for the analysis of elastic large-deflection behaviour are now readily available but apart from the cost, the prospect of having to analyse the numerous cases which occur in most large structures is enough to deter most designers. The object of this book is to provide solutions for a wide range of load cases in a readily accessible form.

In this chapter general design considerations are discussed in order to provide an appreciation of their design significance. Some background is given to design parameters formulated by the Merrison Committee, since they are the only existing guidelines in several areas basic to the explicit design for plate panel imperfections.

## 2.1 Analysis of stiffened plate structures

The steps in the analysis of stiffened plate structures are as follows:

(1) Global analysis

(i) definition of structural geometry including plate and stiffening geometry;
(ii) calculation of global section properties;
(iii) definition of loading for in-service conditions and, if applicable, the erection conditions;
(iv) global analysis to determine global forces and deflections.

(2) Local panel analysis

(i) extraction of in-plane loading from the global analysis;
(ii) calculation of geometric imperfection;
(iii) calculation of residual stresses in case of a welded structure;
(iv) analysis to determine stresses and deformations;
(v) check of stresses and deformations against design criteria.

The above process is repeated with suitable adjustments to plate and stiffener geometry until the design criteria are satisfied. With experience it should not be necessary to repeat the global stage more than once. As noted in the previous chapter, it is not usually necessary to take account of large-deformation effects when computing global section properties. However, after the local panel design parameters are satisfied a final re-run of the global analysis may be required to determine overall deflections.

There will of course be certain applications where plate panels subject to in-plane load are not part of a large stiffened structure. In these cases the process is similar to that given above for local panel analysis, except that panel geometry and loading are defined as part of stage (i).

## 2.2 Limitations of elastic design

Ductile thin plates possess post-yield load-carrying capacity which varies depending on the type of loading and the geometry of the plate, and in this respect plates differ from commonly used beam and column sections. Indeed, it is an accepted fact that local yielding occurs in plate panels in many practical applications as evidenced by the permanent set which can be observed, particularly in ship's plating. Because of this, and because of the trend towards ultimate load design, some structural design engineers question the relevance of elastic design criteria. However, while acknowledging the necessity of estimating the collapse lead, the authors support the continuing application of elastic design because:

(1) Elasto-plastic plate analysis has not yet developed to a stage where design data are readily available in a concise form for the wide range of plate geometry and loading configurations which occur in practice.

(2) Elastic deformation is often a basic design consideration, if only for aesthetic purposes.
(3) First yield, which can be predicted by elastic analysis, represents the onset of permanent set.
(4) Elastic design data are required for fatigue analysis.

## 2.3 Geometry

The results given in this book are for rectangular plates ranging in side ratio (ratio of length *a* to width *b*) from 0·67 to 3. The ratios are generally chosen to cover the significant range in critical buckling load for each particular load configuration. Where particular cases arise having side ratios between the values covered, stresses and deformations can either be taken as those of the nearest side ratio or solutions for the same loading applied to the nearest side ratios for which solutions are given can be linearly interpolated. The solutions are non-dimensionalised so as to be independent of plate thickness.

## 2.4 Initial imperfections

The initial imperfection of a plate panel is the out-of-flatness of the panel in its unloaded condition. These imperfections are an inevitable by-product of the plate rolling, fabrication, transportation and erection process and are the raison d'être for this book. It is to be hoped that a greater appreciation of the design significance and improved production control of these imperfections will lead to simplified design criteria. In the meanwhile engineers who wish to minimise weight by utilising thin plate will have to resort to explicit design methods of the type outlined in this book

Two basic criteria require definition:

(1) Form

The shape or form of the distribution of the initial out-of-plane deflection over the surface area of the plate has a primary influence on the buckling which occurs under load. In most cases deflections and stresses due to applied load will be greatest if the form of the initial imperfection is similar to that of the critical buckling mode.

All the results presented in this book are for initial imperfections in the form of the critical mode corresponding to each particular load case. This will be conservative in many instances, but since there is no guarantee that the critical mode will not occur in practice it must be the governing condition. Figures illustrating load cases and their corresponding critical buckling modes are given in Chapter 4.

(2) Amplitude

In order to enter the design curves given in this book it is necessary to assume an amplitude of the initial imperfection. This design criterion must be related to a fabrication specification and should correspond to tolerances achievable with current shop practices if a cost penalty is to be avoided.

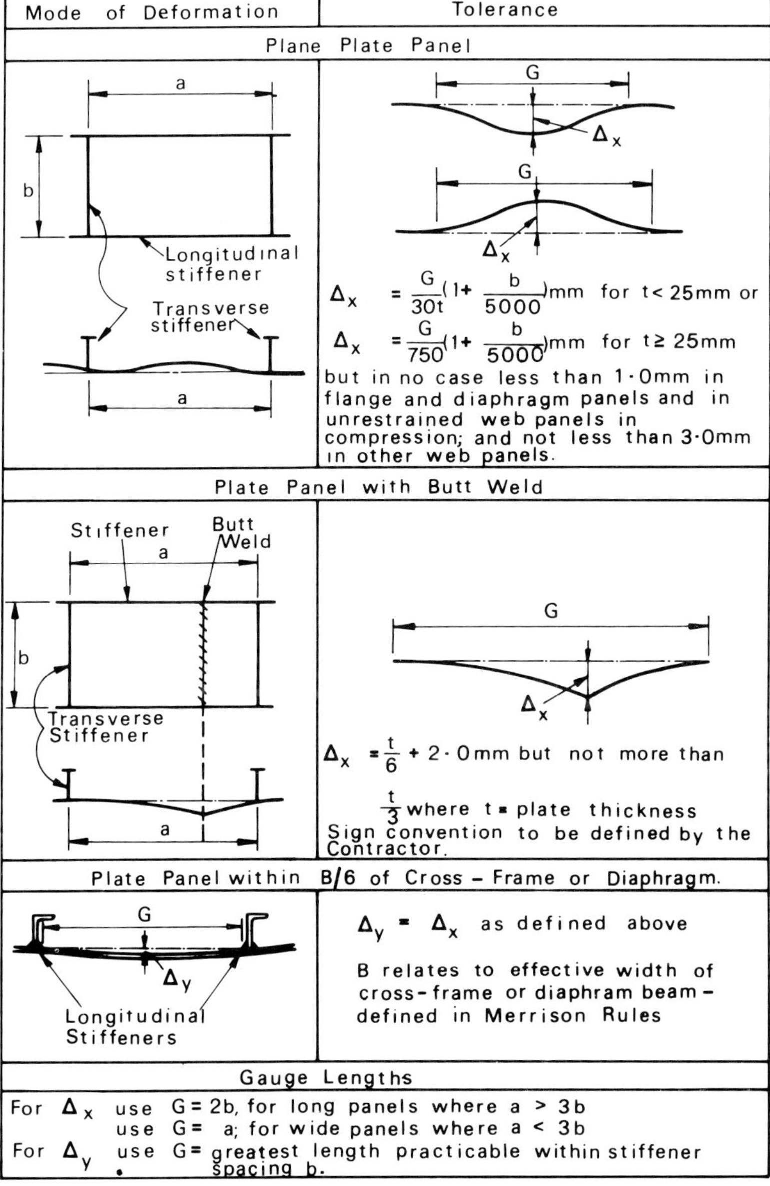


Fig. 2.1 Fabrication tolerances for plate panels in steel box-girder bridges

As part of the background to the Merrison Rules, extensive measurements of imperfections were taken on bridges in England and elsewhere upon which the allowable fabrication tolerances given in Fig. 2.1 were based.

The Merrison Rules specify that the imperfection to be allowed for in design be 1·20 times the fabrication tolerance in the case of discrete panels, and in the case of plate-panel elements in stiffened panels the geometric imperfection should be

$$w_g = [1.2b\Delta_x/G] \sqrt[3]{[1/(N+1)]} \tag{2.1}$$

where $\Delta_x$ is the specified tolerance measured as indicated in Fig. 2.1 over gauge length $G$, and $N$ is the number of longitudinal stiffeners in the stiffened panel under consideration. The above basis for calculating the geometric imperfection is based exclusively on bridge fabrication and can therefore only be used as a guide for other applications.

## 2.5 Plate effectiveness

The plate effectiveness or effective-width concept is a design approximation to take account of the influence of out-of-plane deformation on the in-plane load-carrying capacity of plate panels between stiffeners. Figure 2.2 illustrates

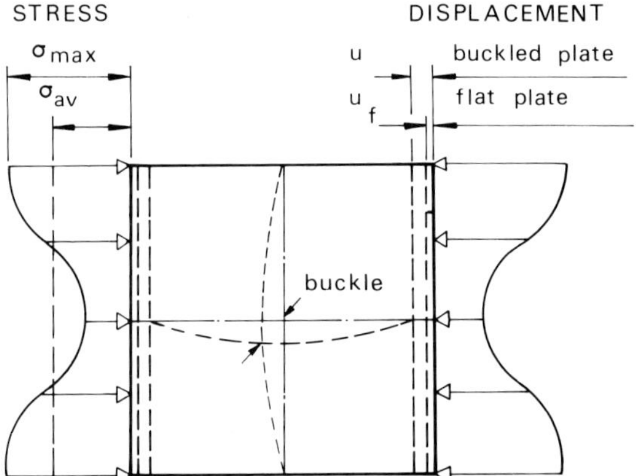


Fig. 2.2 Plate effectiveness parameters

the two aspects of plate behaviour to which this concept can be applied:

$$\begin{aligned} &\text{stress effectiveness: } \lambda_\sigma = \sigma_{av}/\sigma_{max} \\ &\text{stiffness effectiveness: } \lambda_s = u_f/u \end{aligned} \tag{2.2}$$

Values for both the above criteria are given in Chapter 4 for plates subject to uniaxial and biaxial compression and for plates subject to uniaxial and biaxial

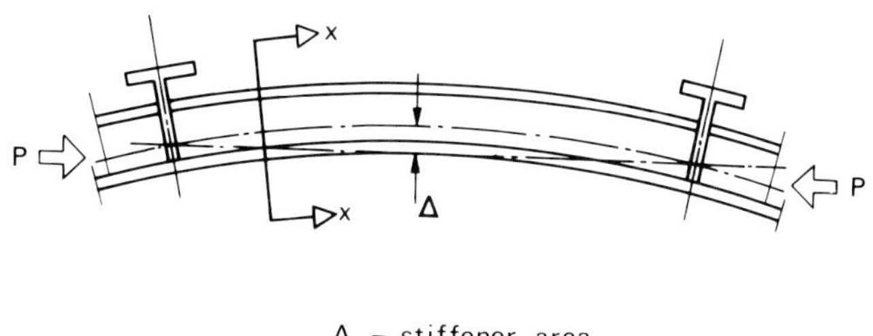


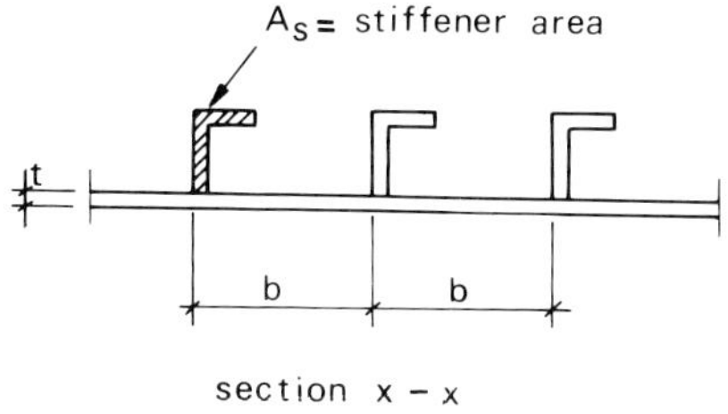


Fig. 2.3 Buckling of stiffened plate

compression combined with shear. The stiffness effectiveness is the value required to assess the influence of loss of plate effectiveness on overall or global deflections and to compute the magnification of stiffened panel imperfections. The stress effectiveness can be used to compute the maximum stress resulting from the magnification of stiffened-panel imperfections.

Figure 2.3 illustrates a section of a typical stiffened plate panel. Stresses induced by magnification of stiffener and plate imperfections can be computed using the expression:

$$\sigma = P/A_e + Pm\Delta/Z_e \tag{2.3}$$

where $\sigma$ = maximum stress

$P$ = applied axial load per unit width

$A_e = (A_s + \lambda_\sigma bt)/b$

$m = 1/(1 - P/P_{cr})$

$P_{cr}$ = critical axial load per unit width of plate–stiffener combination incorporating plate stiffness effectiveness

$\Delta$ = stiffener imperfection

$Z_e$ = section modulus per unit width of plate–stiffener combination incorporating plate stress effectiveness.

## 2.6 Flexural boundary conditions

Figure 2.4 illustrates the buckling of a stiffened panel. In order to isolate a plate panel for the purposes of analysis basic assumptions must be made in regard to the flexural and membrane boundary conditions. So far as flexural conditions are concerned current practice is to assume that all sides are simply sup-

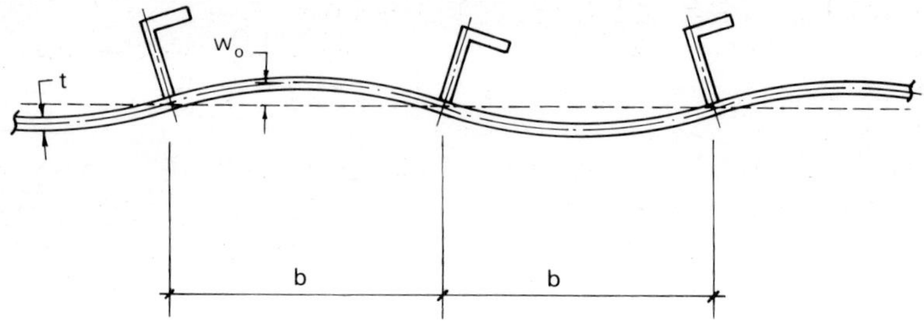


Fig. 2.4 Buckled plate

ported, i.e.

$$\begin{aligned} \text{rigid supports:} \quad & w = 0 \text{ along } x = 0, a \\ & \text{and } y = 0, b \\ \text{rotationally free:} \quad & w_{,xx} = 0 \text{ along } x = 0, a \\ & w_{,yy} = 0 \text{ along } y = 0, b \end{aligned}$$

Although the edge support members are commonly stiffeners, as illustrated in Fig. 2.4, their deflection under load is usually negligible relative to the deflection of the plate panel and hence the assumption of rigid support is realistic.

The assumption that the panel is rotationally-free is conservative, since it is based on antisymmetric buckling across panel boundaries, and completely ignores the torsional rigidity of the stiffeners as well as the restraint afforded by the adjacent panels. Without quantitive data on effective rotational restraint any assumptions other than simple support cannot be justified and all the results given in this book are for this condition. This can be a very conservative assumption and further research is required so that allowance can be made for some degree of constraint.

## 2.7 Membrane boundary conditions

The results given in this book are for two limiting membrane boundary conditions; sides restrained to remain straight and sides free to move in-plane. In the former case the straight sides constrain the buckle, developing edge membrane stresses in the process, and in the latter case the sides 'pull in'. These conditions are illustrated in Fig. 2.5.

The most common case of the restrained condition occurs in plate panel elements in stiffened panels, for 'interior' edges, where compatibility between adjacent panels dictates that they remain straight. It can also occur in 'exterior' or edge panel elements, or single panel elements, where the discontinuous sides are bounded by a stiff framing member. Because of the edge stresses developed in restraining the buckle the applied stress distribution becomes increasingly non-linear as the load increases. Stresses also develop on unloaded restrained sides, which are self equilibrating if the side is free to move while remaining straight, or have a net value if the side is fully restrained. These conditions are illustrated in figures showing the boundary stress and in-plane displacement distributions for each load case in Chapter 4.

The free-side condition commonly occurs along unloaded discontinuous edges of panel elements which are not bounded by stiff framing members. Cases of loaded edges which are free to move are less common, since the load must be able to 'follow' the buckle and hence be applied via a flexible medium. An example is load applied to box girder diaphragms via flexible bearings.

There are of course many cases of intermediate restraint, and cases of involving a mixture of straight and free sides in a single panel element, which are not covered in the results given in this book. These can be appraised by assuming all sides are free or by interpolating between corresponding results for restrained and free sides.

## 2.8 Limiting stress criterion—equivalent stress

Because the state of stress in plates is biaxial, rather than uniaxial as in the case of beams, it is necessary to define a limiting stress criterion which incorporates the three stress components $(\sigma_x, \sigma_y, \tau)$ acting on an element of the plate material. This criterion must provide a reliable measure for determining when the place element passes from an elastic to a plastic state, i.e., when it yields.

There is considerable experimental verification for ductile materials for the following semi-empirical equivalent stress criterion proposed by von Mises:

$$\sigma_e = (\sigma_x^2 + \sigma_y^2 - \sigma_x \sigma_y + 3\tau^2)^{1/2} \tag{2.4}$$

The stresses include the bending and membrane components defined in Chapter 1:

$$\begin{aligned} \sigma_x &= \pm \sigma_{bx} + \sigma_{mx} \\ \sigma_y &= \pm \sigma_{by} + \sigma_{my} \\ \tau &= \pm \tau_b + \tau_m \end{aligned} \tag{2.5}$$

An element is considered to yield when the equivalent stress reaches the yield stress of the material in uniaxial tension, i.e.

$$\sigma_e = \sigma_y \tag{2.6}$$

First yield will always occur at some location on the surface of the plate and this is the basis for the Merrison Rules serviceability limit. The membrane equivalent stress, which can be used as a simple collapse limit state, is given by omitting the bending-stress components in the above expressions. The Merrison Rules give a more complex collapse criterion which is detailed in Section 3.3.

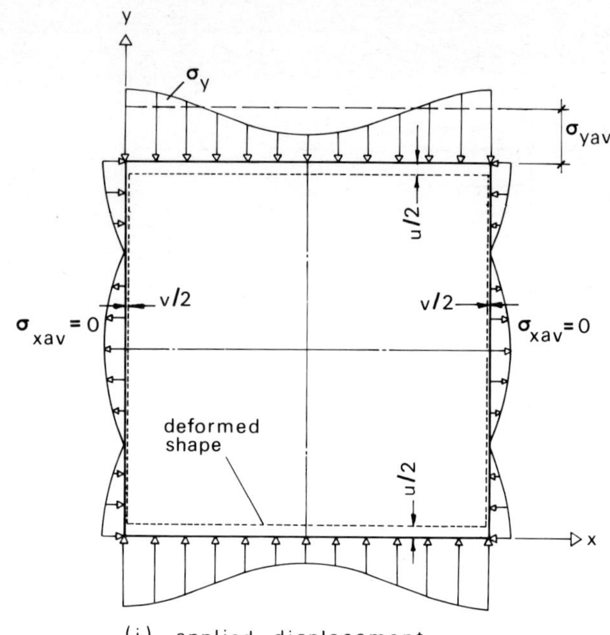


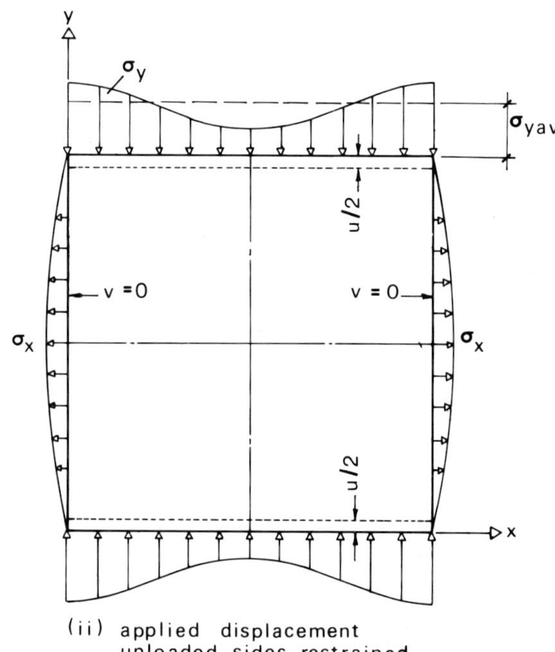


Fig. 2.5 Membrane boundary conditions–illustrated for the case of uniaxial compression in direction of y-axis

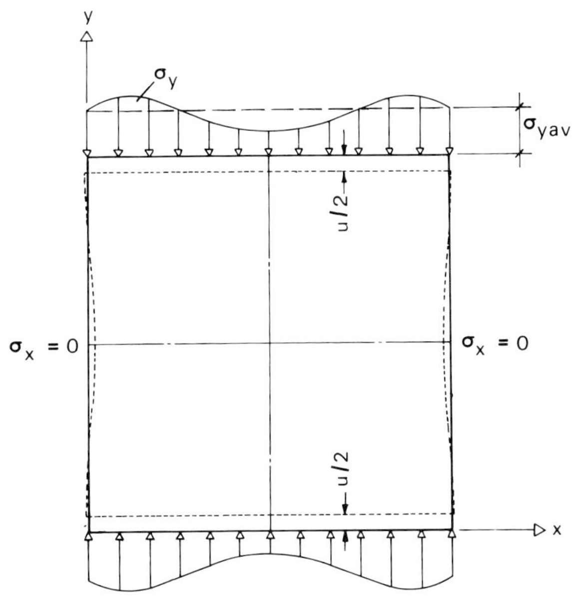


(iii) applied displacement
unloaded sides stress free

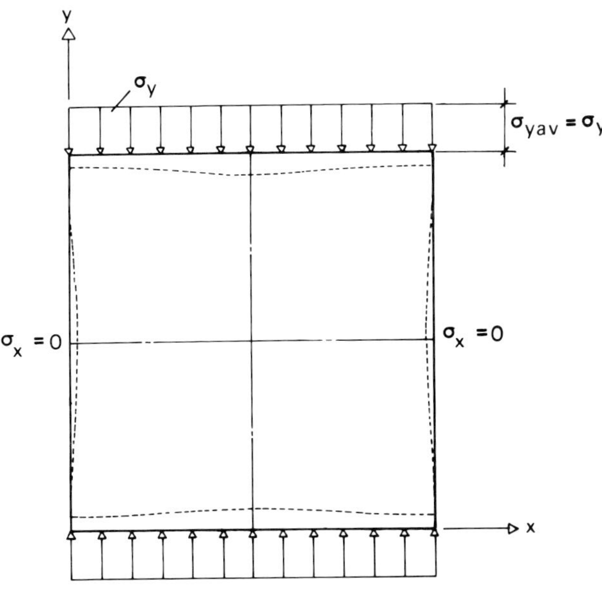


(iv) applied stress
unloaded sides stress free

## 2.9 Surface stress distribution

The location of the maximum surface stress and the gradient of this stress in the region of the maximum have some bearing on its design significance. In general, the maximum is located in the interior of the plate at lower loads and imperfections and moves towards the corners as the load increases. The gradient or slope of the corner stress distribution becomes steeper as load increases. This is illustrated in Fig. 2.6, which shows contours of equivalent surface stress distribution for the case of a square plate subject to uniaxial compression.

The results for equivalent stress plotted in Chapter 4 include the corner value and the value at a location adjacent to the corner. The latter location corresponds to the finite-difference mesh point on the diagonal adjacent to the corner and its actual distance from the corner can be computed from the number of mesh divisions given for each case. By comparing these adjacent values the designer can assess the stress gradient or degree of localisation of maximum stress. It is recommended that some cognisance be taken of stress gradient when it becomes steep.

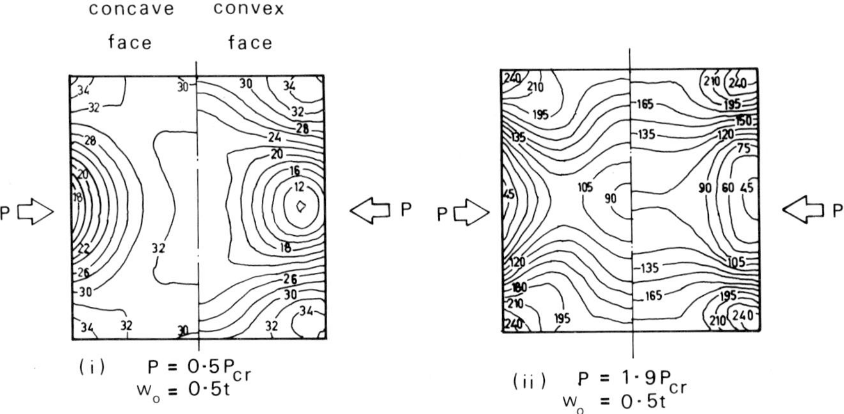


Fig. 2.6 Contours of surface stress for square plate subject to uniaxial compressive displacement, unloaded sides free (to convert contour intervals to dimensional values divide by $Et^2/[12(1-\nu^2)a^2]$)

## 2.10 Residual stresses due to welding

Figure 2.7 illustrates the welding of a typical fillet joint. As the molten metal in the joint cools it shrinks and because this shrinkage is constrained by the adjacent material a self-equilibrating residual stress distribution is set up with the shrinkage zone going into tensile yield. Figure 2.8 shows an idealisation of a typical longitudinal residual stress distribution. The actual shape and extent of this distribution depends on the joint geometry and the weld size, length, number of passes and method of application.

As in the case of geometric imperfections the design significance of residual stresses has been accentuated by the trend toward thin-plate structures. The

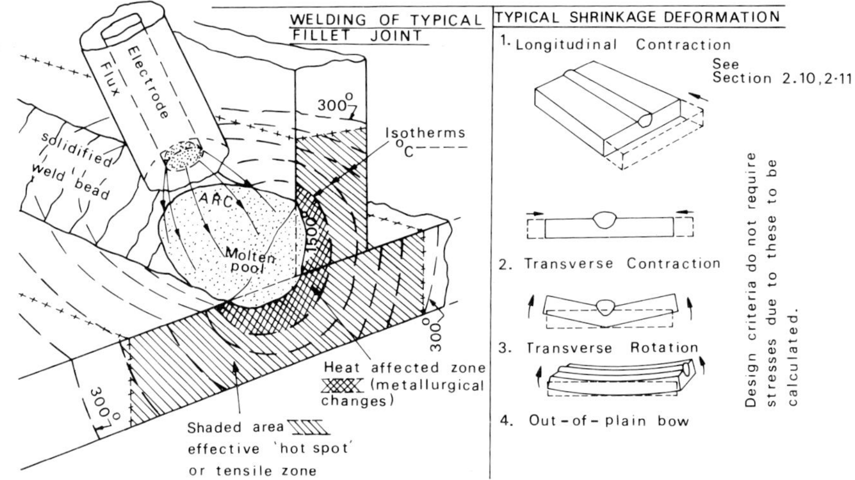


Fig. 2.7 Weld shrinkage

compressive stresses induced in the plating can be of the order of 20–30% of the plate's critical buckling stress, and hence these stresses, the secondary stresses they induce, and the influence they have when combined with stresses induced by subsequent loading cannot be ignored.

Part III of the Merrison Rules provides for the explicit calculation of residual stresses in box-girder bridges. The following extracts are taken from the Rules with the kind permission of HMSO.

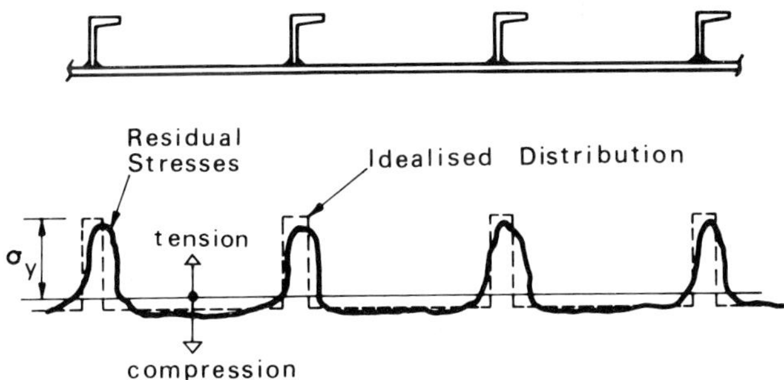


Fig. 2.8 Residual stress idealisation

## 2.11 Allowance for residual stresses in design

### 2.11.1 Calculation of residual stresses

Residual stresses, $\sigma_R$, shall be assessed by considering the shrinkage forces induced by all welds attached to the component concerned calculated in accordance with the expressions given below. The effects of shrinkage forces shall be calculated by assuming the forces to be uniformly distributed over the following areas:

(1) For welds connecting longitudinal stiffeners to plate in a stiffened panel—the gross sectional area of the stiffened panel.
(2) For longitudinal seam and panel junction welds—the total area of cross-section of the box.
(3) For transverse seam welds and connections to transverse stiffeners and diaphragms—an area of a width of plate on each side of the centre line of the seam, stiffener or diaphragm equal to the lesser of:
   (i) Half the distance to the centre line of the adjacent stiffener or diaphragm.
   (ii) The distance to the free edge of the plate.
   (iii) For flanges—one sixth of the width of the flange, or for webs—one sixth of the depth of the web.

### 2.11.2 Shrinkage forces due to welding

General

The following forces, $F_c$ in kilonewtons, due to weld shrinkage shall be assumed to be induced in the direction of each weld, and shall be calculated on the basis of the total nominal cross-sectional area of weld metal in the joint, $A$ (mm$^2$), with due allowance for the total number of weld runs in the joint, $n$, and the mean distance between the welds, $w$. Shrinkage forces transverse to the weld may be ignored.

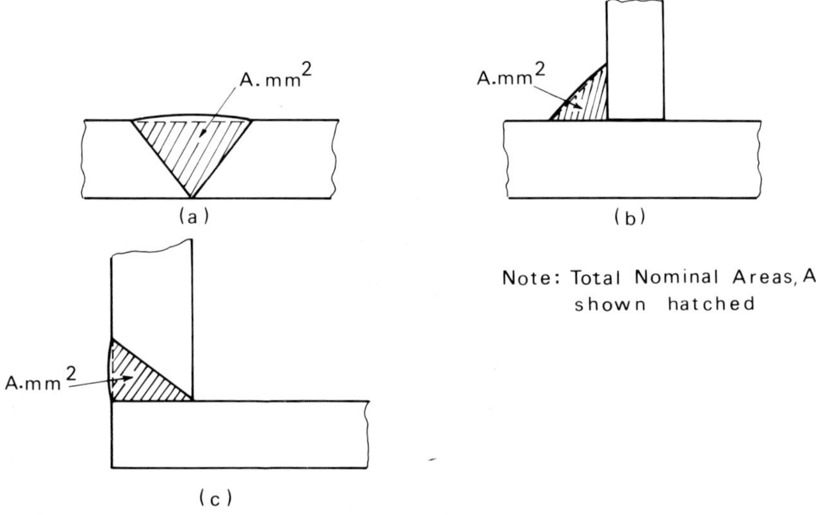


Fig. 2.9 Single-weld joint–single run

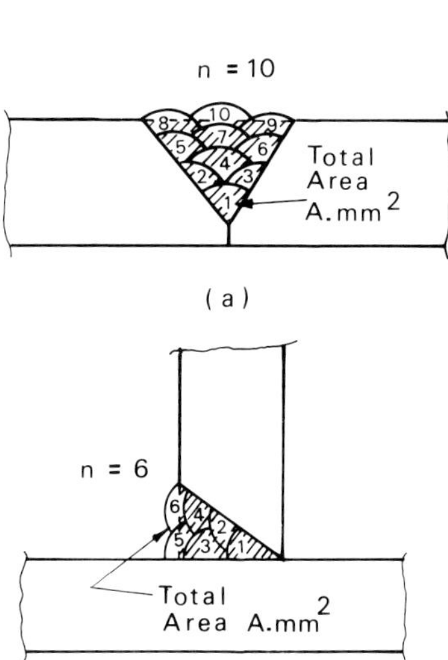


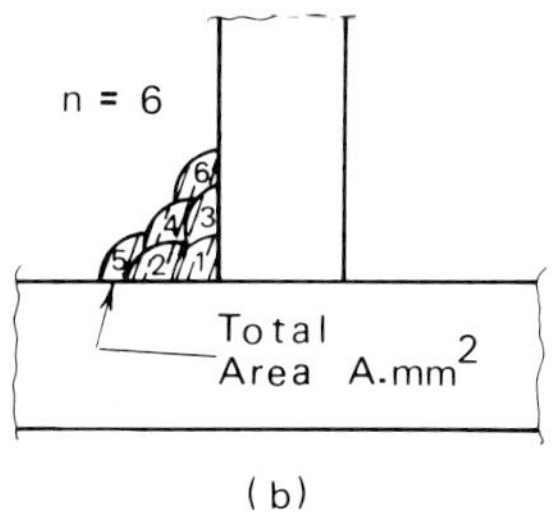


Note: Total Nominal Areas, A, shown hatched

Fig. 2.10 Single-weld joint–multi-run

Single weld joint–single run (Fig. 2.9)

$$F_c = 10{\cdot}0A \tag{2.7}$$

Single weld joint–multi-run (Fig. 2.10)

$$F_c = 10{\cdot}0A/(\sqrt[3]{n})^2 \tag{2.8}$$

Multi-weld joints–runs laid consecutively (Fig. 2.11)

For $w < 60A/\Sigma t(\sqrt[3]{n})^2$

$$F_c = 10{\cdot}0A/(\sqrt[3]{n})^2 + 0{\cdot}1w\Sigma t \tag{2.9}$$

For $w \geqslant 60A/\Sigma t(\sqrt[3]{n})^2$

$$F_c = 16{\cdot}0A/(\sqrt[3]{n})^2 \tag{2.10}$$

Multi-weld joints–runs laid in pairs (Fig. 2.12)

For $w < 60A/\Sigma t(\sqrt[3]{n})^2$

$$F_c = 16{\cdot}0A/(\sqrt[3]{n})^2 + 0{\cdot}15w\Sigma t \tag{2.11}$$

For $w \geqslant 60A/\Sigma t(\sqrt[3]{n})^2$

$$F_c = 25{\cdot}0A/(\sqrt[3]{n})^2 \tag{2.12}$$

Intermittent welds–staggered (Fig. 2.13(a))

$$F_c = 2L_w/(L_m + L_w) \text{ times that due to a continuous weld} \tag{2.13}$$

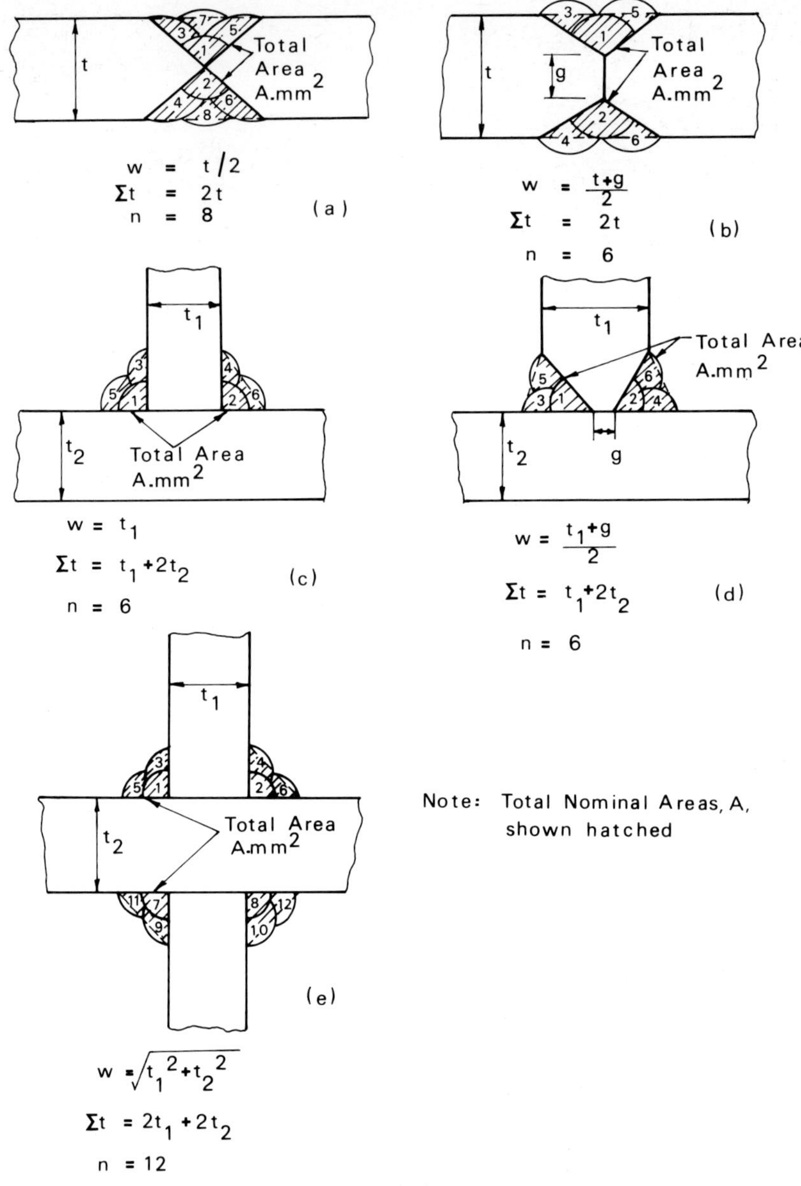


Fig. 2.11 Multi-weld joints–runs laid consecutively

where $L_w$ = full length of each intermittent weld
$L_m$ = clear distance between weld ends on one side.

Intermittent welds–chain (Fig. 2.13(b))

$$F_c = L_w/(L_m + L_w) \text{ times that due to continuous welds on each side} \quad (2.14)$$

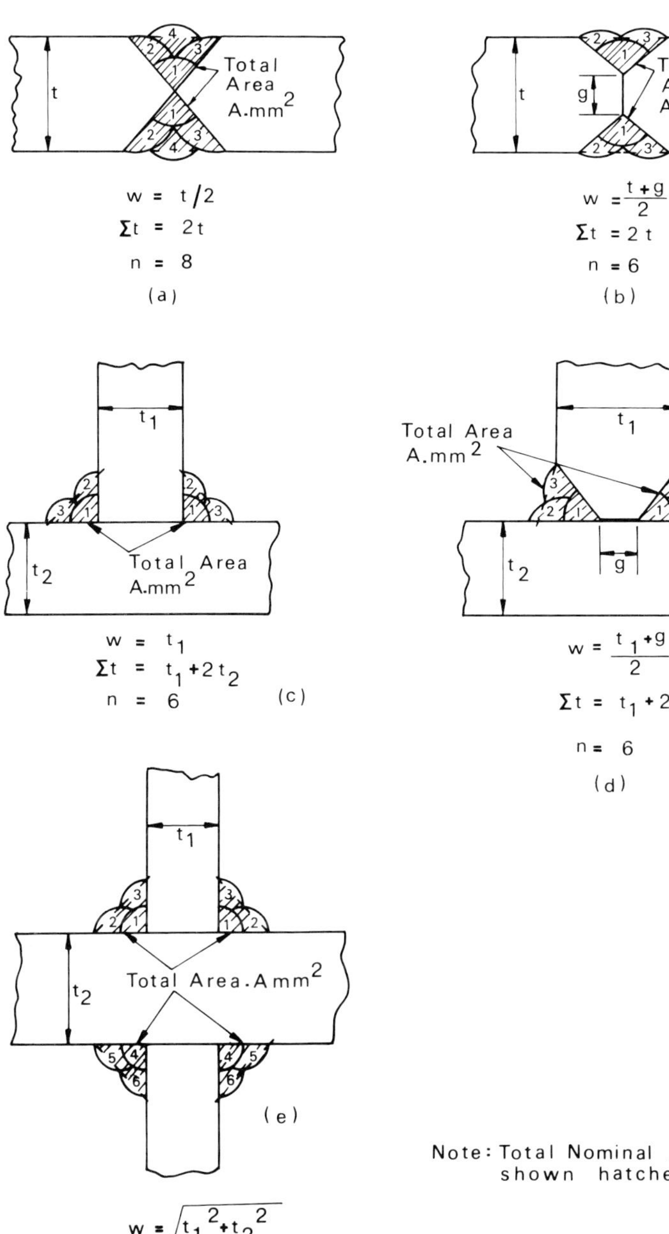


Fig. 2.12 Multi-weld joints–runs laid in pairs

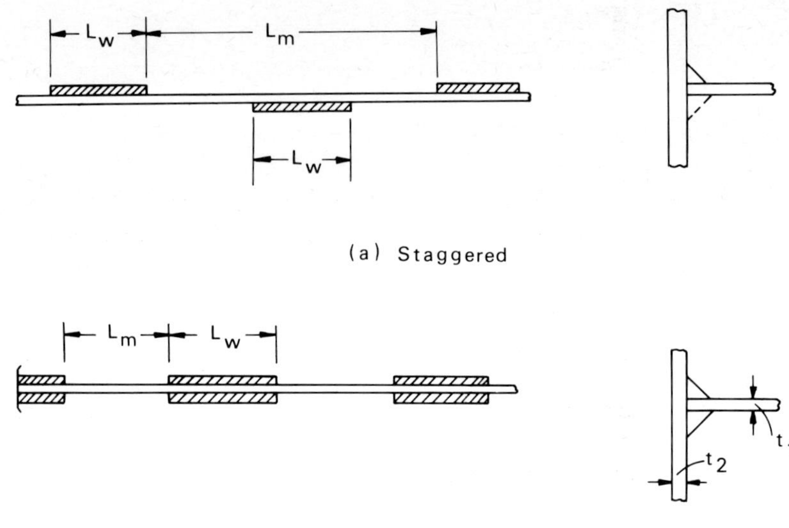


Fig. 2.13 Intermittent welds

### 2.11.3 Effective value of residual stresses in plate panels

In plates in stiffened compression flanges and diaphragms the effective value of residual stress, $\sigma_{\text{Reff}}$, shall be taken as follows:

$$0 < b/t \leqslant 20: \sigma_{\text{Reff}} = 0 \tag{2.15}$$

$$20 < b/t \leqslant 40: \sigma_{\text{Reff}} = \sigma_{\text{Rs}}(b/t - 20)/20 \tag{2.16}$$

$$40 < b/t: \sigma_{\text{Reff}} = \sigma_{\text{Rs}} \tag{2.17}$$

where $\sigma_{\text{Rs}}$ = the residual stress for stiffener attachment welds only
$b$ = the mean spacing of stiffeners
$t$ = the plate thickness.

### 2.11.4 Equivalent imperfection

The Merrison Rules permit designers to analyse residual stresses as either additional loads, or as an equivalent imperfection. If the equivalent imperfection option is chosen, then the geometric imperfection calculated using the expressions given in Section 2.4 must be increased using the following expression.

$$w_0 = (w_g^2 + 8L_x^2 \sigma_{\text{Reff}\,x}/\pi^2 E + 8L_y^2 \sigma_{\text{Reff}\,y}/\pi^2 E)^{1/2} \tag{2.18}$$

where $\sigma_{\text{Reff}\,x}, \sigma_{\text{Reff}\,y}$ = effective residual stresses in $x$- and $y$-directions
$L_x, L_y$ = buckling lengths in $x$- and $y$-directions.

If there is no residual stress in either direction then the term for that direction goes to zero in the above expression. An example illustrating the significance of residual stresses and the alternative methods of analysis is given in Chapter 6.

## 2.12 Elastic properties

The solutions given in this book are for elastic isotropic materials. That is to say the elastic properties of the material are the same in all directions about any given point, and these properties are completely defined by two quantities:

Young's modulus of elasticity; $E$

Poisson's ratio; $\nu$

In fact the solutions given in this book are restrictive in as much as although they are applicable for any value of $E$, they are all for a Poisson's ratio of 0·3. This is the design value for steel and is usually accurate enough for most non-ferrous metals.

Shear modulus is given by

$$G = E/[2(1 + \nu)] \tag{2.19}$$

For materials with values of Poisson's ratio other than 0·3, the corresponding design quantities are to be modified using the following relationships, in which a symbol marked with a star relates to the actual Poisson's ratio, $\nu^*$.

For deflections

$$W^* = 1{\cdot}1(1 - \nu^{*2})W \tag{2.20}$$

For bending stresses

$$\begin{aligned} \sigma^*_{bx} &= 1{\cdot}1\ [(1 - 0{\cdot}3\nu^*)\sigma_{bx} + (\nu^* - 0{\cdot}3)\sigma_{by}] \\ \sigma^*_{by} &= 1{\cdot}1\ [(1 - 0{\cdot}3\nu^*)\sigma_{by} + (\nu^* - 0{\cdot}3)\sigma_{bx}] \end{aligned} \tag{2.21}$$

For membrane stresses

$$\begin{aligned} \sigma^*_{mx} &\simeq \sigma_{mx} \\ \sigma^*_{my} &\simeq \sigma_{my} \end{aligned} \tag{2.22}$$

For the maximum equivalent stresses given, it is not possible to evaluate their corresponding accurate values for Poisson's ratio other than 0·3. However, an upper limit to the maximum deviation from the given maximum equivalent stresses may be obtained by making the following assumptions:

(1) The contribution of membrane stresses is zero.
(2) For values of Poisson's ratio less than 0·3, the stresses are taken as uniaxial, and for Poisson's ratio greater than 0·3 the state of stress is biaxial with stresses equal in both directions.

The above assumptions are not of practical significance, but are required for the evaluation of the maximum possible values of $\sigma_{e\,max}$ and $\sigma_{max}$ for Poisson's ratio different from 0·3. Thus:

for $\nu^* < 0{\cdot}3$

$$\begin{aligned} \sigma^*_{e\,max} &= 1{\cdot}1(1 - 0{\cdot}3\nu^*)\sigma_{e\,max} \\ \sigma^*_{max} &= 1{\cdot}1(1 - 0{\cdot}3\nu^*)\sigma_{max} \end{aligned} \qquad (2.23)$$

for $\nu^* > 0{\cdot}3$

$$\begin{aligned} \sigma^*_{e\,max} &= 0{\cdot}77(1 + \nu^*)\sigma_{e\,max} \\ \sigma^*_{max} &= 0{\cdot}77(1 + \nu^*)\sigma_{max} \end{aligned} \qquad (2.24)$$

The expressions (2.23), (2.24) may be used as a guide in design. The actual values of stresses for a Poisson's ratio other than 0·3 are in fact closer to the values given in the curves than the adjusted values obtained from the preceding examples.

## 2.13 Fatigue

In appraising a structure the designer must consider the effect of cyclic or repeated loading which can result in the formation and propagation of cracks. These cracks are usually initiated by stress raisers such as small defects or discontinuities and they can occur under applied stresses which are less than the material yield stress. Welded joints are a prime source of discontinuities and defects, and as noted in the Section on residual stresses, the material in the vicinity of the weld is stressed to yield because of the constraint against shrinkage during the cooling process.

Designers will be familiar with the concept of weld detail classification which reflects the potential for crack initiation. For example, the fatigue strength of a fillet-welded lap joint is much lower than that of a fillet-welded flange-to-web connection.

The standard method of performing fatigue calculations is to derive the number of cycles ($N$) to produce failure for a particular maximum stress level ($S$) from an experimentally derived $S$–$N$ curve. The ratio of $n$ (the actual number of cycles) to $N$ is termed the fatigue damage ratio for the stress level $S$. For spectrum loading the summation of the individual damage ratios, $\Sigma(n/N)$, ignoring stress levels below a specified cut-off level, is termed the cumulative damage or Miner's summation which corresponds to failure when $\Sigma(n/N) = 1$.

Weld classifications and $S$–$N$ curves are readily available in design literature. The Merrison Rules have recommended some variations to standard practice in regard to failure calculations, the most basic of which is explicit allowance for residual welding stresses. The use of the more readily calculated equivalent stress value rather than principal stress value is also allowed under certain conditions.

The Merrison Rules recommend two levels of fatigue criteria. These are referred to as 'serviceability' and 'collapse' although they are quite dissimilar from the two levels of load factors used to assess static loads. Essentially, they are simply two levels of severity, the more severe (or 'collapse') criterion, being required unless the following requirements are all satisfied.

(1) Propagated cracking will not result in catastrophic collapse of a bridge.
(2) Propagated cracking will not result in sudden impairment of bridge performance.
(3) Cracking when repaired would restore the bridge to full user service.
(4) Cracking would be detectable by normal visual inspection and would be accessible for repair.
(5) Repetitive cracking of a like kind in similar members could be repaired without prohibitive cost and without complete traffic closure.

CHAPTER THREE

# Design Parameters, Procedure and Criteria

The preceding chapters outline the basic mechanics of large-deflection behaviour of thin plates subject to in-plane load and provide guidelines for establishing the boundary conditions, loading, initial imperfection, residual stresses etc., appropriate to a particular case.

Design data are presented in graphical form in Chapter 4 and in tabular form in Appendix 1. The various parameters are plotted and tabulated as functions of the applied load expressed as a fraction of the corresponding critical buckling load. In most instances the graphical data will provide all the information required to appraise a plate for a given load situation. The tabular data are supplementary to that given in the graphs and provide additional data, in particular with reference to uniaxial stress components, which may be required for particular checks such as fatigue calculations.

## 3.1 Design parameters

The various parameters for which design data are provided are described in Table 3.1. These parameters are presented in non-dimensional form as follows:

### (1) In-plane loading

The in-plane loading in each case is defined by $P/P_{cr}$, where $P$ is the applied load and $P_{cr}$ is the corresponding critical buckling load. The critical buckling stress is given by

$$\left.\begin{aligned} &\sigma_{cr} \text{ or } \tau_{cr} = k\sigma_b \\ \text{and} \quad &P/P_{cr} = \sigma_{av}/\sigma_{cr} \text{ or } \tau_{av}/\tau_{cr} \end{aligned}\right\} \tag{3.1}$$

Table 3.1 **Design parameters**

| Symbol | Quantity | Location |
|---|---|---|
| $w^{*,\dagger}$ | Out-of-plane deflection | Maximum (can vary with load) |
| $\sigma_{es}{}^{*}$ | Equivalent surface stress | Various locations as noted (one of which is maximum) |
| $\sigma_{em}{}^{*}$ | Equivalent membrane stress | Maximum (not plotted when maximum is applied edge value) |
| $\lambda_{\sigma x}, \lambda_{\sigma y}{}^{*}$ | Stress effectiveness | In direction of applied uniform compression |
| $\lambda_{sx}, \lambda_{sy}{}^{*}$ | Stiffness effectiveness | In direction of applied uniform compression |
| $\sigma_{xav}, \sigma_{yav}{}^{*}$ | Average edge restraining stress | Along edges restrained against in-plane movement |
| $u, v^{\dagger}$ | In-plane displacement | Along edges subject to uniform edge displacement |
| $\sigma_{mx}, \sigma_{my}{}^{\dagger}$ | Axial membrane stress | Various locations as noted |
| $M_x, M_y{}^{\dagger}$ | Bending moment | At point of maximum out-of-plane deflection |
| $M_{xy}{}^{\dagger}$ | Twisting moment | At corners and also at point of maximum out-of-plane deflection in cases of applied shear |

*See graphs
†See tables

where $k$ = critical load factor (see respective graphs)

$$\sigma_b = \pi^2 E/[12(1 - \nu^2)(b/t)^2] \tag{3.2}$$

Where applicable, suffixes refer to co-ordinate axes and locations of the reference critical stress. In cases of combined and/or variable applied load one component and/or location is defined as the reference stress.

### (2) Out-of-plane deflections

Maximum out-of-plane deflection $w$ is given by

$$W = w/t \tag{3.3}$$

Numerical suffixes refer to location on the plate. Where arrows are shown next to a location, the maximum moves in the direction of the arrows as load increases.

(3) In-plane displacements

In-plane axial displacements $u$ and $v$ in the $x$- and $y$-directions are defined, for cases of applied displacement, as a function of $u_f$ and $v_f$, the corresponding displacements which would occur if the plate were perfectly flat, thus:

$$U = u/u_f \tag{3.4}$$

$$V = v/v_f \tag{3.5}$$

$u_f$ and $v_f$ are defined on the figures illustrating typical boundary stress and in-plane displacement distributions for each load case.

In-plane shear displacement $s$ is defined, for cases of applied shear, as a function of $s_f$ the corresponding displacement which would occur if the plate were perfectly flat, thus:

$$S = s/s_f \tag{3.6}$$

(4) Plate effectiveness

For cases of compression and combined compression and shear loading, stress and stiffness (where applicable) plate effectiveness are given, as follows:
stress effectiveness:

$$\left.\begin{aligned} \lambda_{\sigma x} &= \sigma_{x\text{av}}/\sigma_{x\text{max}} \\ \lambda_{\sigma y} &= \sigma_{y\text{av}}/\sigma_{y\text{max}} \end{aligned}\right\} \tag{3.7}$$

stiffness effectiveness:

$$\left.\begin{aligned} \lambda_{sx} &= u_f/u \\ \lambda_{sy} &= v_f/v \end{aligned}\right\} \tag{3.8}$$

The above stresses and in-plane displacements are defined on the figures illustrating typical boundary stress and in-plane displacement distributions for each load case.

(5) Equivalent stress

The equivalent stress at various locations on the plate is given by the non-dimensional parameters $\bar{\sigma}_{es}$ and $\bar{\sigma}_{em}$ for surface and membrane equivalent stress respectively. These values relate to the corresponding actual equivalent stresses $\sigma_{es}$ and $\sigma_{em}$ as follows:

$$\bar{\sigma}_{es} = \sigma_{es}/\sigma_{eR} \tag{3.9}$$

$$\bar{\sigma}_{em} = \sigma_{em}/\sigma_{eR} \tag{3.10}$$

where $\sigma_{eR}$ is a reference equivalent membrane stress. It is a stress which would apply if the plate were perfectly flat and is defined on the figures illustrating typical boundary stress and in-plane displacement distributions for each load case.

Numerical suffixes refer to location on the plate.

(6) Axial membrane stress

Axial membrane stress $\sigma_{mx}$ and $\sigma_{my}$ in the $x$- and $y$-directions are given in the tables only. They are defined in non-dimensional form as follows:

$$\left.\begin{aligned}\bar{\sigma}_{mx} &= \sigma_{mx}/\sigma_{Rx}\\ \bar{\sigma}_{my} &= \sigma_{my}/\sigma_{Ry}\end{aligned}\right\} \tag{3.11}$$

where $\sigma_{Rx}$ and $\sigma_{Ry}$ are reference axial stresses defined on the figures illustrating typical boundary stress and in-plane displacement distributions for each load case. Numerical suffixes refer to location on the plate.

Shear membrane stress $\tau_m$ is given in the tables only. It is defined in non-dimensional form as follows:

$$\bar{\tau}_m = \tau_m/\tau_R \tag{3.12}$$

Note that in cases of applied shear only, the non-dimensional divisors $\sigma_{Rx}$ and $\sigma_{Ry}$ equal $\tau_{av}$.

(7) Moments

Axial bending moments $M_x$ and $M_y$ in the $x$- and $y$-directions, and twisting moment $M_{xy}$, are given in the tables only. They are defined in non-dimensional form as follows:

$$\left.\begin{aligned}\bar{M}_x &= M_x(b^2/Et^4)\\ \bar{M}_y &= M_y(b^2/Et^4)\\ \bar{M}_{xy} &= M_{xy}(b^2/Et^4)\end{aligned}\right\} \tag{3.13}$$

## 3.2 Design procedure

The usual procedure is to check or appraise a chosen plate configuration for applied forces which have been determined from a global analysis incorporating stiffness properties based on assumed plate thicknesses. The plate appraisal involves the calculation of deflection and stresses, which must conform to design criteria such as those outlined in Section 3.3.

In cases involving uniform compression the plate effectiveness must be checked. If the value assumed in the global analysis is not sufficiently accurate or if effectiveness was ignored in the initial analysis it will be necessary to repeat the global analysis in order to correctly define global deflections.

In cases of restrained plate sides, average side forces can be calculated. These may be required in a check of side-framing members.

For a specified plate geometry, material properties, loading and welding process the plate appraisal procedure is as follows:

Calculate:

(1) Residual stresses
(2) Initial imperfection, $w_0/t$ (which can incorporate the effect of residual stresses)
(3) Loading, $P/P_{cr}$ (which can incorporate residual stresses if not already included in the initial imperfection)

Enter the design curves for the computed values of $P/P_{cr}$ and $w_0/t$ to determine deflection and stresses, which are reduced to dimensional form as follows:

$$w = W \times t \qquad \text{mm (in)} \qquad (3.14)$$

$$\sigma_{es} = \bar{\sigma}_{es} \times \sigma_{eR} \qquad \text{N/mm}^2 \text{ (psi)} \qquad (3.15)$$

$$\sigma_{em} = \bar{\sigma}_{em} \times \sigma_{eR} \qquad \text{N/mm}^2 \text{ (psi)} \qquad (3.16)$$

In cases involving uniform compression the values of stress and/or stiffness effectiveness are read directly from the curves and in cases of restrained edges the average edge restraining stress is given by

$$\sigma_{av} = \bar{\sigma}_{av} \times \sigma_R \qquad \text{N/mm}^2 \text{ (psi)} \qquad (3.17)$$

Values of $\sigma_{eR}$ and $\sigma_R$ are defined on the figures illustrating typical boundary stress and in-plane displacement for each load case.

## 3.3 Design criteria

The global analysis of the structure is carried out for loads which are factored on a statistical basis to ensure adequate improbability of occurrence during the prescribed design life, and the performance of components is then checked to ensure that they do not contravene specified design criteria when subject to forces derived in the global analysis, together with local loads if they occur.

The load factors and load combinations for various structures are specified in appropriate Codes of Practice. Some Codes also specify factors to be applied to material properties in conjunction with particular loading conditions. Discussion of these aspects is outside the scope of this book. The reader is referred to the Merrison Rules for load factors applicable to steel box-girder bridges.

The performance of components, in particular plates, can be described in terms of the maintenance of serviceability, and the provision of strength, as follows:

(1) Serviceability criteria

*(a) Control of elastic out-of-plane deflection*
In the case of plates subject to in-plane load the designer is primarily interested in the stresses arising as a result of out-of-plane deflections rather than the deflection per se. However, there may be cases where the actual magnitude of deflection under working conditions is of interest and in such cases it is usually specified as a function of the plate thickness and/or span depending on the particular application. An example is the limitation of web panel deflections in certain box girder applications because of aesthetic considerations. There may also be applications where considerations in regard to cracking of surface protective coating imposes a deflection limitation.

*(b) Control of initial yielding*
If it is desired to avoid permanent set in a plate panel, as a limit of serviceability, then material yielding must not take place anywhere on the plate under working conditions. Yielding will usually initiate on the surface of the plate and for a ductile material the equivalent stress defined in Section 2.8 is the recommended yield criterion. The design curves in Chapter 4 give values of surface equivalent stress at various locations. Since the location of the maximum moves as the load varies, each location value must be checked and compared to establish the maximum, which will correspond to the serviceability limit when

$$\sigma_{es} = \sigma_y \tag{3.18}$$

In many applications yield initiates in the corners of the plate and is very localised, and in such cases it may be unrealistic as well as uneconomic to specify the onset of yield as the serviceability limit. As discussed in Section 2.9 a limited spread of yield or 'overstress' is often acceptable, and in order to provide a guide to the designer in such cases the curves in Chapter 4 include values of equivalent surface stress at the node adjacent to the corner of the plate.

(2) Strength criteria (collapse)

*(a) Control of membrane yield*
The attainment of membrane yield or yield through the thickness of the plate generally corresponds quite closely to collapse load for a plate subject to in-plane load. Since elasto-plastic solutions are not yet generally available in concise design form, some approximate method must be used to establish the load at which membrane yield occurs.

Membrane yield stress derived from a purely elastic analysis is only approximate, since it does not take account of surface yielding, which results in the shedding of subsequent load to lower stressed regions of the plate, giving rise to a more uniform stress distribution. In the absence of more rigorous data it may therefore be assumed that collapse load based on membrane yield derived from an elastic analysis is conservative. The design curves in Chapter 4 include values of maximum equivalent membrane stress, and a safe approximation to the collapse load is reached when

$$\sigma_{em} = \sigma_y \tag{3.19}$$

The Merrison Rules for box-girder bridges provide empirical expressions for estimating the collapse load which take account of surface yielding, as follows:

(i) Load cases not involving shear:

For restrained panels and unrestrained panels in which $\sigma_{cr} > \sigma_y/2$, the applied stress to cause collapse is the greater of:

$$\sigma_{ec} = \sigma_y(1 - \sigma_y^2/(8\sigma_{esy}\sigma_{cr})) \tag{3.20}$$

$$\sigma_{ec} = \sigma_{esy} \tag{3.21}$$

where $\sigma_{esy}$ is the load to cause $\sigma_{es} = \sigma_y$ and corresponds to the serviceability limit.

For unrestrained panels in which $\sigma_{cr} \leqslant \sigma_y/2$, eq. (3.21) applies.

(ii) Load cases involving shear:

If $\sigma_{cr} > \sigma_y/2$, the applied stress to cause collapse is the greater of

$$\sigma_{ec} = \sigma_y(1 - \sigma_y^2/(8\sigma_{esy}\sigma_{cr})) \tag{3.22}$$

$$\sigma_{ec} = (\alpha^2\sigma_y^2 - 3\alpha^2(\tau_y')^2 + 3(\tau_y')^2(1 - \sigma_y^2/(8\sigma_{esy}\sigma_{cr}))^{1/2} \tag{3.23}$$

If $\sigma_{cr} \leqslant \sigma_y/2$ the applied stress to cause collapse is:

$$\sigma_{ec} = (\alpha^2\sigma_y^2 - 3\alpha^2(\tau_y')^2 + 3(\tau_y')^2(2 - 1/2\alpha)^2)^{1/2} \tag{3.24}$$

where $\alpha = \sigma_{esy}/\sigma_y$
$\tau_y' = \tau_{av}(\sigma_y/\sigma_{esy})$
$\tau_{av}$ = average applied shear stress

*(b) Control of fatigue*

The equivalent surface stresses plotted in the design curves in Chapter 4 give a basis for computing fatigue on the basis outlined in Section 2.13. If a uniaxial criteria is preferred the values given in the tables in Appendix 1 can be used.

CHAPTER FOUR

# Design Data

Design data are presented in this chapter in graphical form. Supplementary data are presented in Appendix 1 in tabular form. The data are grouped in sections, each relating to a particular type of loading, and each section includes a figure or figures illustrating the boundary stresses and in-plane displacements together with details of the form of initial imperfection used in obtaining the solution and some introductory notes. The form of imperfection is in all cases the critical buckling mode.

The non-dimensional factors, required to reduce the data given in the graphs to dimensional form, are given on the figures illustrating each load case. The non-dimensional relationships are given in Section 3.1 and the procedure for using the graphs to evaluate design criteria is described in Chapter 3. The figure illustrating each load case also includes details of the side ratios for which solutions are given, together with the finite-difference mesh divisions used in each case. The plate geometry, showing finite-difference mesh divisions, is illustrated in Fig. 4.1.

As discussed in Section 3.3(1(*b*)), it is safe and more economic to allow for localised yield in the corner regions under serviceability conditions in certain applications. To facilitate this procedure results are given for equivalent surface stress at the corner and at the diagonal node adjacent to the corner.

Obviously design judgement is required in deciding on an acceptable extent of yield, and this procedure should be used with caution. It has particular application where the stress gradient is very steep, but if in doubt the designer should not allow the corner stress to exceed yield. It should be noted of course that the maximum stress does not necessarily occur at the corners and results are given for other locations on the plate where this is the case, usually in the lower loading range and at lower values of imperfection.

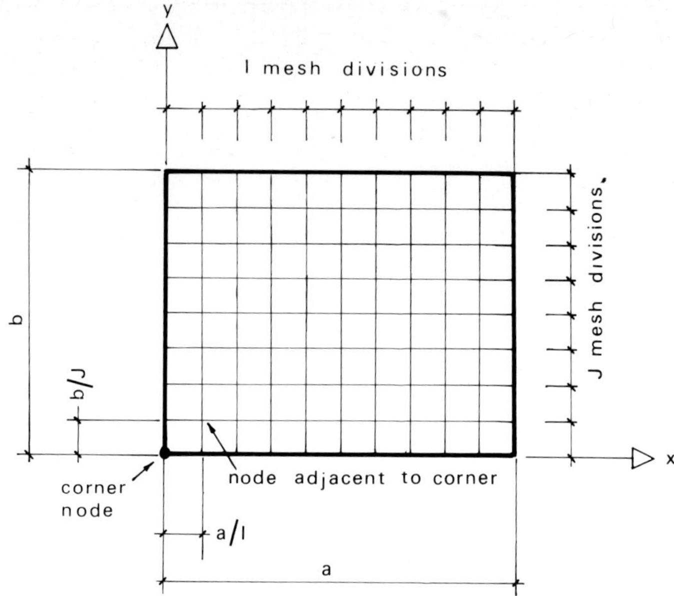


Fig. 4.1 Plate geometry showing finite-difference mesh

The graph or graphs of design data for each case incorporate a diagrammatic representation of the loading, showing the locations to which the suffixes given on the variables refer. The value of the critical buckling load is also given on the graphs in terms of a critical load factor ($k$) and a reference stress ($\sigma_b$) which is a function of the plate geometry and modulus of elasticity (see Section 3.1(1)). For all cases except combined bearing and bending, values of the critical load factor are taken from the literature. In the case of combined bearing and bending, factors were derived using the finite-difference program as described in Section 1.7.2.

## 4.1 Uniform compression

### 4.1.1 Uniform stress

Solutions are given for uniform uniaxial and biaxial compressive stress applied to square plates ($a/b = 1$). This condition arises when load is applied to the side of the plate via a flexible medium, such that the side is free to deform in-plane and 'follow' the buckle.

Uniform stress is not nearly as common a loading condition as uniform displacement, but it can occur in practice and is a more severe condition in terms of deflection and stress at comparable load levels. A soft bearing under a diaphragm is an example of a situation giving rise to this type of loading condition, although the other edges will usually be subject to a condition closer to uniform displacement, in which case the uniform stress results will be conservative. Stiffness effectiveness is not plotted for these cases since they will not generally arise in situations where loss of plate stiffness is a design consideration.

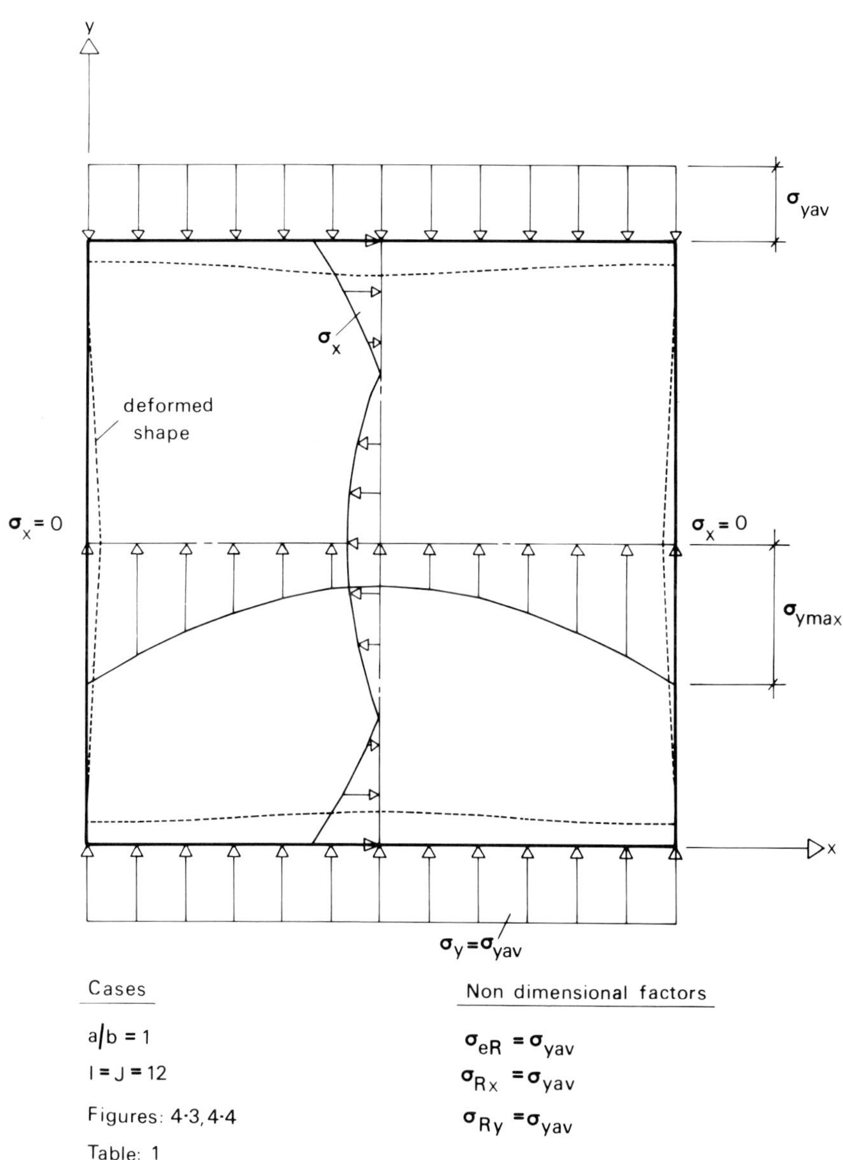


Fig. 4.2 Uniform uniaxial compressive stress; unloaded sides stress free

(a) Unaxial

Typical boundary stresses and in-plane displacements: Fig. 4.2
Initial imperfections: $w_i = w_0 \sin(\pi x/a) \sin(\pi y/b)$
Design data: Figs. 4.3, 4.4
Reference equivalent stress: $\sigma_{eR} = \sigma_{yav}$

As the load is applied a self-equilibrating axial-stress distribution develops in

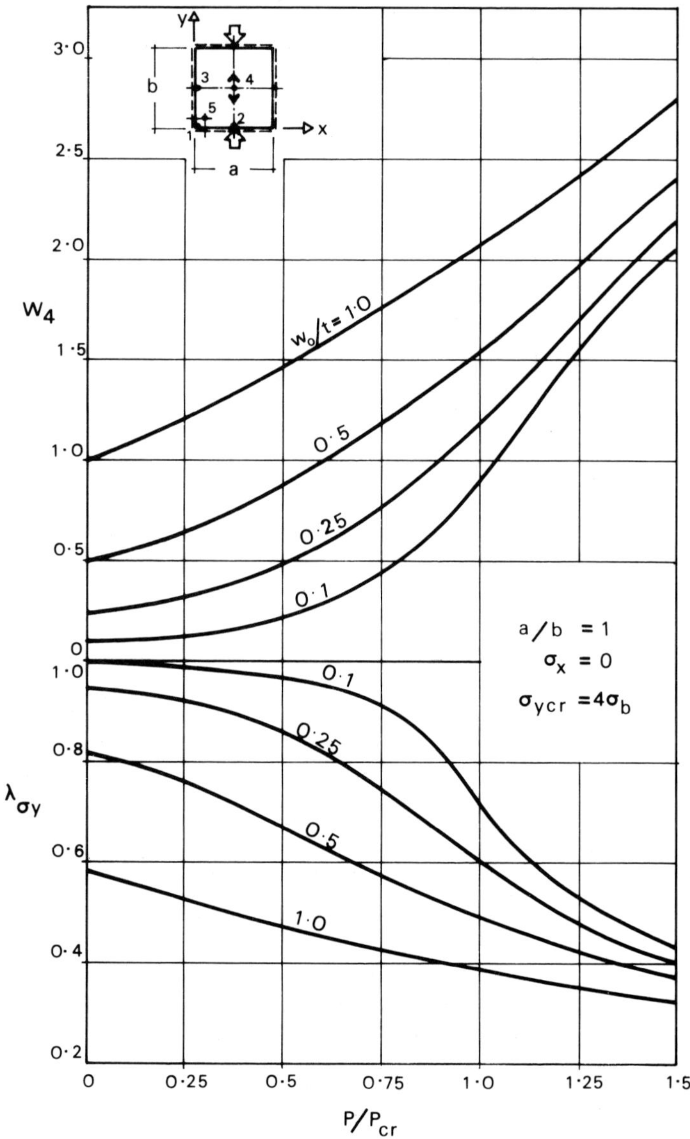


Fig. 4.3 Square ($a/b$ = 1) plate subjected to uniform uniaxial compressive stress; unloaded sides stress free. Deflection and effectiveness

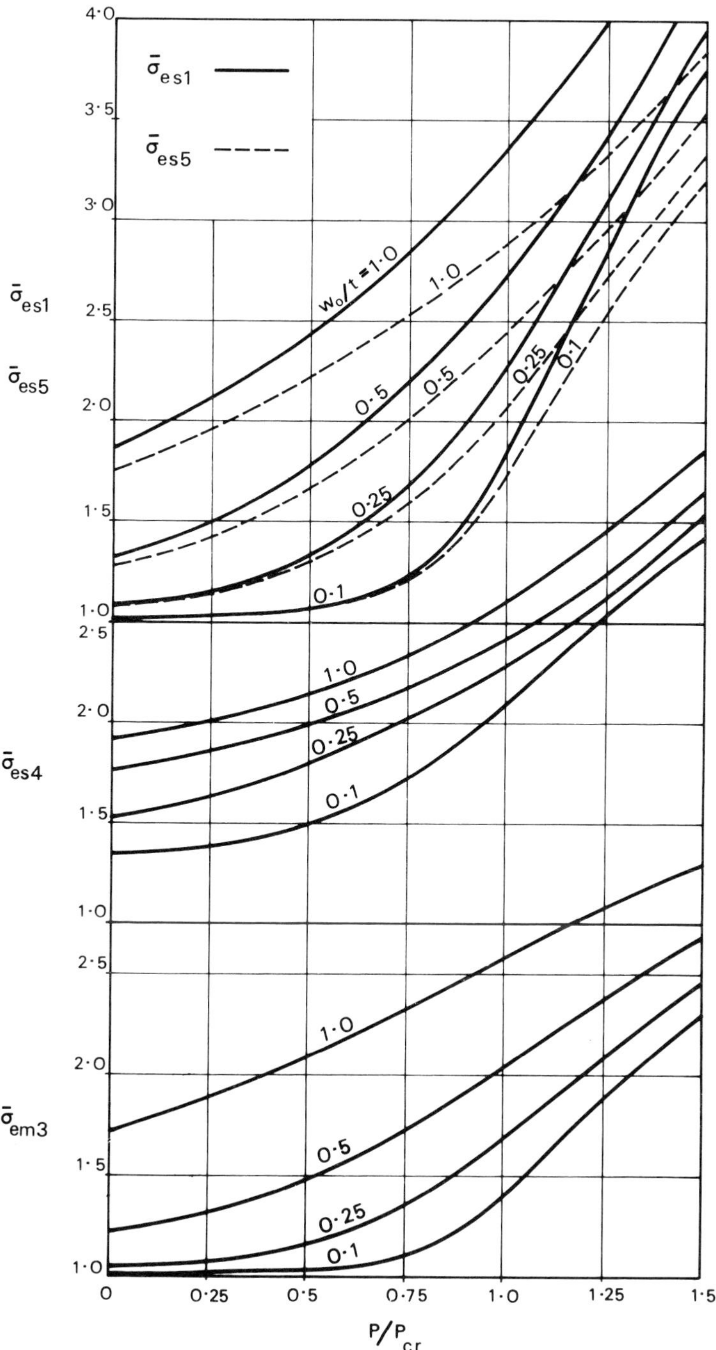


Fig. 4.4 Square (*a*/*b* = 1) plate subjected to uniform uniaxial compressive stress; unloaded sides stress free. Equivalent membrane and surface stresses

the interior of the plate in the unloaded direction. Due to shedding of load to the stiffer side regions in the interior of the plate as the buckle develops, the maximum axial and equivalent membrane stress occurs at mid-side (location 3) on the unloaded edges. Maximum equivalent surface stress initiates in the centre of the plate for all cases plotted and moves to the corners as load increases. The transition to the corners is more rapid for higher values of $w_0/t$.

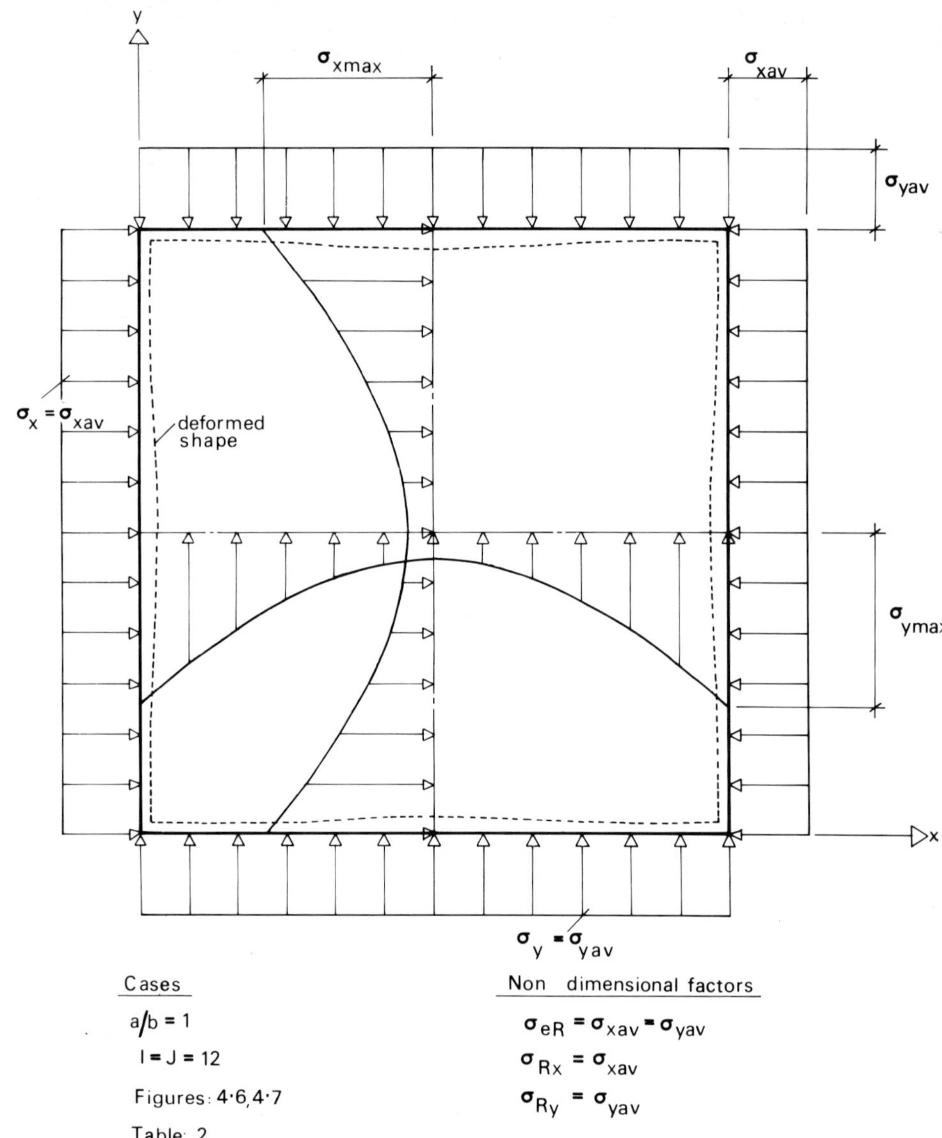


Fig. 4.5 Uniform equal biaxial compressive stress ($\sigma_x = \sigma_y$)

(b) Biaxial

Typical boundary stresses and in-plane displacements: Fig. 4.5
Initial imperfections: $w_i = w_0 \sin(\pi x/a) \sin(\pi y/b)$
Design data: Figs. 4.6, 4.7
Reference equivalent stress: $\sigma_{eR} = \sigma_{xav} = \sigma_{yav}$

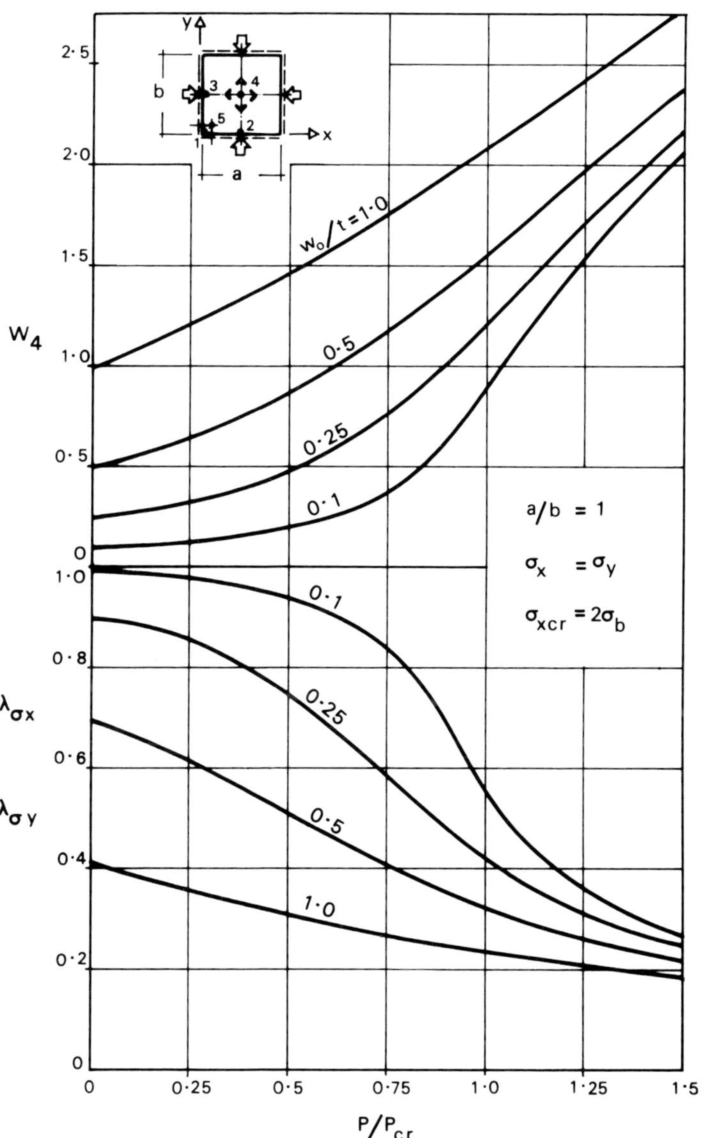


Fig. 4.6 Square ($a/b = 1$) plate subjected to uniform equal biaxial compressive stress ($\sigma_x = \sigma_y$). Deflection and effectiveness

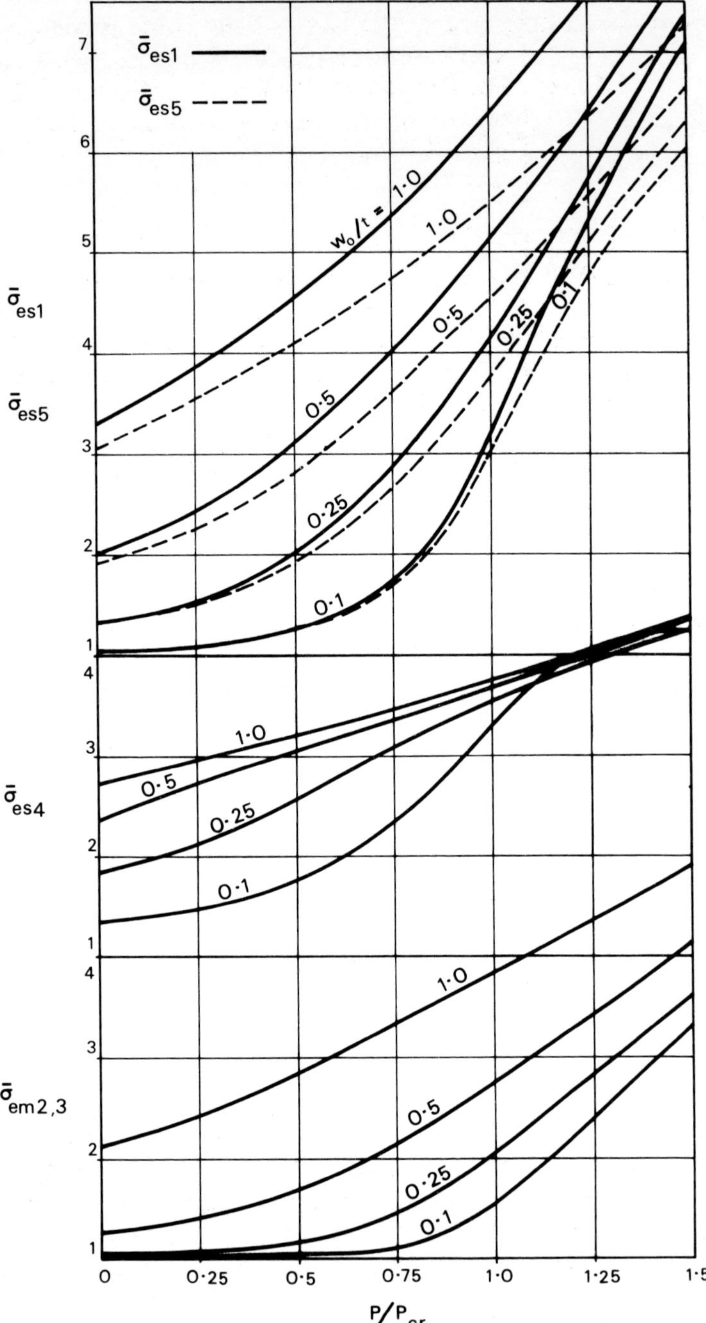


Fig. 4.7 Square ($a/b = 1$) plate subjected to uniform equal biaxial compressive stress ($\sigma_x = \sigma_y$). Equivalent membrane and surface stresses

Maximum axial and equivalent membrane stress occurs at mid-sides (locations 2 and 3) and is more severe than in the case of uniaxial stress at equal proportions of the respective critical loads. Also, surface stresses are higher and the transition of the maximum from centre plate to the corners as load increases is more rapid in the case of biaxial stress.

### 4.1.2 Uniform displacement

Solutions are given for uniaxial and biaxial uniform compressive displacement applied to plates with side ratios of $a/b = 1, 2$ and 3. In the uniaxial cases results are given for unloaded sides stress-free and for unloaded sides straight, where the load is applied along the longer side for $a/b = 2$ and 3. For rectangular plates ($a/b \neq 1$) where load is applied to the short side the designer should use the results for $a/b = 1$. The biaxial results are for average stresses in the ratio of 1 and 2, where in the latter case the higher load is in the direction of the longer axis for $a/b = 2$ and 3.

The most common case of uniform displacement occurs in interior panels in an array of panels where compatibility between adjacent sides dictates that the sides remain straight. The boundary condition along the discontinuous side of an edge panel depends on the stiffness of the side framing, which will usually be flexible enough to give rise to the stress-free condition. There will be cases of directly applied load where the medium through which the load is applied is sufficiently stiff to maintain the straight-edge condition.

#### (a) Uniaxial

Typical boundary stresses and in-plane displacements:
(i) Unloaded sides stress-free: Fig. 4.8
(ii) Unloaded sides straight: Fig. 4.15
Initial imperfections: $w_i = w_0 \sin(\pi x/a) \sin(\pi y/b)$
Design data:
(i) Unloaded sides stress-free: Figs. 4.9–14
(ii) Unloaded sides straight: Figs. 4.16–21
Reference equivalent stress: $\sigma_{eR} = \sigma_{yav}$

In the case of stress-free unloaded sides, the free sides 'follow' the buckle and a self-equilibrating axial stress distribution develops in the unloaded direction in the interior of the plate. Where unloaded sides are constrained to remain straight but are free to move this self-equilibrating stress distribution extends to the sides, and in cases of edge panels becomes a loading on the side framing. Panels with stress-free unloaded sides are more flexible than panels with straight unloaded sides resulting in higher stresses at comparable loads.

Maximum axial and equivalent membrane stress occurs midway along the unloaded sides (location 3) in the case of stress-free unloaded sides. In the case of straight unloaded sides maximum axial membrane stress occurs at the corner (location 1), but the tensile stresses developed in restraining the buckle give rise to maximum equivalent stress midway along the unloaded sides (location 3).

For square plates maximum equivalent surface stress initiates at centre panel (location 4) for low values of $w_0/t$, but as load increases, and from the onset of loading for higher values of $w_0/t$, it initiates at or near the corners (location 1).

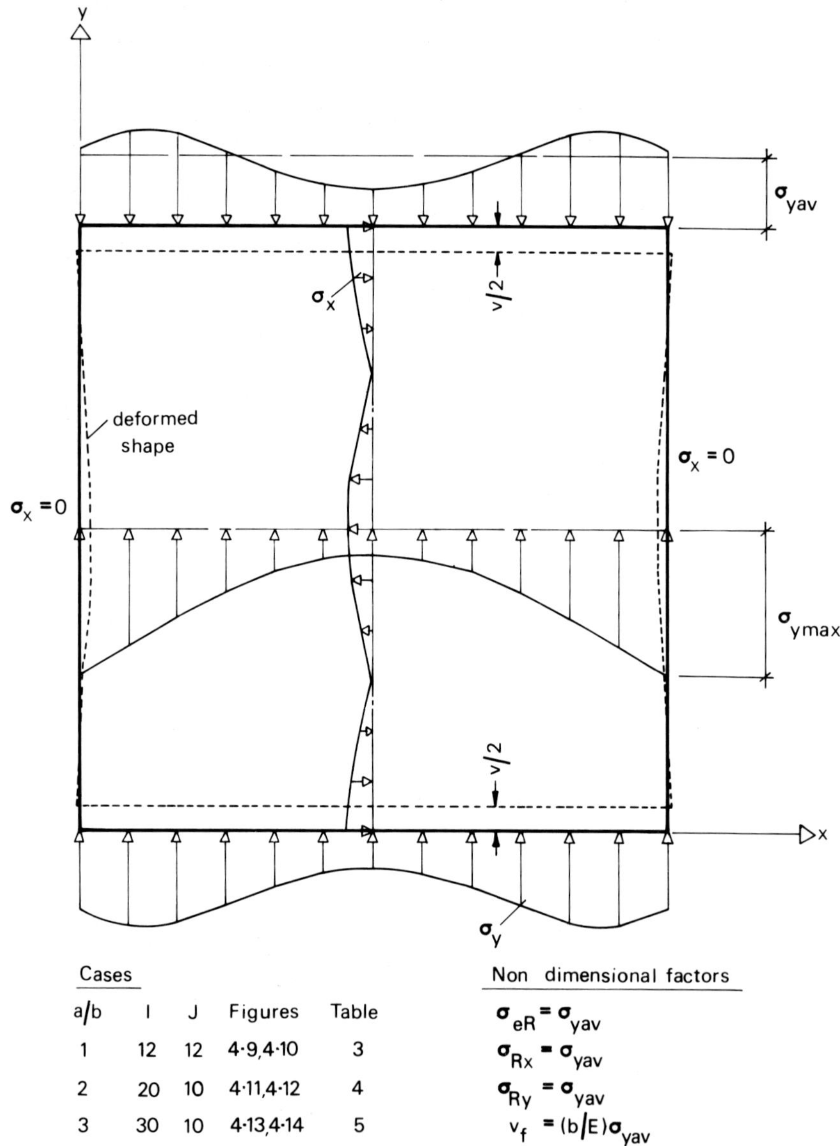


Cases

| a/b | I | J | Figures | Table |
|---|---|---|---|---|
| 1 | 12 | 12 | 4·9,4·10 | 3 |
| 2 | 20 | 10 | 4·11,4·12 | 4 |
| 3 | 30 | 10 | 4·13,4·14 | 5 |

Non dimensional factors

$\sigma_{eR} = \sigma_{yav}$

$\sigma_{Rx} = \sigma_{yav}$

$\sigma_{Ry} = \sigma_{yav}$

$v_f = (b/E)\sigma_{yav}$

Fig. 4.8 Uniform uniaxial compressive displacement; unloaded sides stress-free

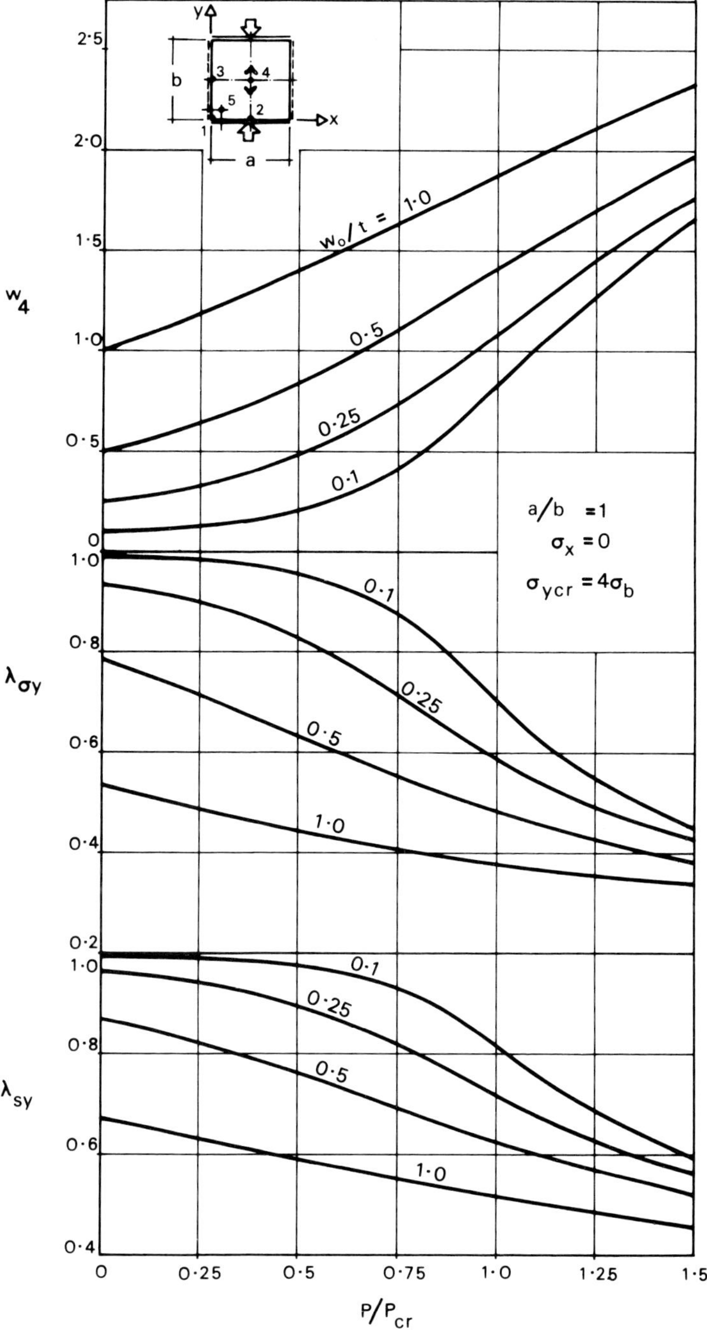


Fig. 4.9 Square ($a/b = 1$) plate subjected to uniform uniaxial compressive displacement; unloaded sides stress-free. Deflection and effectiveness

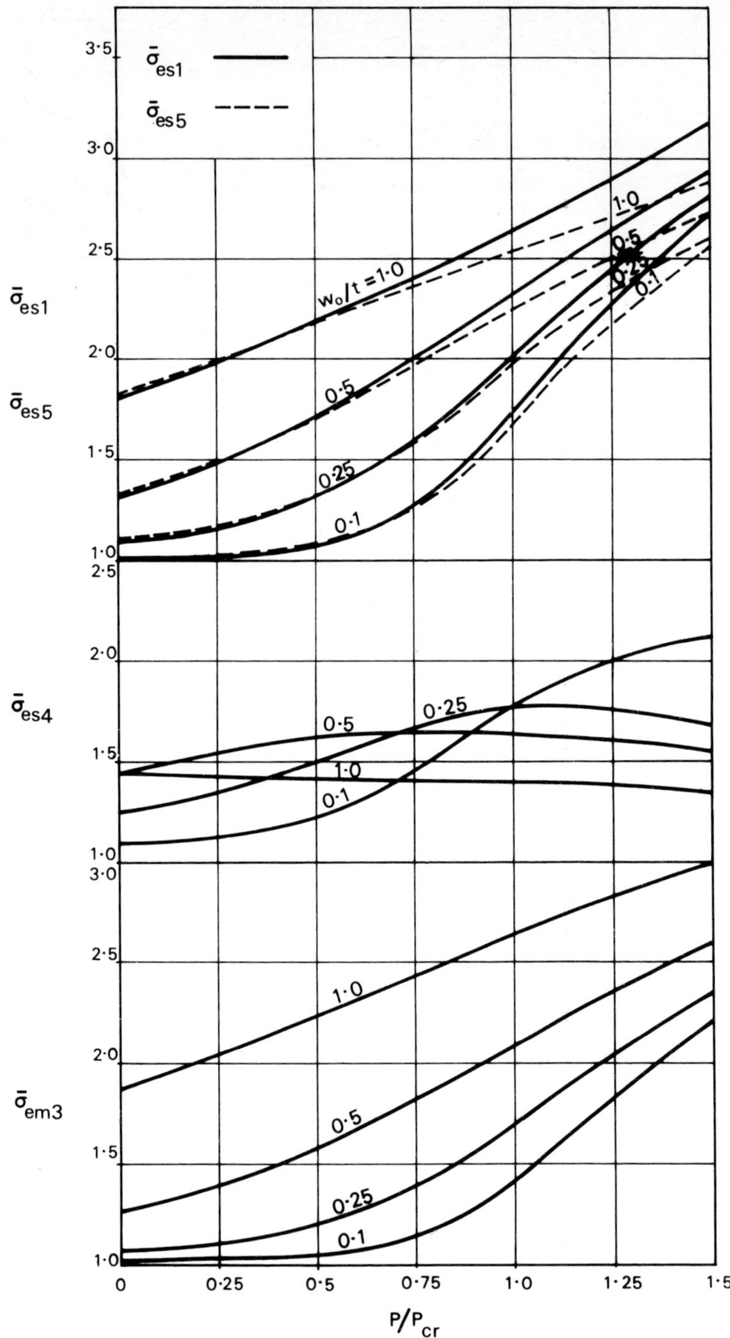


Fig. 4.10 Square ($a/b = 1$) plate subjected to uniform uniaxial compressive displacement; unloaded sides stress-free. Equivalent membrane and surface stresses

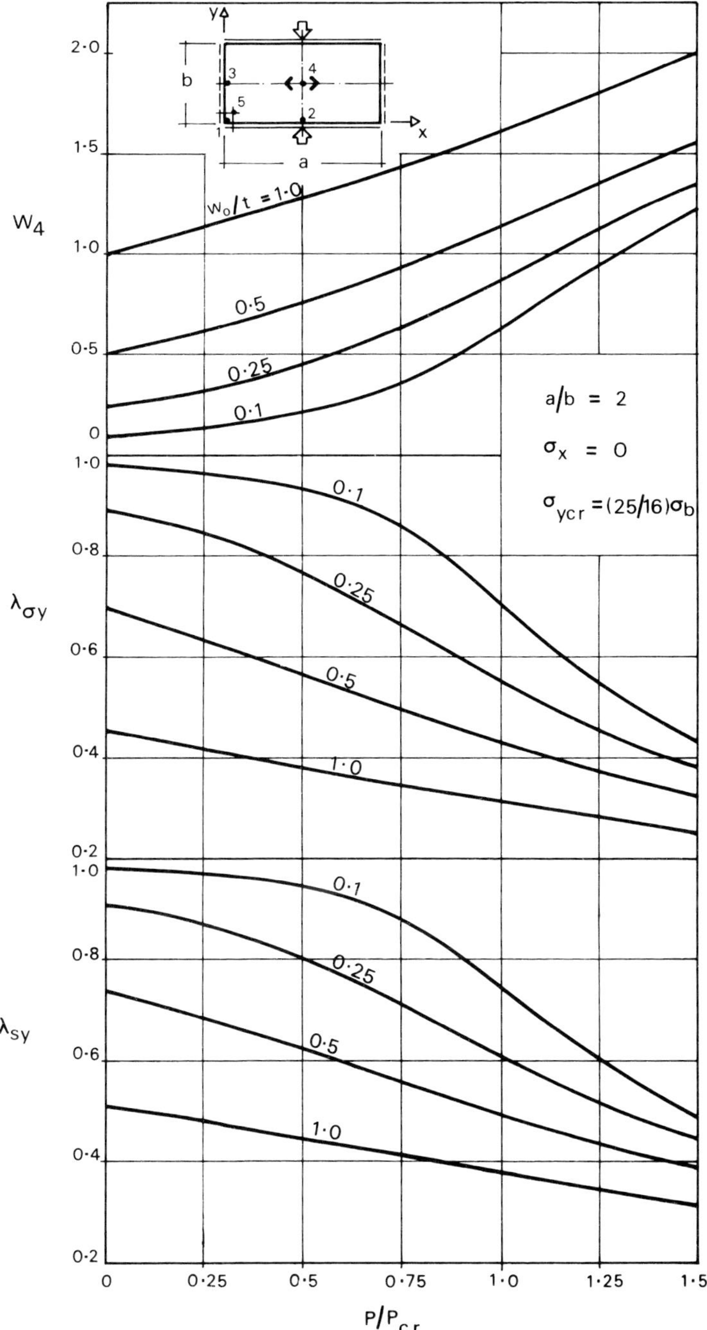


Fig. 4.11 Rectangular ($a/b = 2$) plate subjected to uniform uniaxial (short axis) compressive displacement; unloaded sides-stress free. Deflection and effectiveness

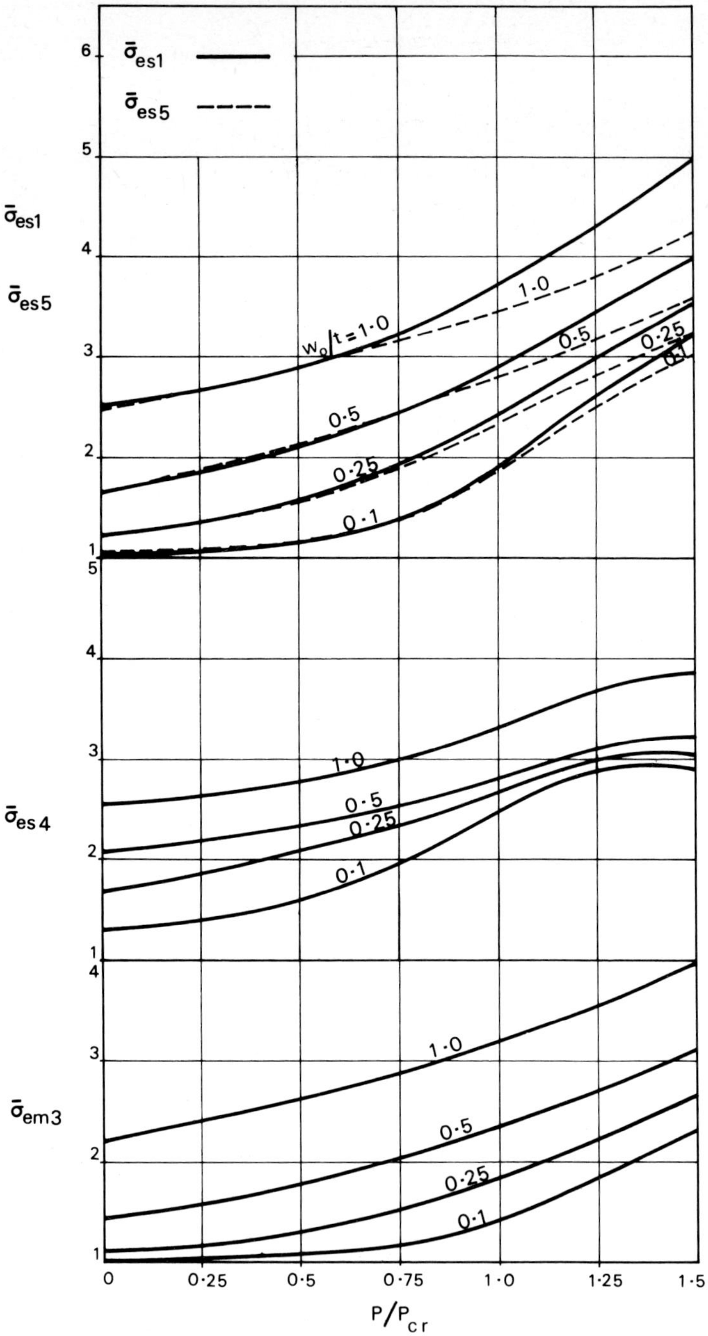


Fig. 4.12 Rectangular ($a/b = 2$) plate subjected to uniform uniaxial (short axis) compressive displacement; unloaded sides stress-free. Equivalent membrane and surface stresses

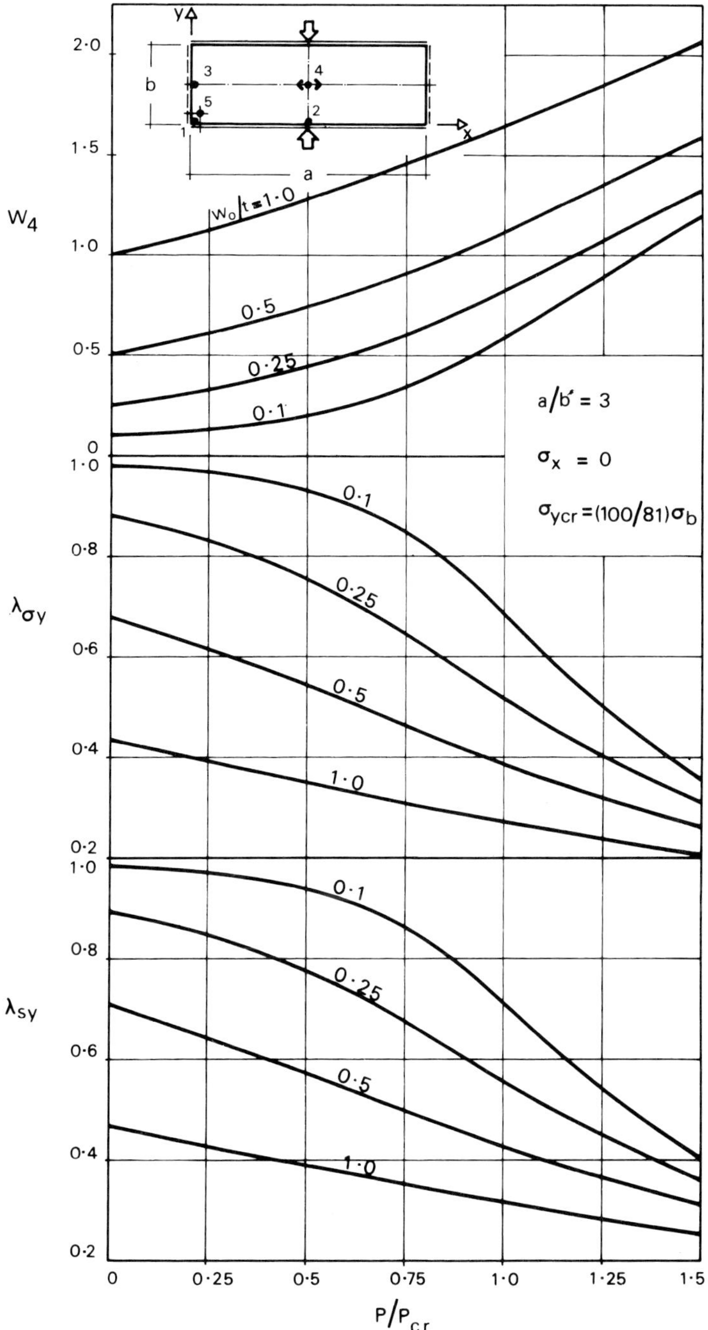


Fig. 4.13 Rectangular ($a/b = 3$) plate subjected to uniform uniaxial (short axis) compressive displacement; unloaded sides stress-free. Deflection and effectiveness

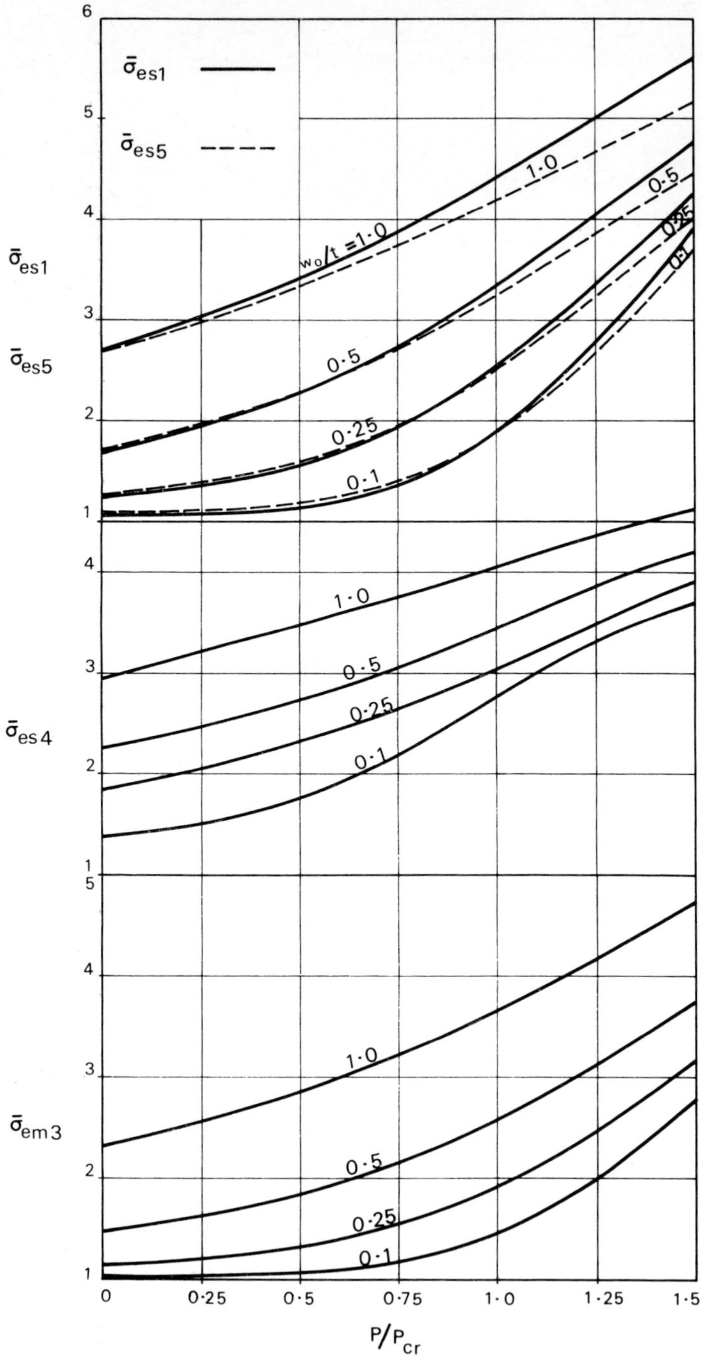


Fig. 4.14 Rectangular ($a/b = 3$) plate subjected to uniform uniaxial (short axis) compressive displacement; unloaded sides stress-free. Equivalent membrane and surface stresses

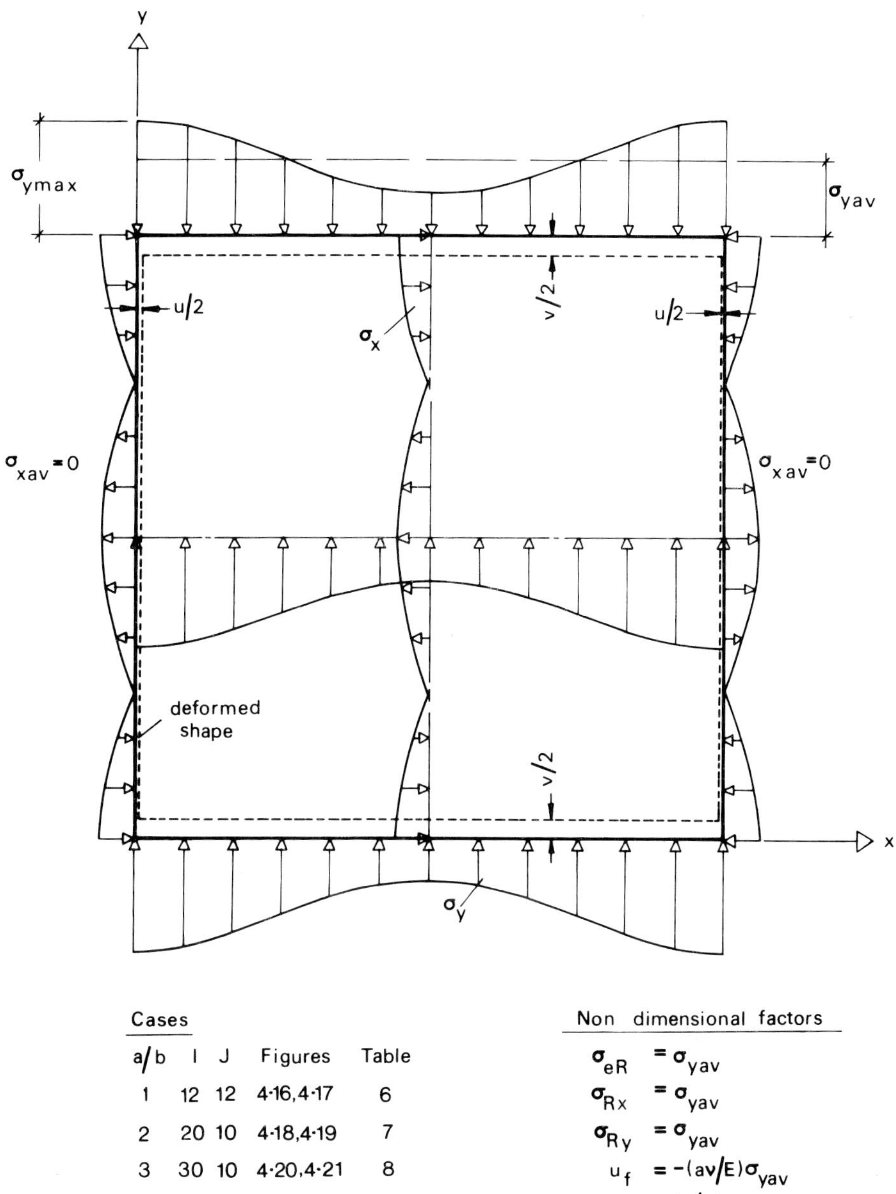


Cases

| a/b | I | J | Figures | Table |
|---|---|---|---|---|
| 1 | 12 | 12 | 4·16, 4·17 | 6 |
| 2 | 20 | 10 | 4·18, 4·19 | 7 |
| 3 | 30 | 10 | 4·20, 4·21 | 8 |

Non dimensional factors

$\sigma_{eR} = \sigma_{yav}$

$\sigma_{Rx} = \sigma_{yav}$

$\sigma_{Ry} = \sigma_{yav}$

$u_f = -(av/E)\sigma_{yav}$

$v_f = (b/E)\sigma_{yav}$

Fig. 4.15 Uniform uniaxial compressive displacement; unloaded sides straight

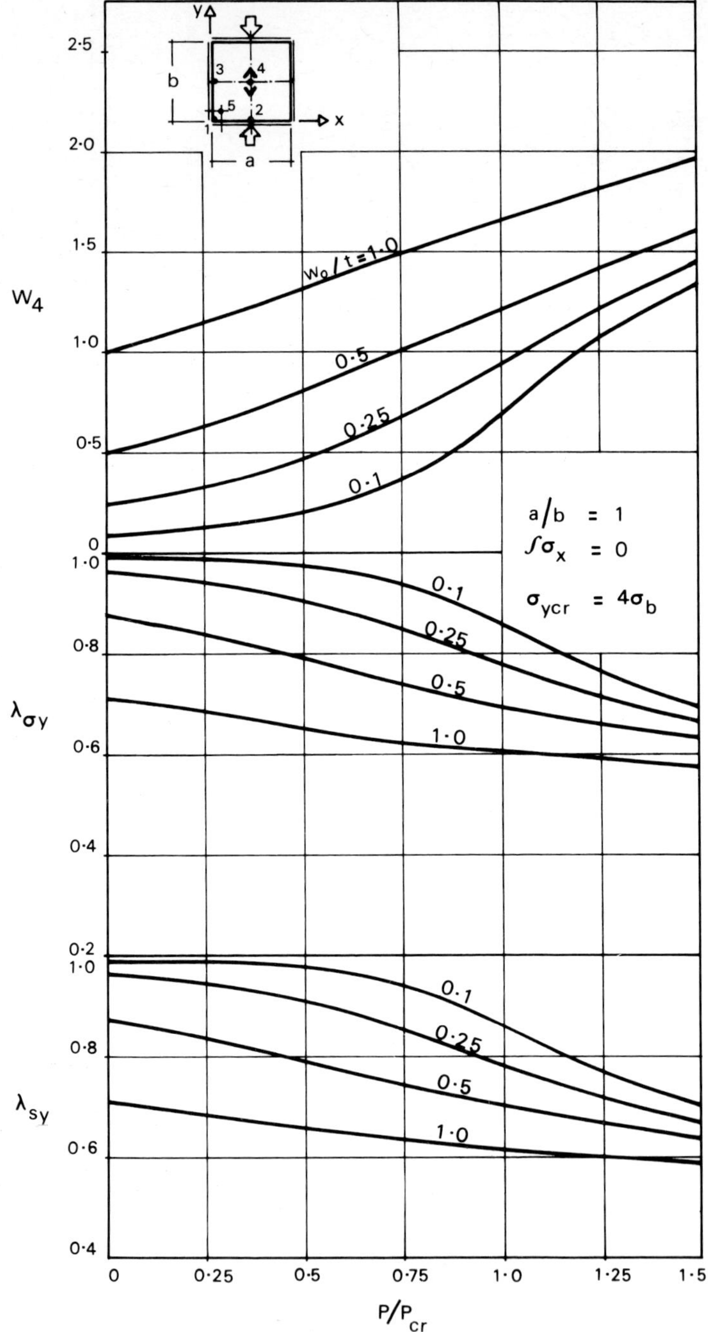


Fig. 4.16 Square ($a/b = 1$) plate subjected to uniform uniaxial compressive displacement; unloaded sides straight. Deflection and effectiveness

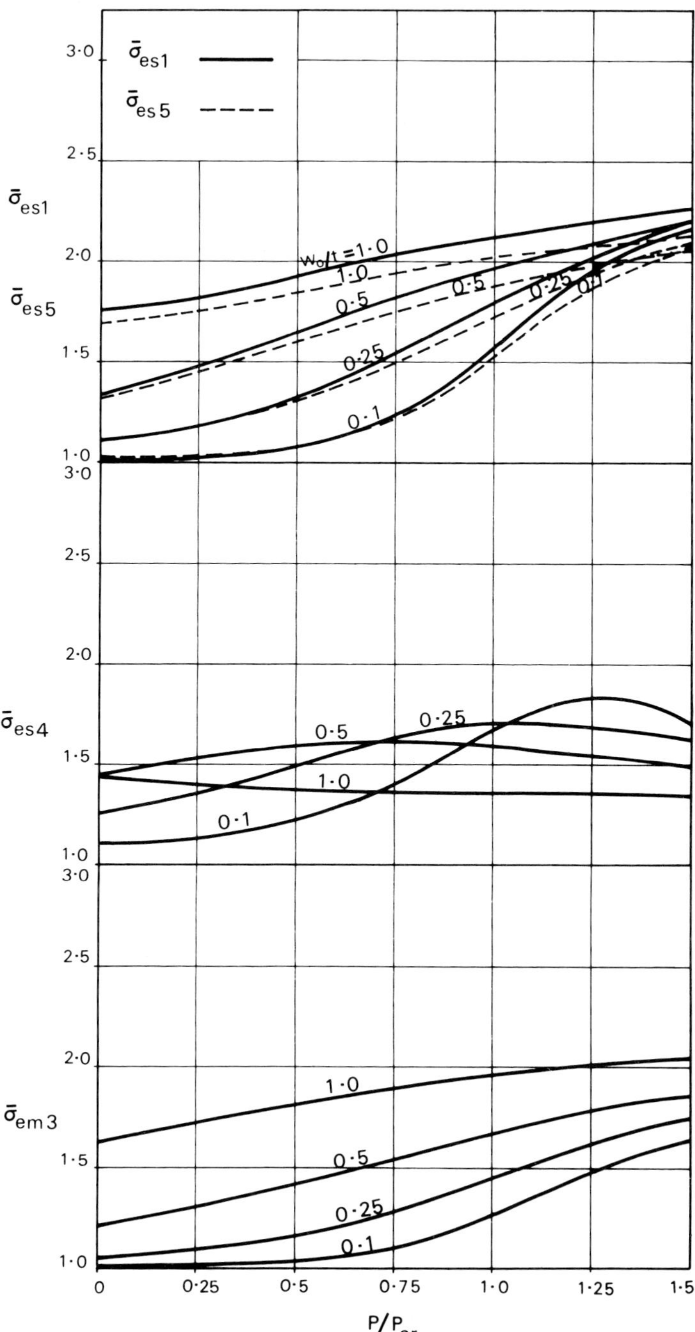


Fig. 4.17 Square (*a*/*b* = 1) plate subjected to uniform uniaxial compressive displacement; unloaded sides straight. Equivalent membrane and surface stresses

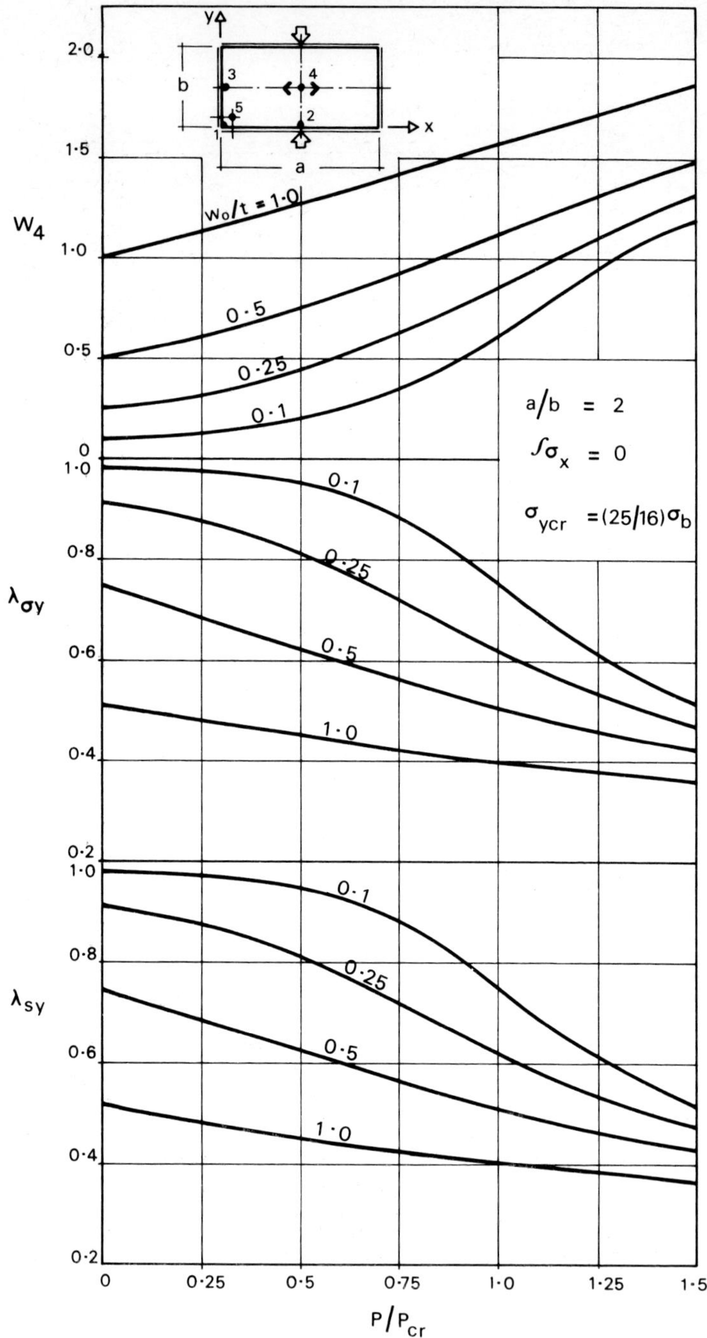


Fig. 4.18 Rectangular ($a/b = 2$) plate subjected to uniform uniaxial (short axis) compressive displacement; unloaded sides straight. Deflection and effectiveness

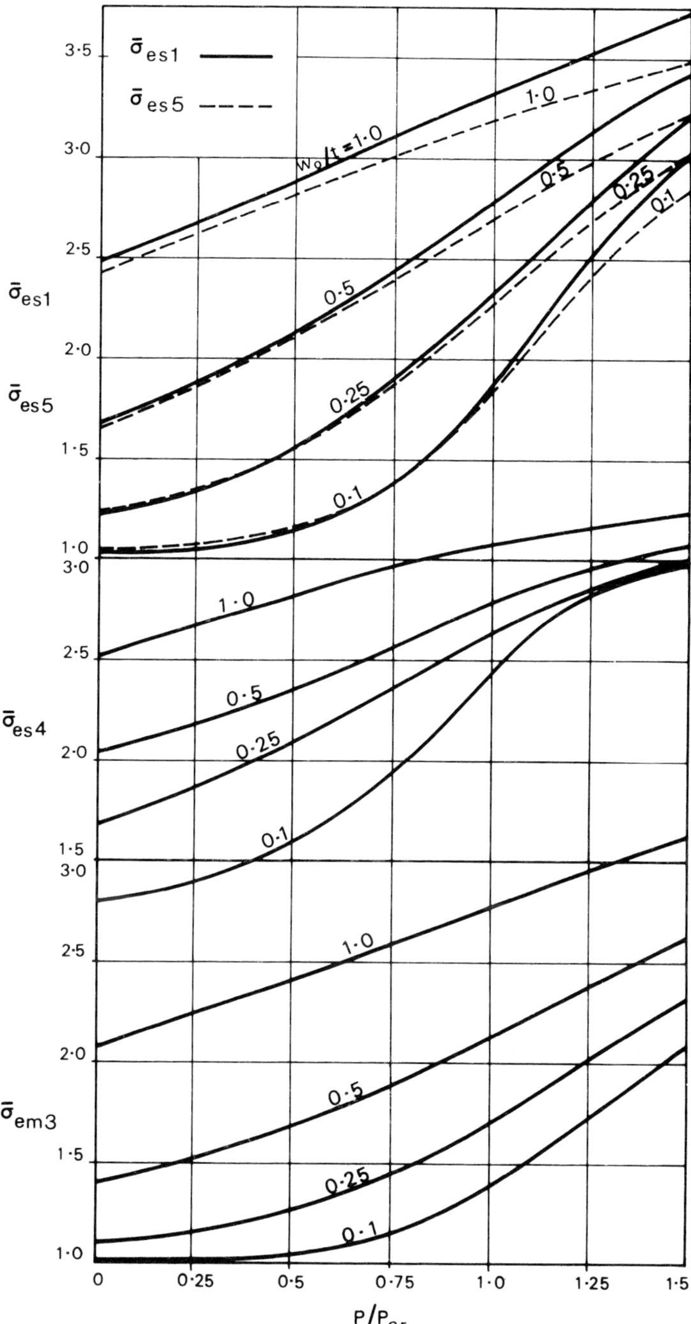


Fig. 4.19 Rectangular ($a/b = 2$) plate subjected to uniform uniaxial (short axis) compressive displacement; unloaded sides straight. Equivalent membrane and surface stresses

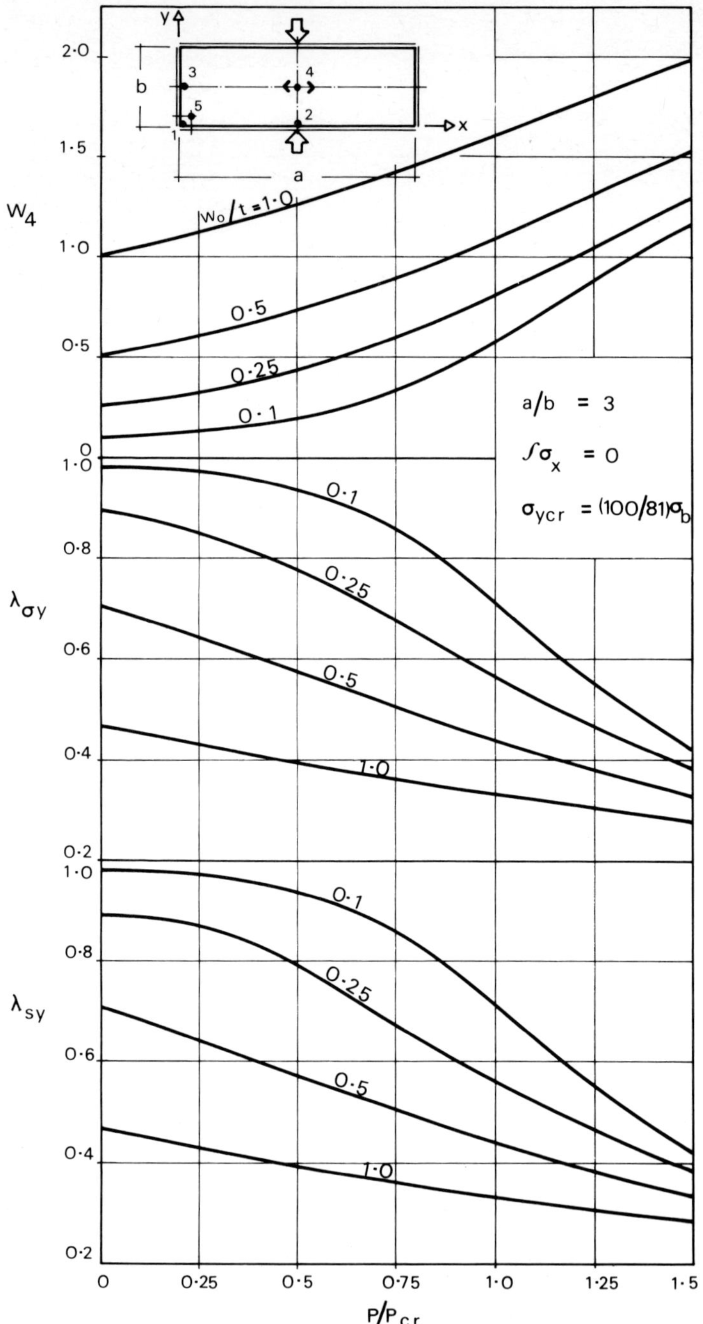


Fig. 4.20 Rectangular ($a/b = 3$) plate subjected to uniform uniaxial (short axis) compressive displacement; unloaded sides straight. Deflection and effectiveness

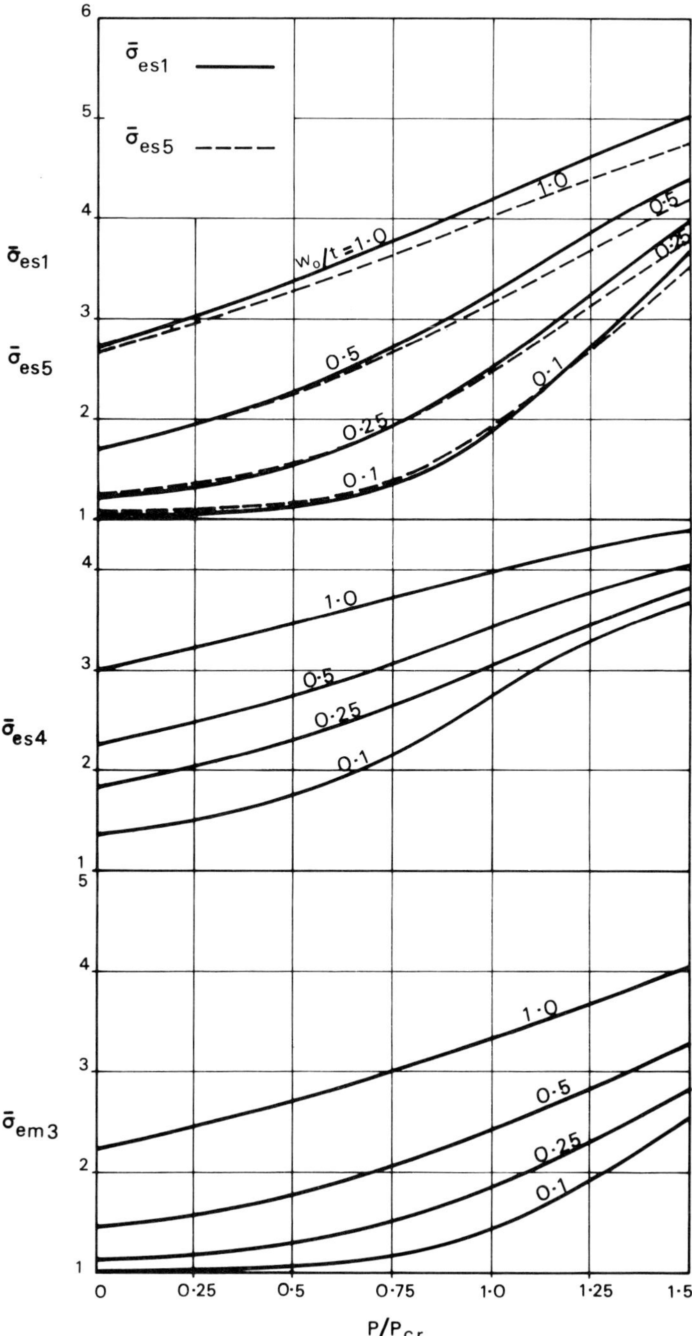


Fig. 4.21 Rectangular ($a/b = 3$) plate subjected to uniform uniaxial (short axis) compressive displacement; unloaded sides straight. Equivalent membrane and surface stresses

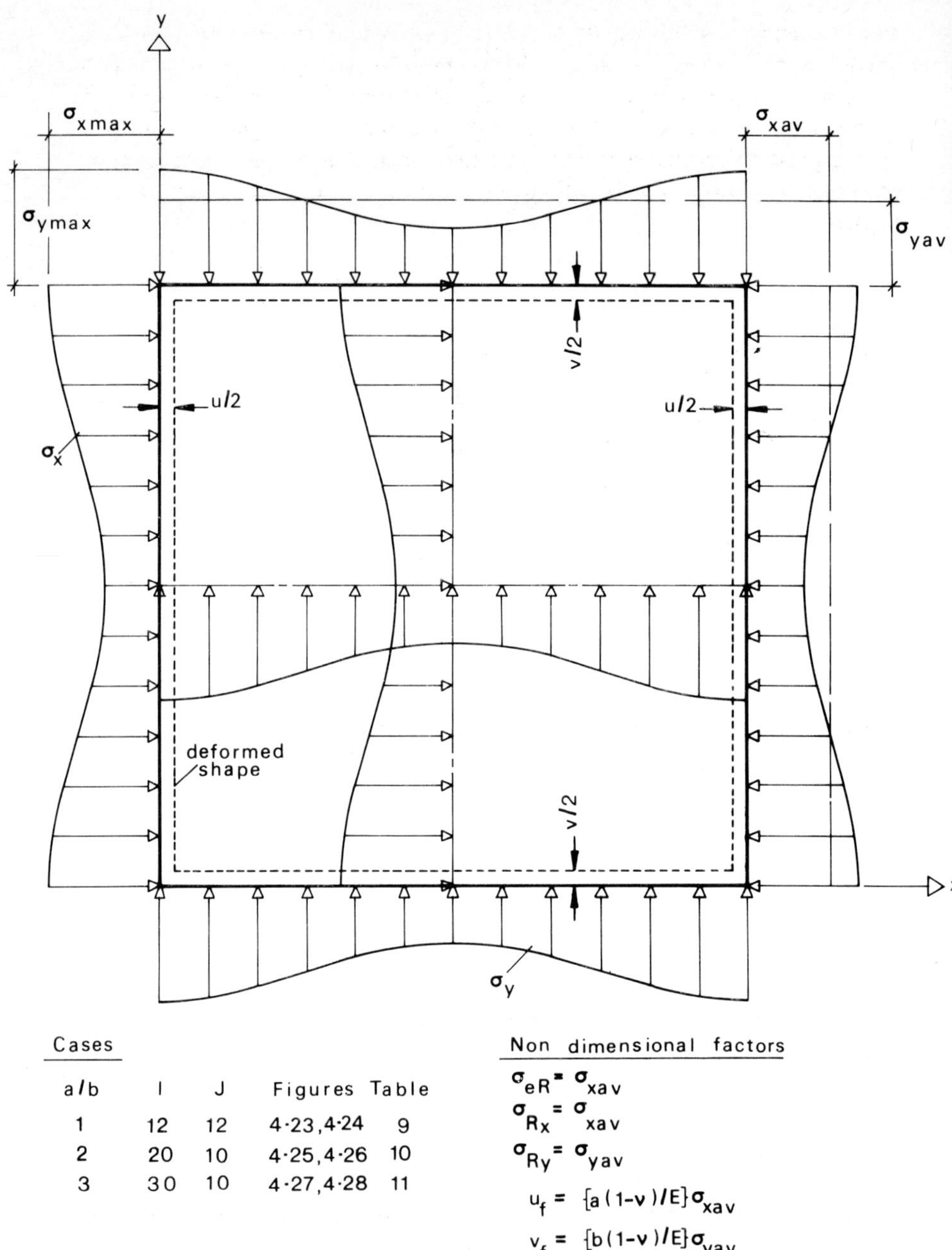


Fig. 4.22 Uniform equal biaxial compressive displacement ($\sigma_{xav} = \sigma_{yav}$)

For rectangular plates ($a/b \neq 1$), due to behaviour which becomes increasingly similar to simple column action as side ratio increases, the tendency for the maximum stress to initiate and remain at centre panel (location 4) becomes more pronounced. Because of this 'column type' behaviour, rectangular plates subject to compression in the short-axis direction (including biaxially loaded plates) have relatively lower reserves of strength and higher margins of safety should be applied.

(b) Biaxial

Typical boundary stresses and in-plane displacements:
(i) $\sigma_{x\mathrm{av}} = \sigma_{y\mathrm{av}}$: Fig. 4.22
(ii) $\sigma_{x\mathrm{av}} = 2\sigma_{y\mathrm{av}}$: Fig. 4.29
Initial imperfections: $w_\mathrm{i} = w_0 \sin(\pi x/a) \sin(\pi y/b)$
Design data:
(i) $\sigma_{x\mathrm{av}} = \sigma_{y\mathrm{av}}$: Figs. 4.23–28
(ii) $\sigma_{x\mathrm{av}} = 2\sigma_{y\mathrm{av}}$: Figs. 4.30–35
Reference equivalent stress:
(i) $\sigma_{x\mathrm{av}} = \sigma_{y\mathrm{av}}$: $\sigma_{\mathrm{eR}} = \sigma_{x\mathrm{av}}$
(ii) $\sigma_{x\mathrm{av}} = 2\sigma_{y\mathrm{av}}$: $\sigma_{\mathrm{eR}} = (\sigma_{x\mathrm{av}}^2 + \sigma_{y\mathrm{av}}^2 - \sigma_{x\mathrm{av}}\sigma_{y\mathrm{av}})^{1/2}$
$= (3/4)^{1/2}\sigma_{x\mathrm{av}}$

The redistribution of loading as the buckle develops results in variation of location of maximum equivalent membrane stress in some cases. Note the significantly lower values of effectiveness, at equal proportions of the critical load, for biaxial as compared to uniaxial loading.

## 4.2 Shear

### 4.2.1 Uniform stress

Typical boundary stresses and in-plane displacements: Fig. 4.36
Initial imperfections: Fig. 4.39(i)
Design data: Fig. 4.37
Reference equivalent stress: $\sigma_{\mathrm{eR}} = (3\tau_{\mathrm{av}}^2)^{1/2} = (3)^{1/2}\tau_{\mathrm{av}}$

Solutions are given for uniform shear stress applied to square plates ($a/b = 1$). This condition can arise on one or more sides where the panel is not continuous and where the discontinuous sides are not restrained by a stiff framing member. As in the case of uniform compression this is a more severe condition than that of restrained sides, since the sides can 'follow' the buckle.

Although the sides are free of normal stresses, a self-equilibrating normal stress distribution develops in the interior of the plate. Maximum membrane stress occurs at centre panel (location 4). Maximum equivalent surface stress occurs at the corners (location 1) from the onset of loading for all values of $w_0/t$

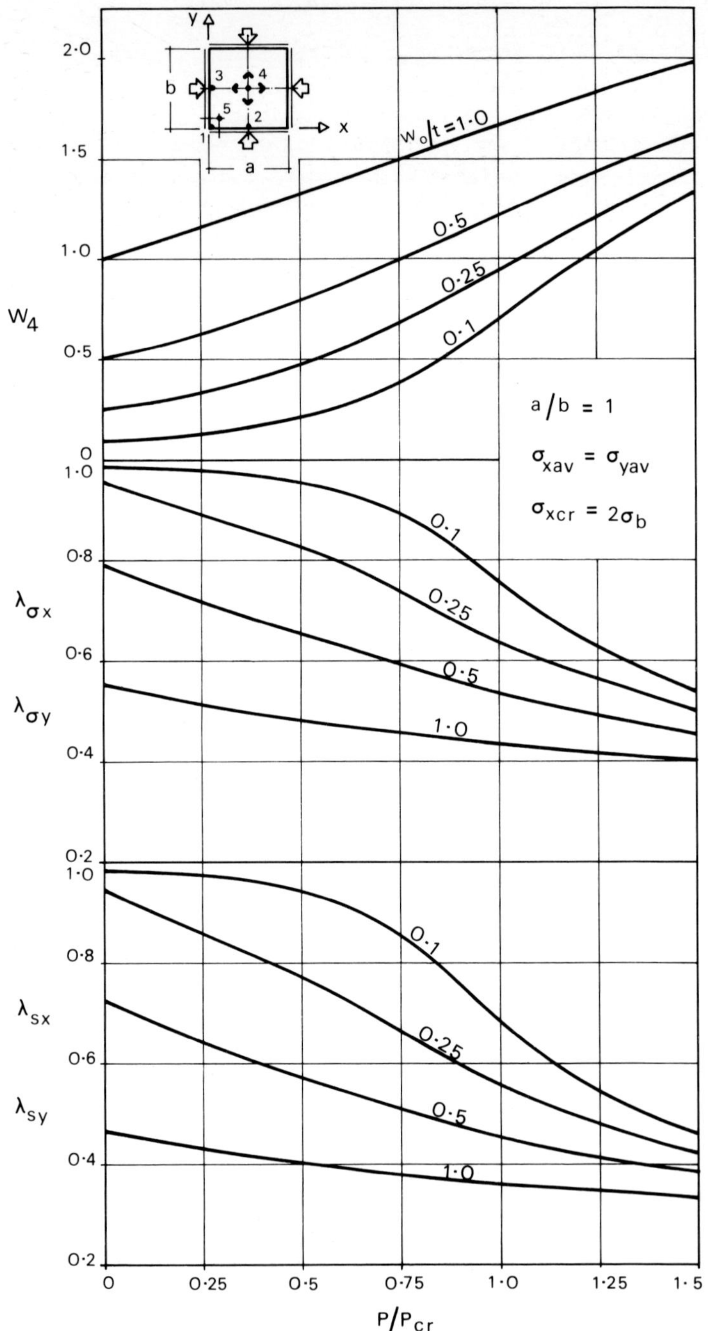


Fig. 4.23 Square ($a/b = 1$) plate subjected to uniform equal biaxial compressive displacement ($\sigma_{xav} = \sigma_{yav}$). Deflection and effectiveness

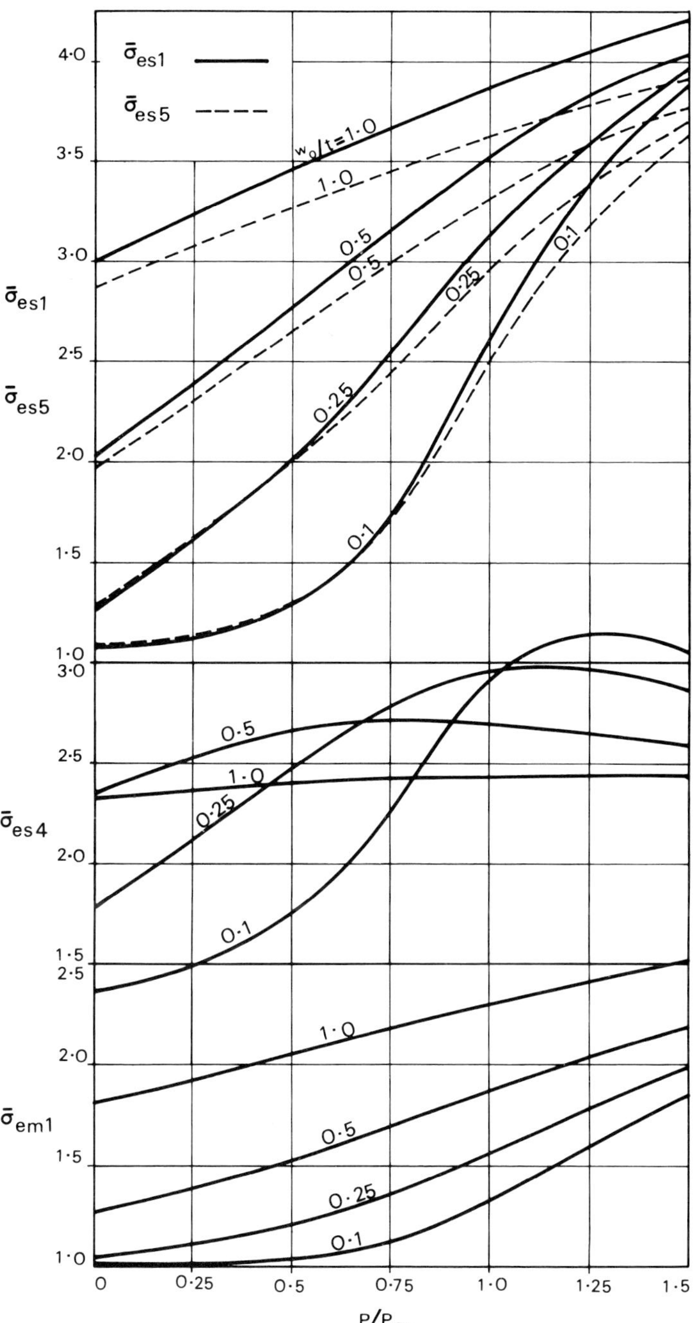


Fig. 4.24 Square ($a/b$ = 1) plate subjected to uniform equal biaxial compressive displacement ($\sigma_{xav} = \sigma_{yav}$). Equivalent membrane and surface stresses

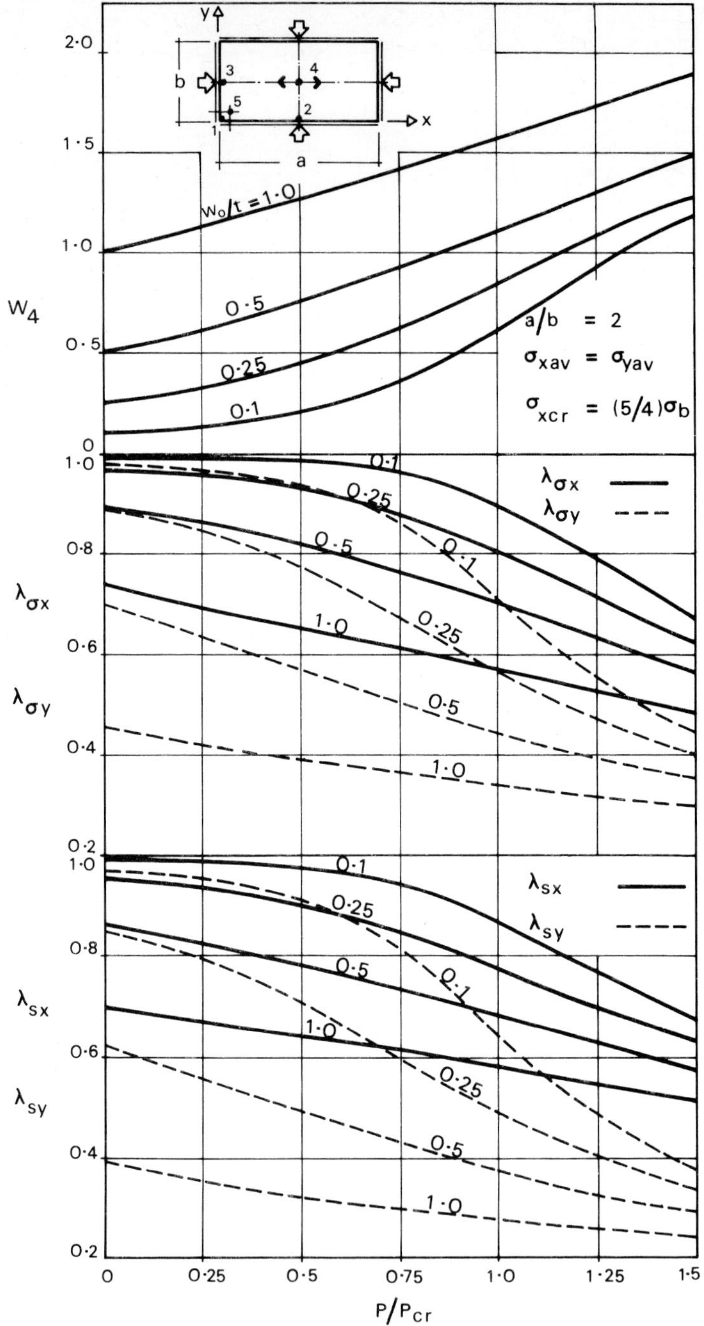


Fig. 4.25 Rectangular ($a/b = 2$) plate subjected to uniform equal biaxial compressive displacement ($\sigma_{xav} = \sigma_{yav}$). Deflection and effectiveness

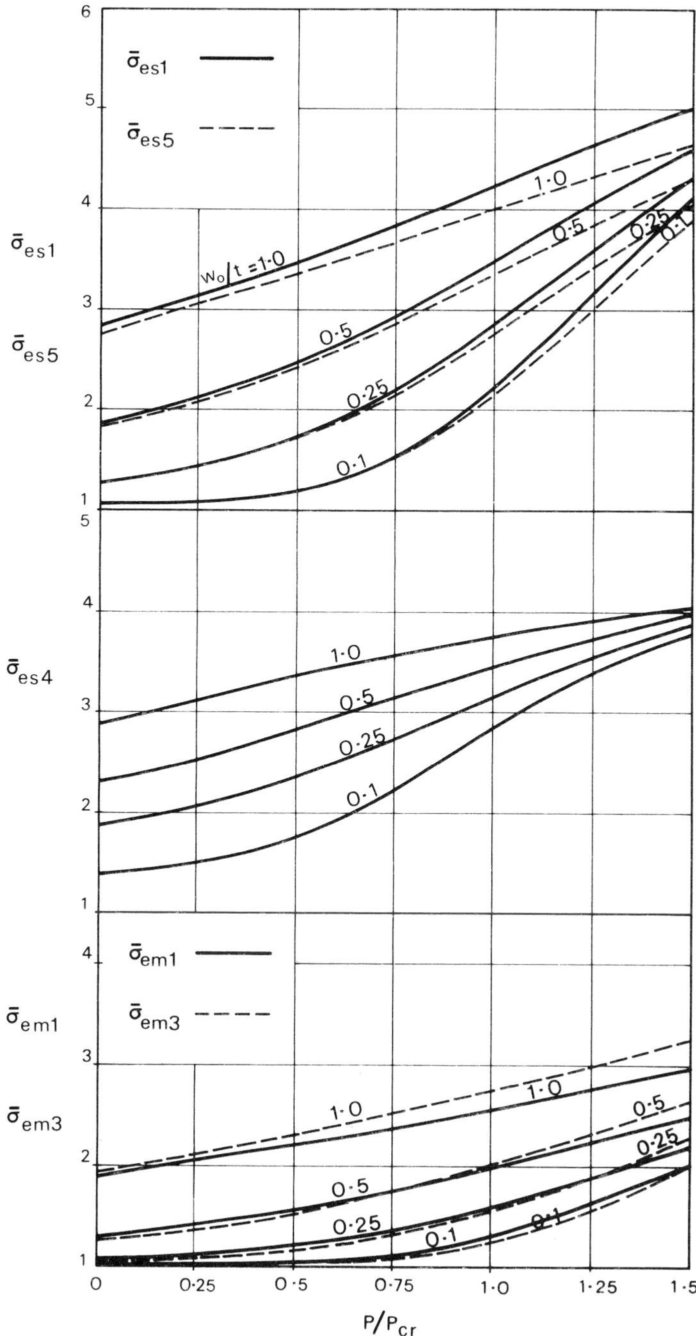


Fig. 4.26 Rectangular ($a/b = 2$) plate subjected to uniform equal biaxial compressive displacement ($\sigma_{xav} = \sigma_{yav}$). Equivalent membrane and surface stresses

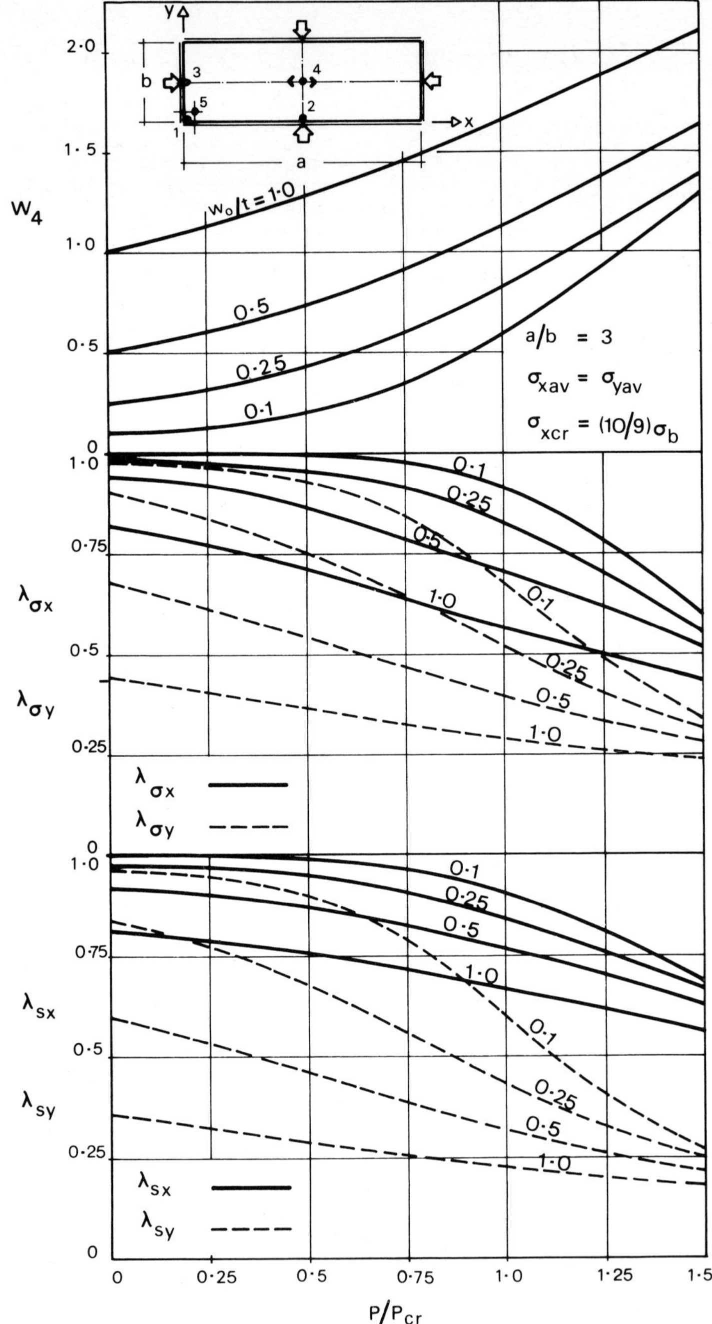


Fig. 4.27 Rectangular ($a/b = 3$) plate subjected to uniform equal biaxial compressive displacement ($\sigma_{xav} = \sigma_{yav}$). Deflection and effectiveness

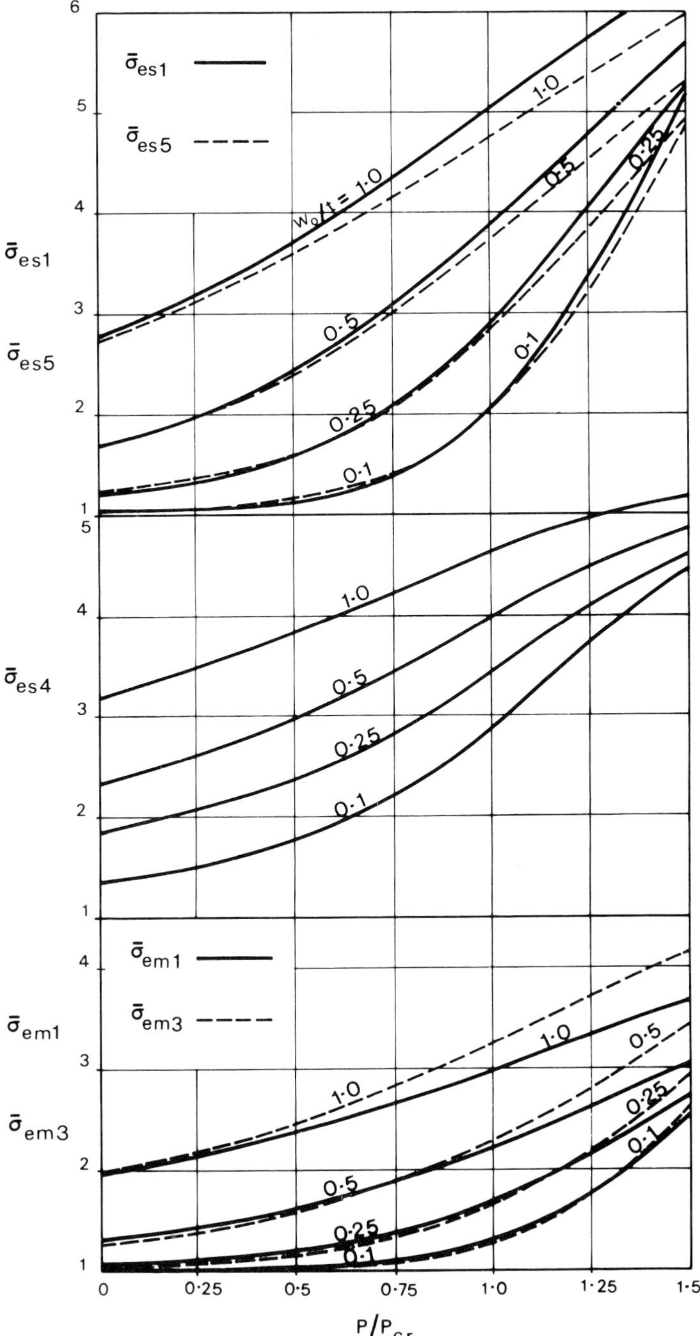


Fig. 4.28 Rectangular ($a/b = 3$) plate subjected to uniform equal biaxial compressive displacement ($\sigma_{xav} = \sigma_{yav}$). Equivalent membrane and surface stresses

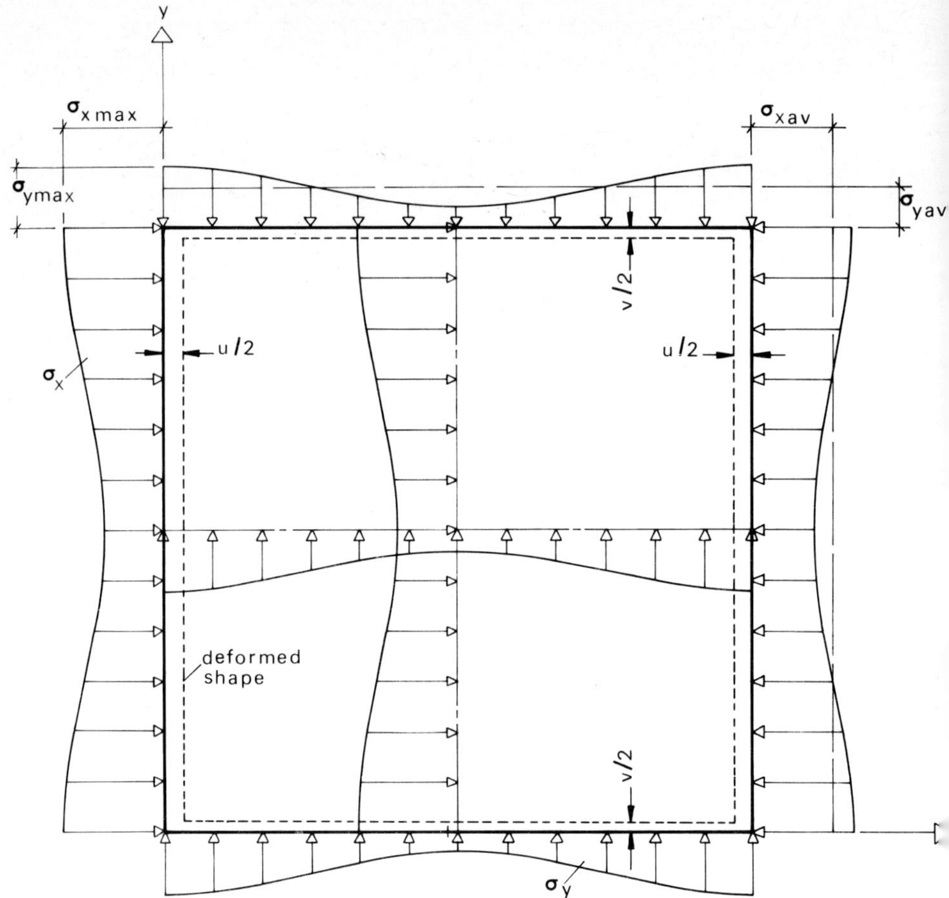


**Cases**

| a/b | I | J | Figures | Table |
|---|---|---|---|---|
| 1 | 12 | 12 | 4·30,4·31 | 12 |
| 2 | 20 | 10 | 4·32,4·33 | 13 |
| 3 | 30 | 10 | 4·34,4·35 | 14 |

**Non dimensional factors**

$$\sigma_{eR} = (3/4)^{1/2}\sigma_{xav}$$

$$\sigma_{Rx} = \sigma_{xav}$$

$$\sigma_{Ry} = \sigma_{yav}$$

$$u_f = \{a(1-\nu/2)/E\}\sigma_{xav}$$

$$v_f = \{b(1-2\nu)/E\}\sigma_{yav}$$

Fig. 4.29 Uniform unequal biaxial compressive displacement ($\sigma_{xav} = 2\sigma_{yav}$)

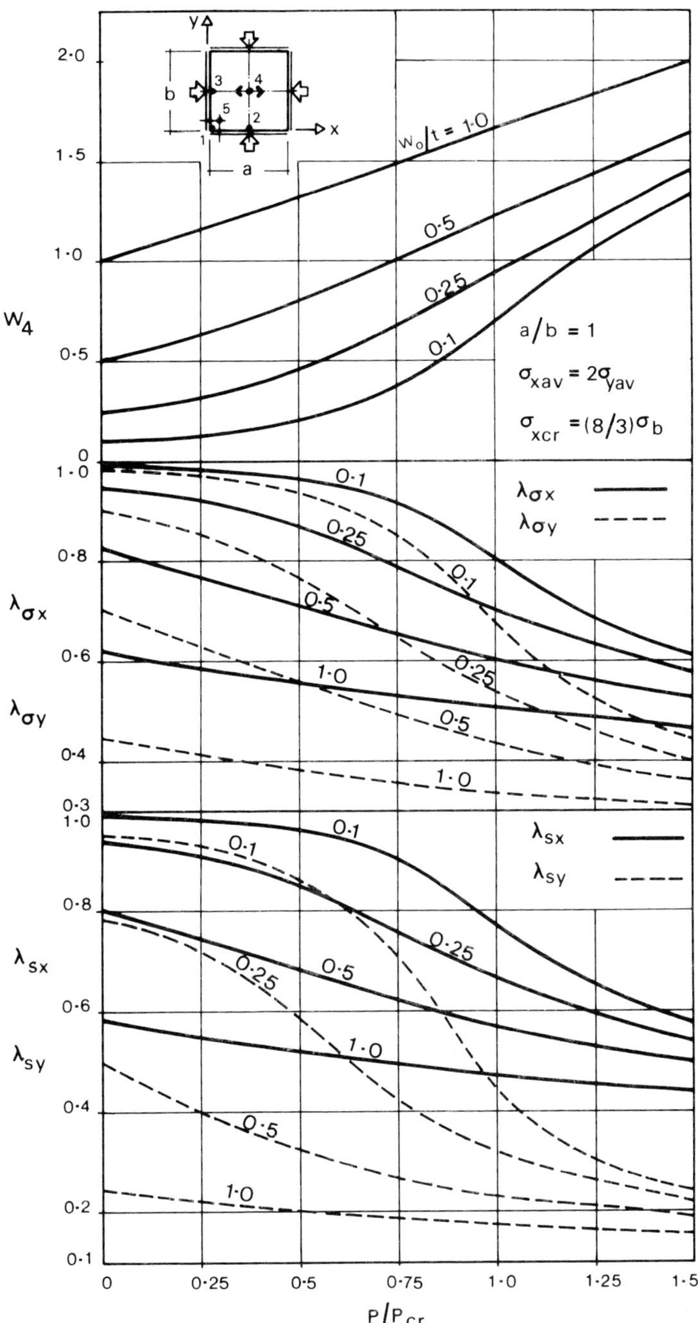


Fig. 4.30 Square ($a/b = 1$) plate subjected to uniform unequal biaxial compressive displacement ($\sigma_{xav} = 2\sigma_{yav}$). Deflection and effectiveness

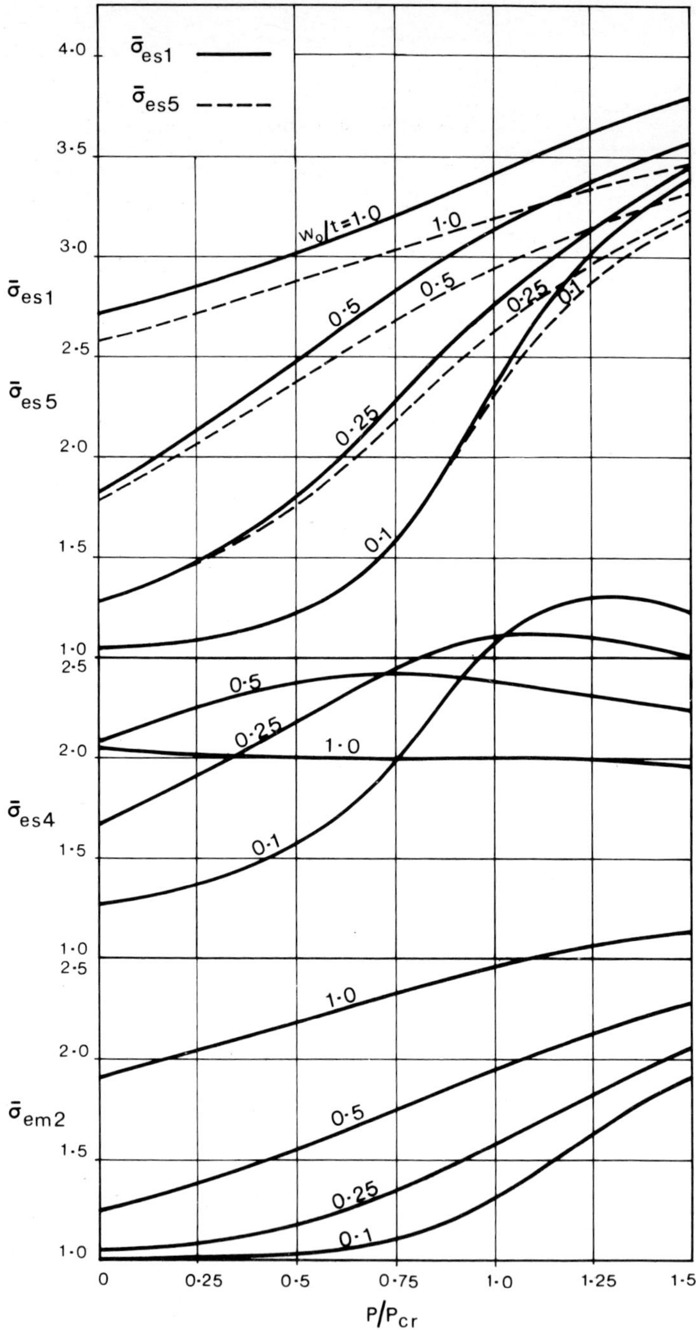


Fig. 4.31 Square ($a/b = 1$) plate subjected to uniform unequal biaxial compressive displacement ($\sigma_{xav} = 2\sigma_{yav}$). Equivalent membrane and surface stresses

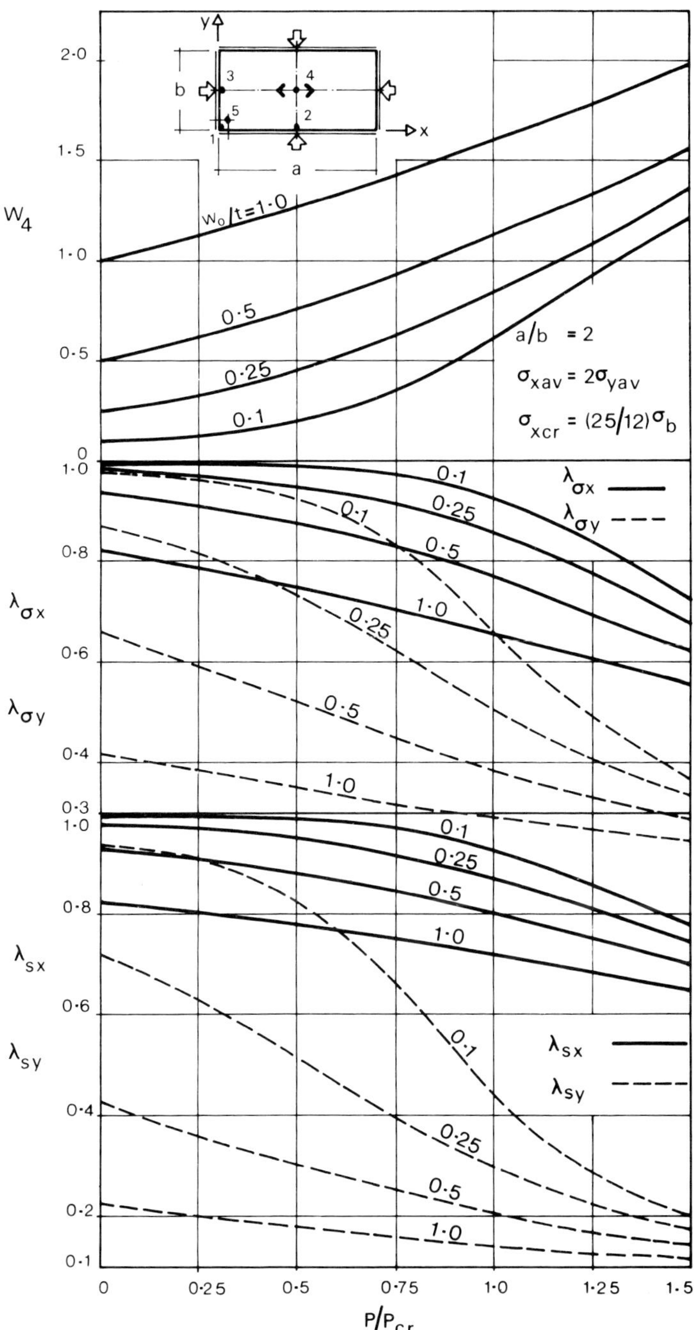


Fig. 4.32 Rectangular ($a/b = 2$) plate subjected to uniform unequal biaxial compressive displacement ($\sigma_{xav} = 2\sigma_{yav}$). Deflection and effectiveness

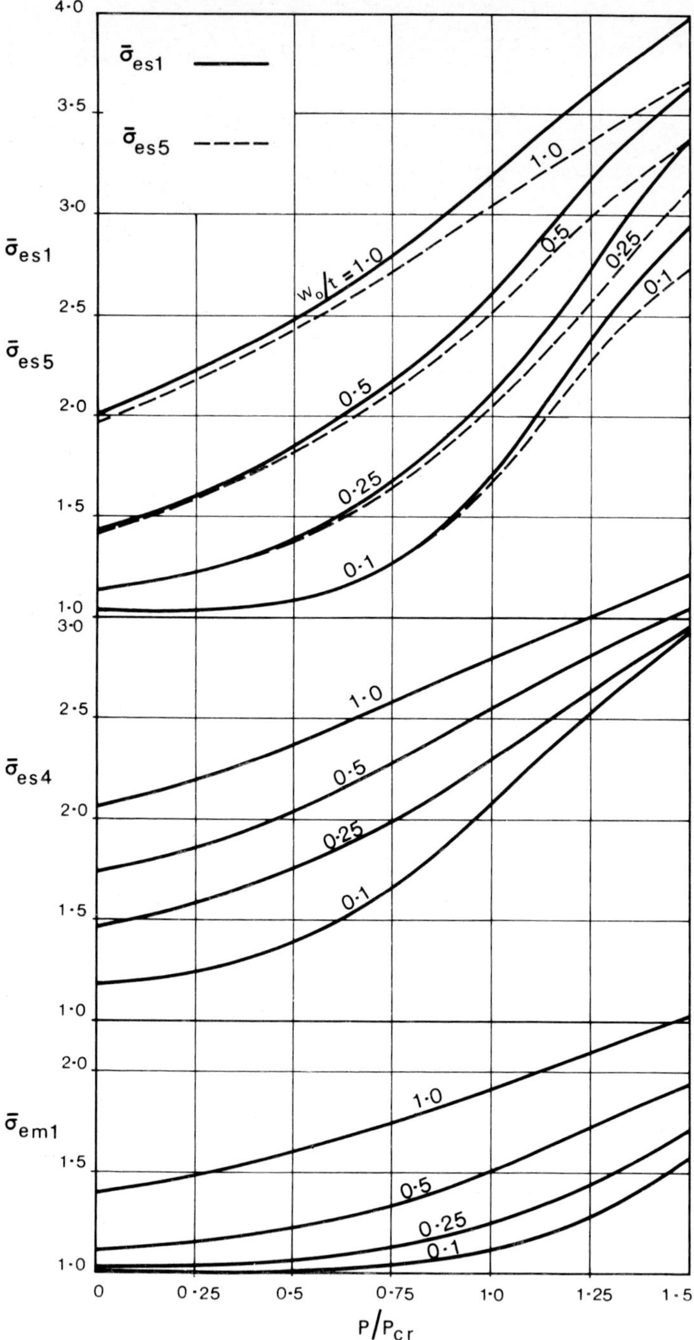


Fig. 4.33 Rectangular ($a/b = 2$) plate subjected to uniform unequal biaxial compressive displacement ($\sigma_{xav} = 2\sigma_{yav}$). Equivalent membrane and surface stresses

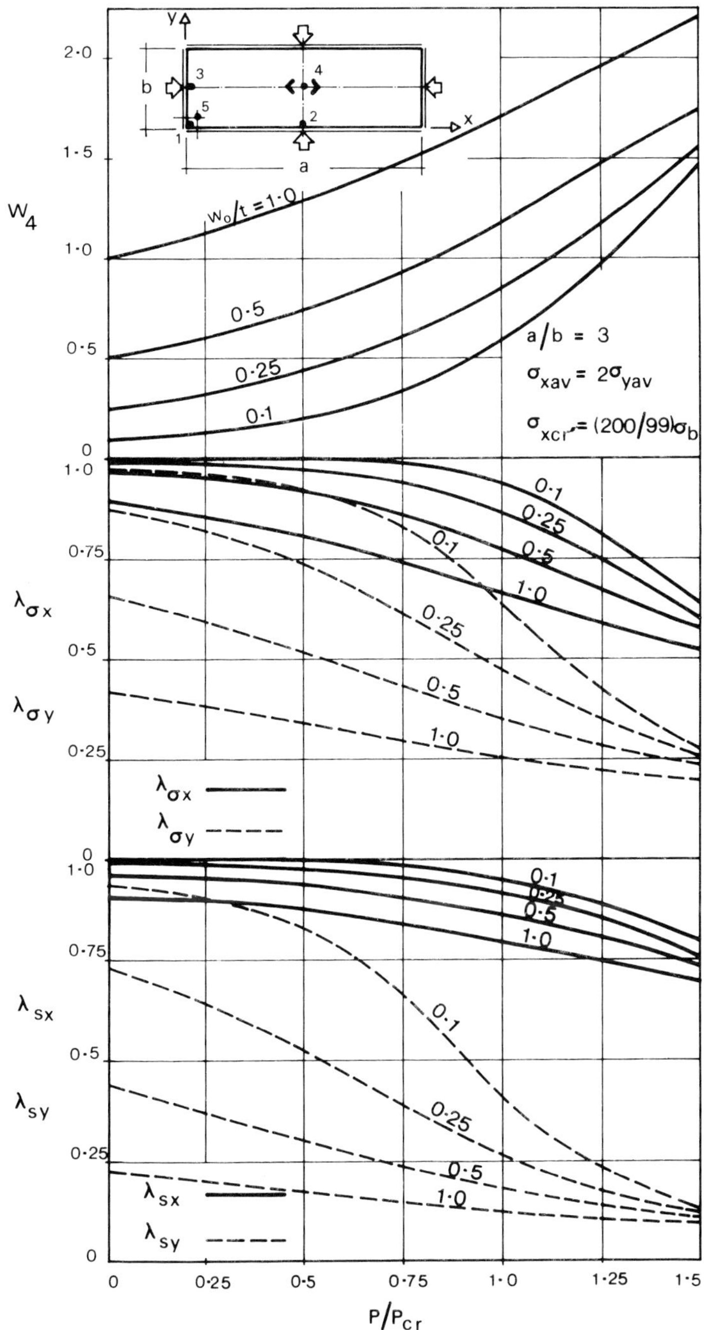


Fig. 4.34 Rectangular ($a/b = 3$) plate subjected to uniform unequal biaxial compressive displacement ($\sigma_{xav} = 2\sigma_{yav}$). Deflection and effectiveness

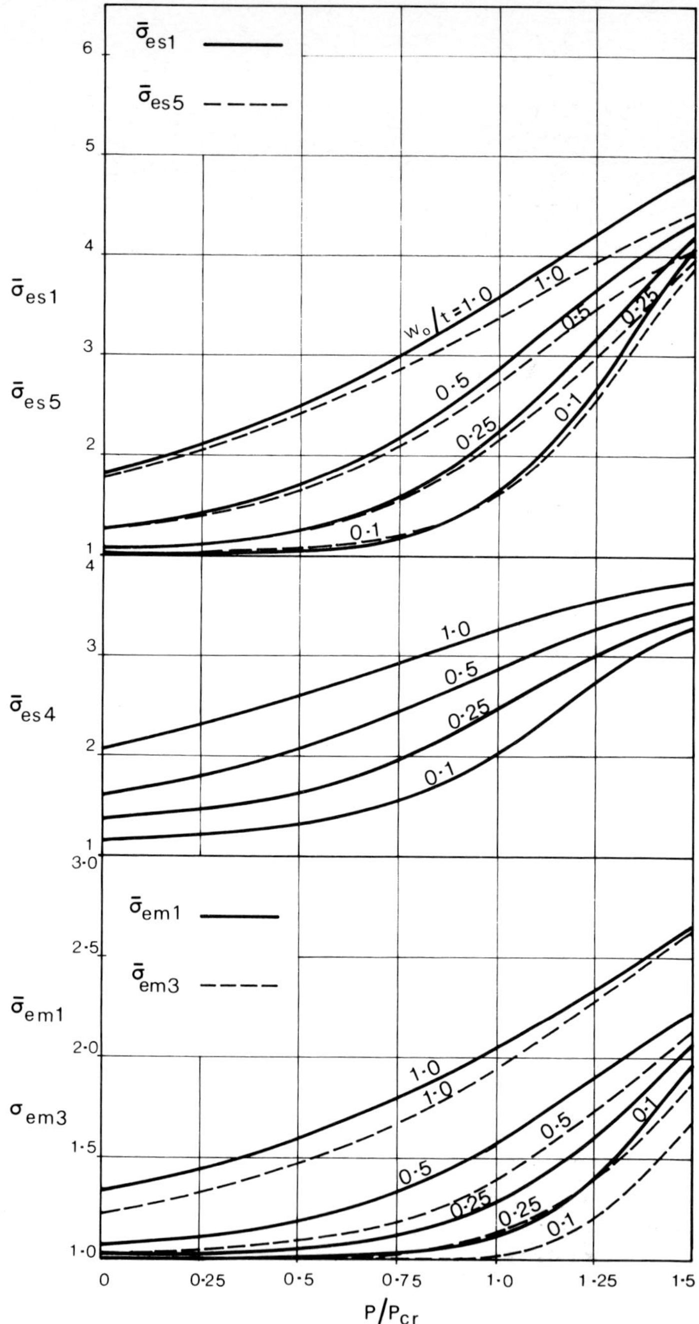


Fig. 4.35 Rectangular ($a/b = 3$) plate subjected to uniform unequal biaxial compressive displacement ($\sigma_{xav} = 2\sigma_{yav}$). Equivalent membrane and surface stresses

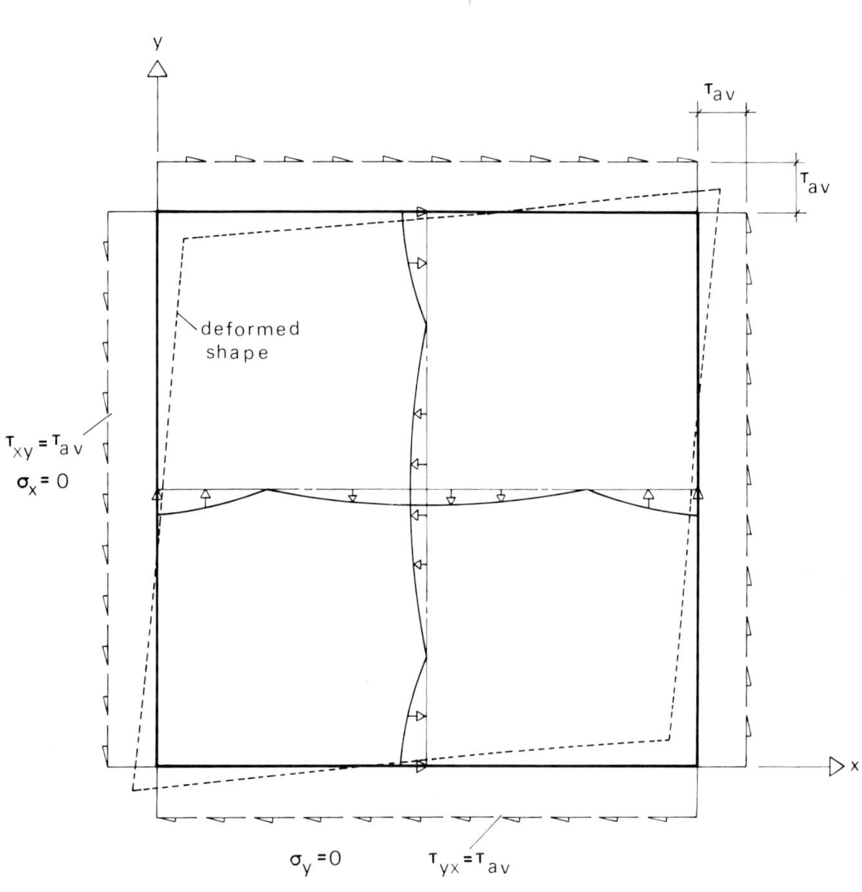


<u>Cases</u>

a/b = 1

I = J = 11

Figure: 4·37

Table : 15

<u>Non dimensional factors</u>

$\sigma_{eR} = (3)^{1/2}\tau_{av}$

$\sigma_{Rx} = \tau_{av}$

$\sigma_{Ry} = \tau_{av}$

$\tau_R = \tau_{av}$

$s_f = (a\tau_{av})/G$

Fig. 4.36 Uniform shear stress

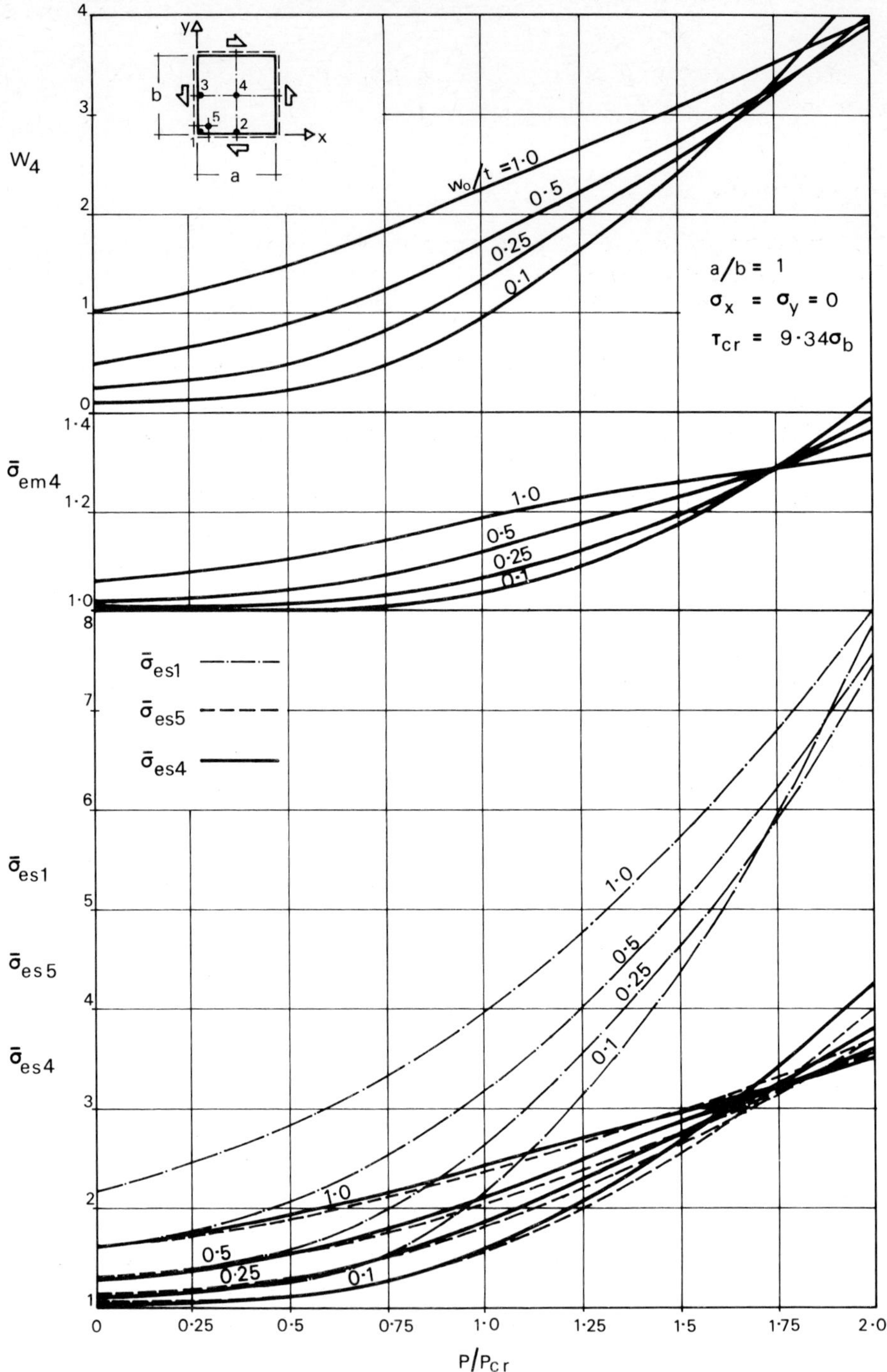


Fig. 4.37 Square ($a/b = 1$) plate subjected to uniform shear stress. Deflection, equivalent membrane and surface stresses

and is very localised. Recourse to local yielding is therefore necessary to avoid specifying unnecessarily thick plate.

### 4.2.2 Uniform displacement

Typical boundary stresses and in-plane displacements: Fig. 4.38
Initial imperfections: Fig. 4.39
Design data: Figs. 4.40–43
Reference equivalent stress: $\sigma_{eR} = (3\tau_{av}^2)^{1/2} = (3)^{1/2}\tau_{av}$

Solutions are given for uniform shear displacement with sides restrained against net in-plane movement for side ratios of $a/b = 1$, 1·5, 2 and 3. The restraint against in-plane movement results in the development of tensile stresses along the sides. This condition commonly arises in web panels, in regions of zero bending moment or adjacent to the neutral axis, for all types of girders where compatibility of adjacent panel sides dictates straight immobile sides.

Where a panel is bounded by a stiff side framing member the normal stresses must be resisted by that member. Values of average normal side stresses are plotted for the purpose of checking stresses in framing members and maximum values are given in the tables in Appendix 1 if required.

Maximum equivalent membrane stress occurs at the corners (location 1) but is not plotted because it only marginally exceeds the applied stress. If required, it can be computed from the results given in the tables in Appendix 1, by substituting into the expression for equivalent stress at location 1:

$$\sigma_{em1} = (3\tau_{m1}^2)^{1/2} = (3)^{1/2}\tau_{m1}$$

The location of the maximum equivalent surface stress varies depending on side ratio. The tendency for surface stresses at, and near, the corners (location 1) to reverse in longer plates as load increases is due to the development of higher buckling modes. As in the case of uniform shear stress, economic design requires an allowance for restricted yield in the case of square or near square plates.

## 4.3 Combined uniform compressive and shear displacement

Solutions are given for uniaxial and biaxial compression combined with shear for side ratios of $a/b = 1$ and 2. All sides are constrained to remain straight (in the uniaxial cases the unloaded sides are free to move in-plane) corresponding to interior panels or panels bounded by stiff framing members. In the case of uniaxial loading applied to plates of side ratio $a/b = 2$ the loading is applied to the shorter sides. The uniaxial results are for equal average axial and shear stress and the biaxial results are for average axial stress equal to half the average shear stress.

These loading conditions arise in all major structures although very few solutions are available elsewhere in the literature. Plate stiffness effectiveness is significantly reduced by the presence of shear, a factor requiring consideration in assessing global deflections.

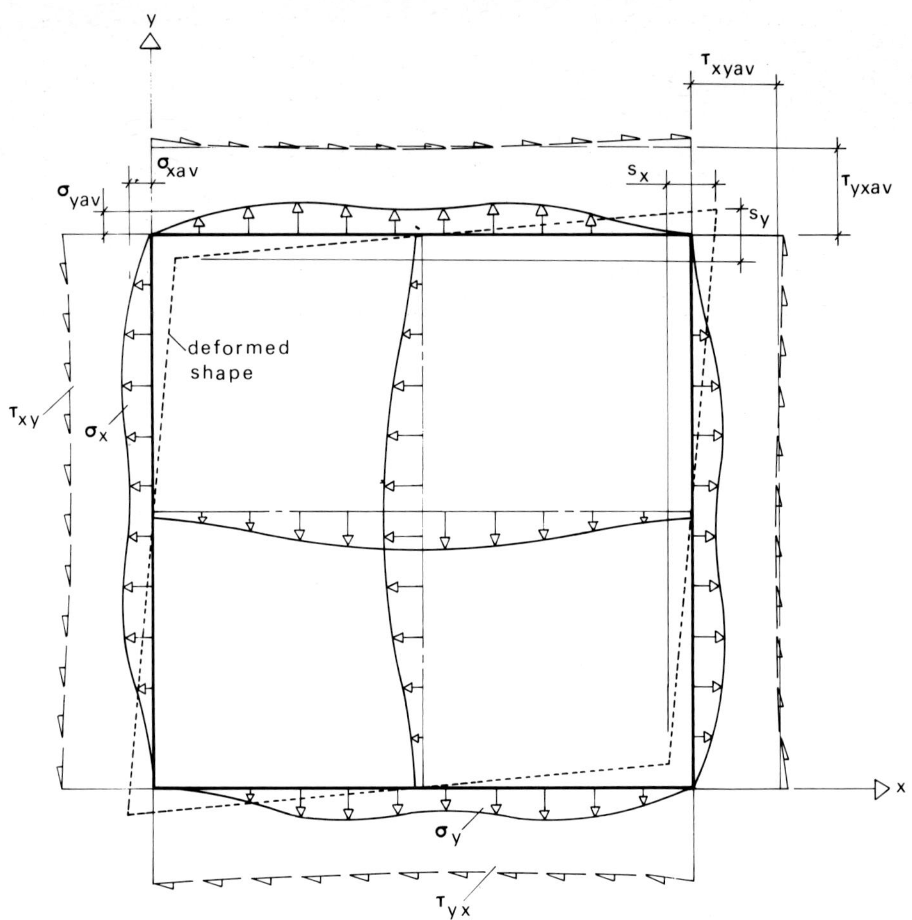


$s = s_x + s_y, \quad \tau_{av} = (\tau_{xyav} + \tau_{yxav})/2$

| Cases | | | | |
|---|---|---|---|---|
| a/b | I | J | Figure | Table |
| 1 | 11 | 11 | 4·40 | 16 |
| 1·5 | 15 | 11 | 4·41 | 17 |
| 2 | 22 | 11 | 4·42 | 18 |
| 3 | 30 | 10 | 4·43 | 19 |

Non dimensional factors

$\sigma_{eR} = (3)^{1/2} \tau_{av}$

$\sigma_{Rx} = \tau_{av}$

$\sigma_{Ry} = \tau_{av}$

$\tau_R = \tau_{av}$

$s_f = (a/G)\tau_{av}$

Fig. 4.38 Uniform shear displacement

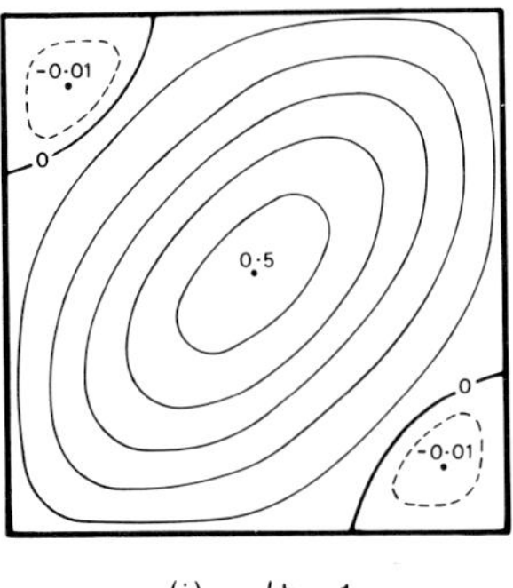


(i) a / b = 1

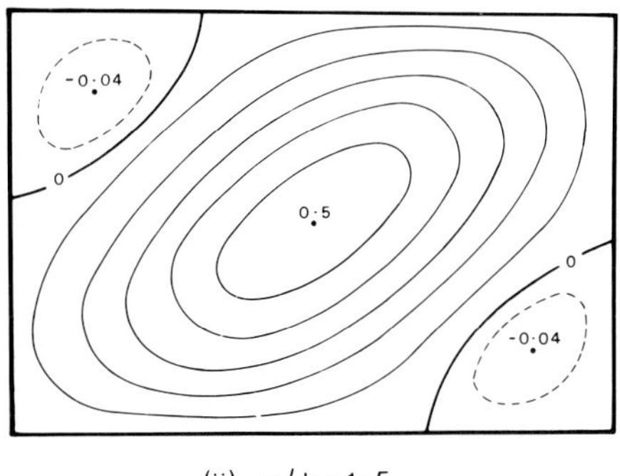


(ii) a / b = 1·5

(iii) a / b = 2

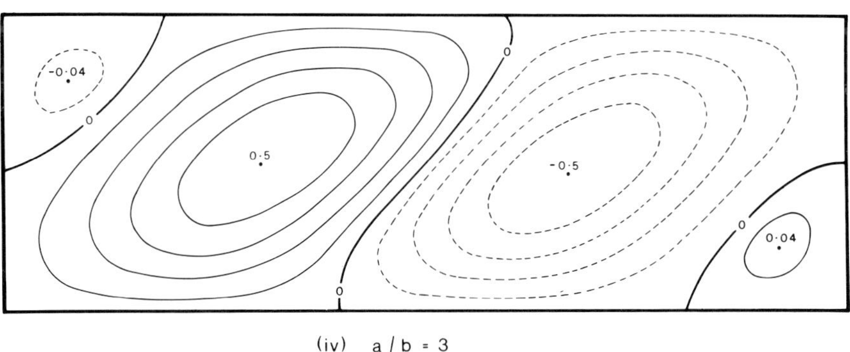


(iv) a / b = 3

Fig. 4.39 Initial imperfection contours ($w_0/t = 0{\cdot}5$). Uniform shear

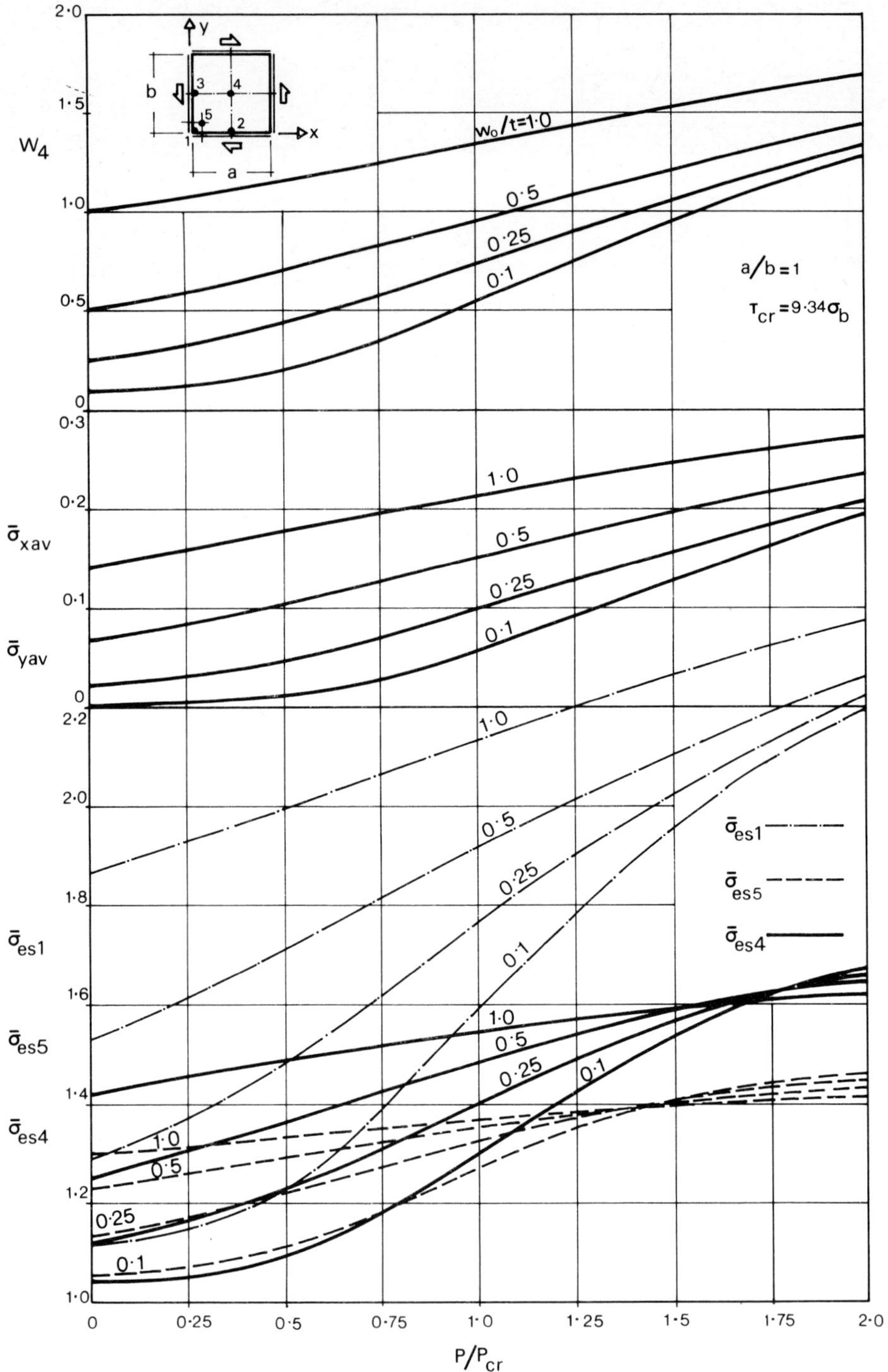


Fig. 4.40 Square ($a/b = 1$) plate subjected to uniform shear displacement. Deflection, average edge restraining stresses and equivalent surface stresses

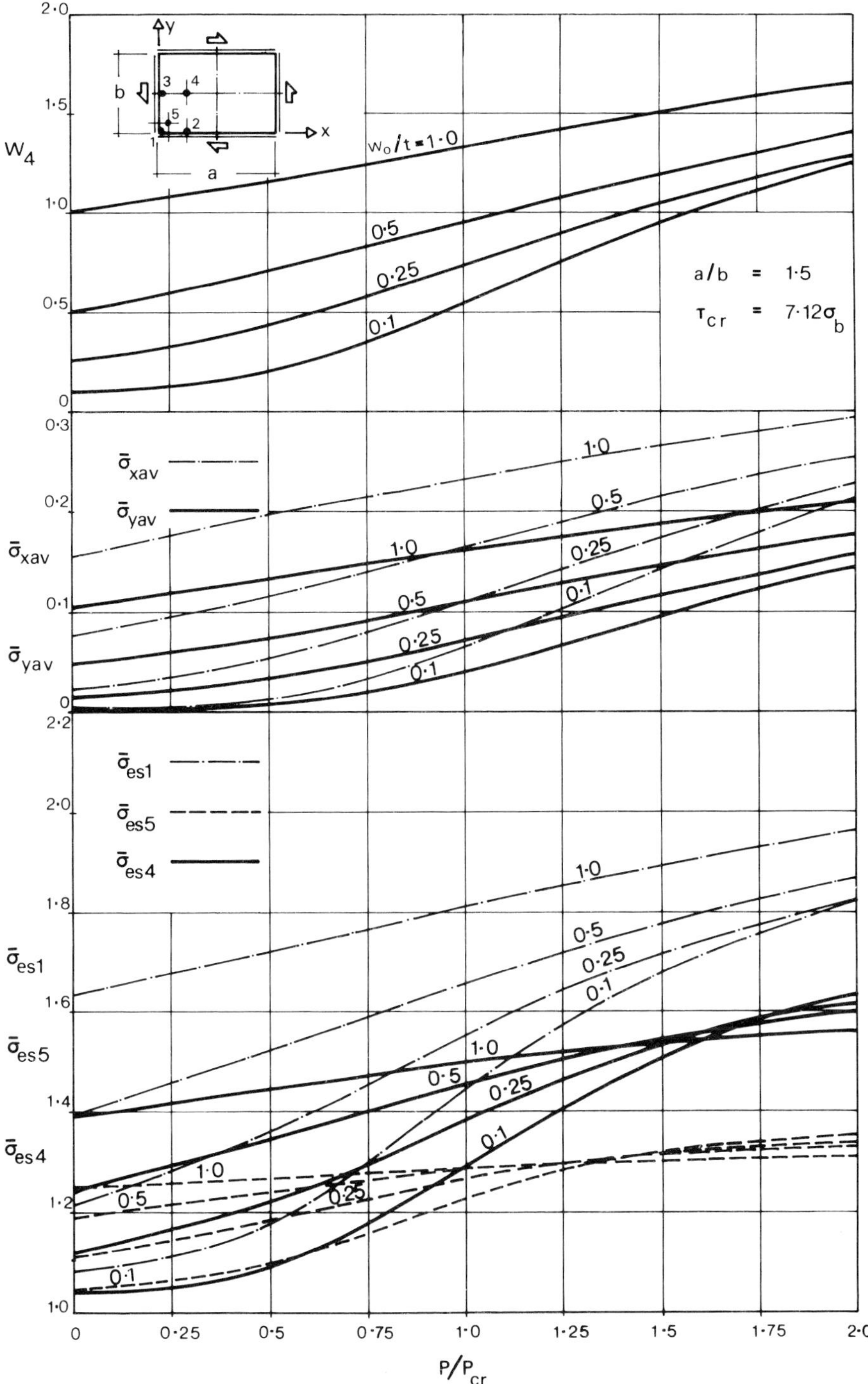


Fig. 4.41 Rectangular ($a/b = 1{\cdot}5$) plate subjected to uniform shear displacement. Deflection, average edge restraining stresses and equivalent surface stresses

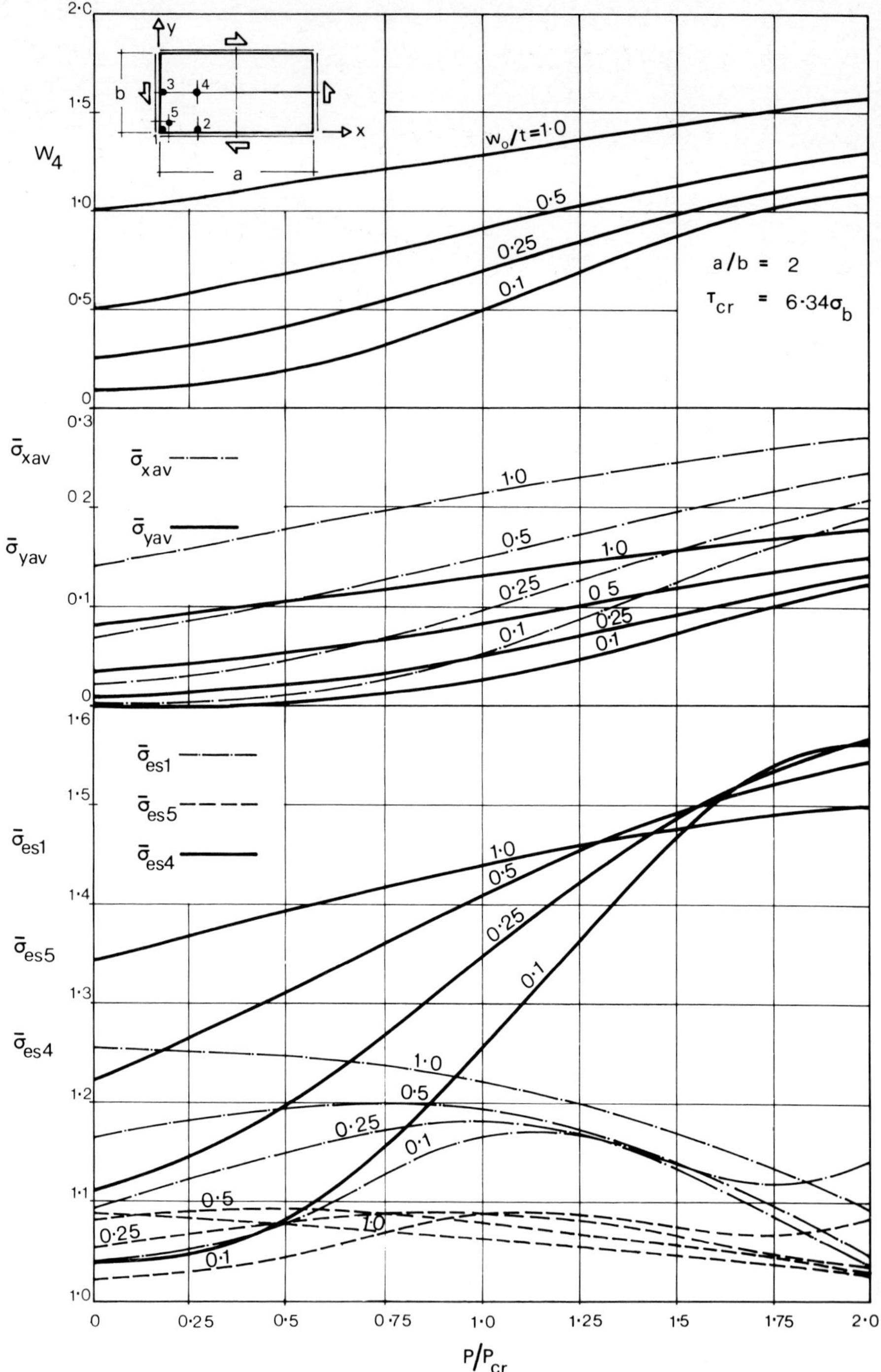


Fig. 4.42 Rectangular ($a/b = 2$) plate subjected to uniform shear displacement. Deflection, average edge restraining stresses and equivalent surface stresses

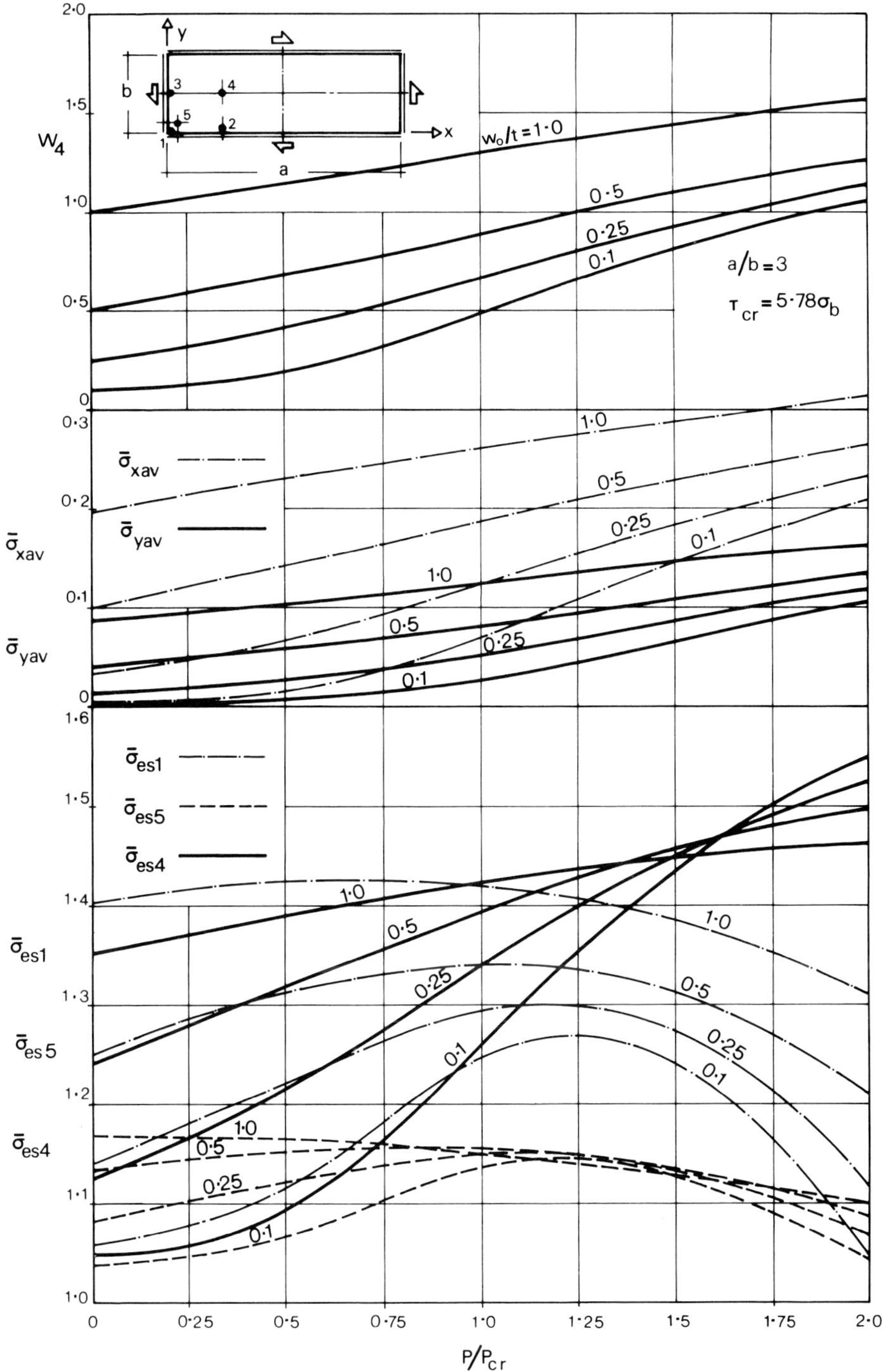


Fig. 4.43 Rectangular ($a/b = 3$) plate subjected to uniform shear displacement. Deflection, average edge restraining stresses and equivalent surface stresses

### 4.3.1 Uniaxial compression and shear

Typical boundary stresses and in-plane displacements: Fig. 4.44
Initial imperfections: Fig. 4.45
Design data: Figs. 4.46–49
Reference equivalent stress: $\sigma_{eR} = (\sigma_{xav}^2 + 3\tau_{av}^2)^{1/2}$

$$= 2\sigma_{xav}$$

In restraining the buckle, self-equilibrating stress distributions develop along the unloaded sides. Maximum axial and equivalent membrane stresses and maximum equivalent surface stress occur at the corners (location 1).

The surface stress gradient is very steep in the corners and allowance should therefore be made for local yield.

### 4.3.2 Biaxial compression and shear

Typical boundary stresses and in-plane displacements: Fig. 4.50
Initial imperfections: Fig. 4.51
Design data: Figs. 4.52–55
Reference equivalent stress:

$$\sigma_{eR} = (\sigma_{xav}^2 + \sigma_{yav}^2 - \sigma_{xav}\sigma_{yav} + 3\tau_{av}^2)^{1/2}$$
$$= (13)^{1/2}\sigma_{xav}$$

Maximum axial and equivalent membrane stresses and maximum equivalent surface stresses occur at the corners (location 1). The steep surface stress gradient in the corner regions necessitates design for local yield.

## 4.4 Varying uniaxial compressive displacement

For a flat plate this loading would give rise to a triangular stress distribution. For an imperfect plate this triangular distribution becomes increasingly non-linear as the buckle develops due to shedding of load to the stiffer side regions. Solutions are given for side ratios of 0·667, 1, 1·5 and 2 for unloaded sides restrained and for unloaded sides unrestrained. This type of loading occurs in stiffened web panels where the zero-stress side lies along the neutral axis and will be effectively constrained and the side in line with maximum stress will also usually be constrained except in the case of an edge panel bounded by a flexible framing member.

### 4.4.1 Unloaded sides restrained

Typical boundary stresses and in-plane displacements: Fig. 4.56
Initial imperfections: Fig. 4.57
Design data: Figs. 4.58–61
Reference equivalent stress: $\sigma_{eR} = \sigma_{x1}$

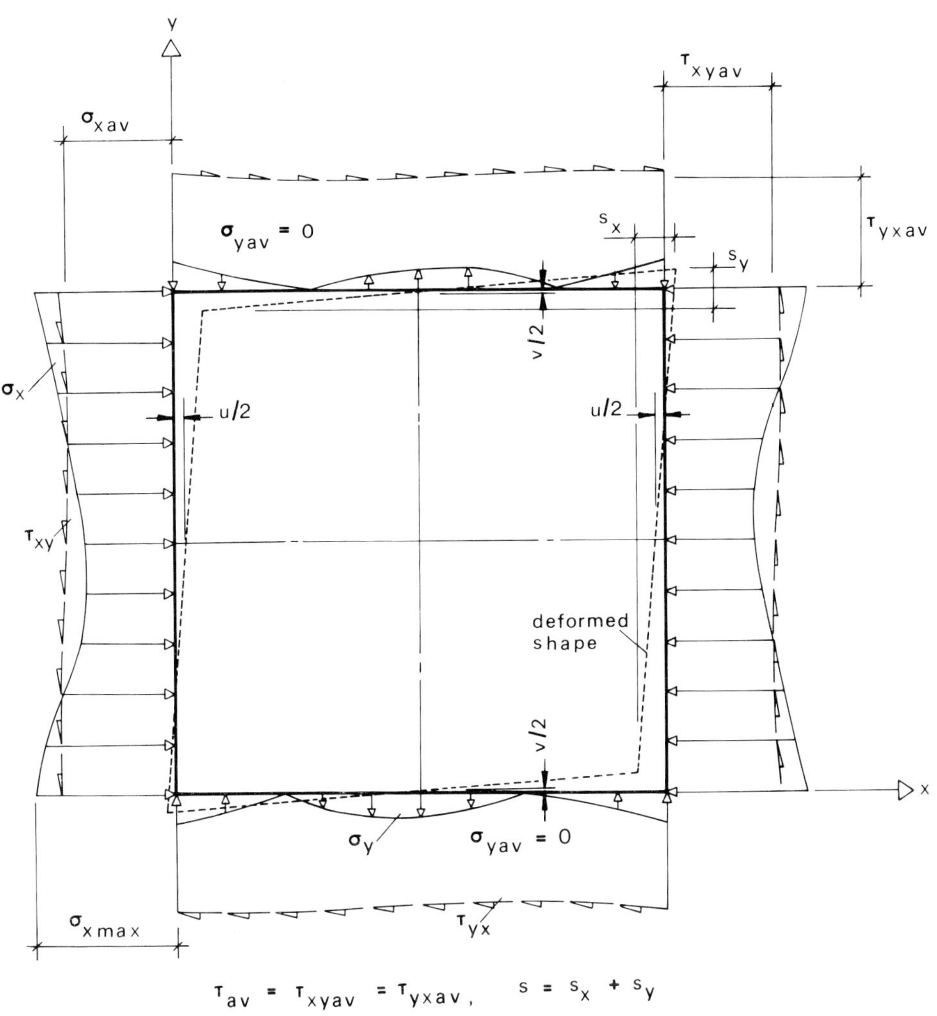


**Cases**

| a/b | I | J | Figures | Table |
|---|---|---|---|---|
| 1 | 10 | 10 | 4·46,4·47 | 20 |
| 2 | 20 | 10 | 4·48,4·49 | 21 |

**Non dimensional factors**

$\sigma_{eR} = 2\sigma_{xav}$

$\sigma_{Rx} = \sigma_{xav}$

$\sigma_{Ry} = \sigma_{xav}$

$\tau_R = \tau_{av}$

$u_f = (a/E)\sigma_{xav}$

$v_f = -(b\nu/E)\sigma_{xav}$

$s_f = (a/G)\tau_{av}$

Fig. 4.44 Combined uniform uniaxial compressive and shear displacement ($\sigma_{xav} = \tau_{av}$)

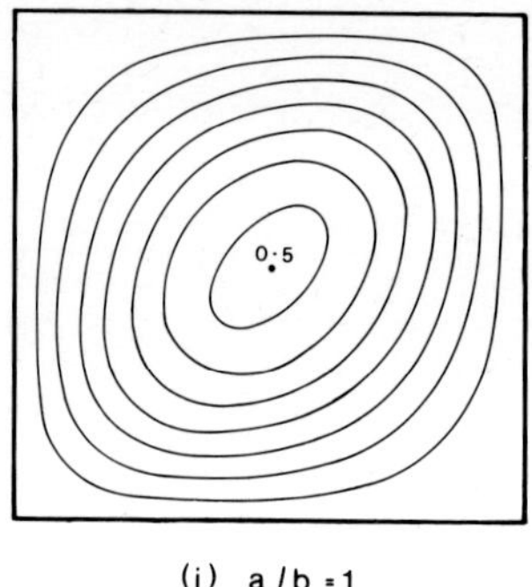


(i) a / b = 1

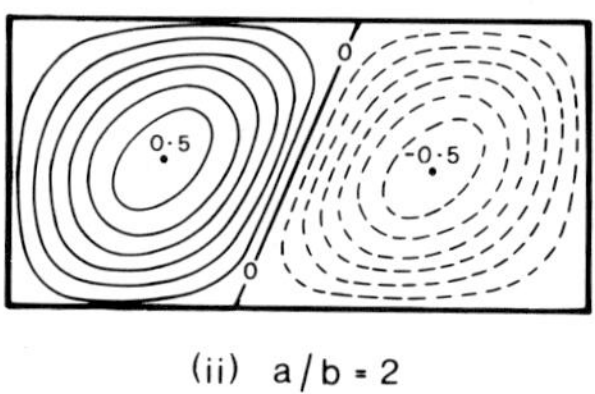


(ii) a / b = 2

Fig. 4.45 Initial imperfection contours ($w_0/t = 0{\cdot}5$). Combined uniaxial compression and shear ($\sigma_{xav} = \tau_{av}$)

The non-linearity in the applied loading distribution corresponds to a loss of plate effectiveness, but in the case of girders this can be ignored in assessing global deformations, and the applied stress $\sigma_{x1}$ can be taken as the value given by the global analysis on the assumption that the web is flat. Where an unloaded side is bounded by a framing member the average stresses plotted can be used in checking that member. Maximum edge restraining stresses are provided in the tables in Appendix 1 if required.

Maximum equivalent membrane stress occurs partway along the unloaded side (location 2). It is not plotted but can be computed from the results given in the tables in Appendix 1 by substituting into the following expression for equivalent stress at location 2:

$$\sigma_{em2} = (\sigma_{mx2}^2 + \sigma_{my2}^2 - \sigma_{mx2}\sigma_{my2})^{1/2}$$

where

$$\sigma_{mx2} = \sigma_{mx1}$$

Maximum equivalent surface stress occurs at the corner (location 1) for all cases plotted. Because of the nature of the loading combined with the restrained edge conditions, the corner stresses are relatively low and it will be noted that the adjacent diagonal stresses do not or only just exceed the applied membrane

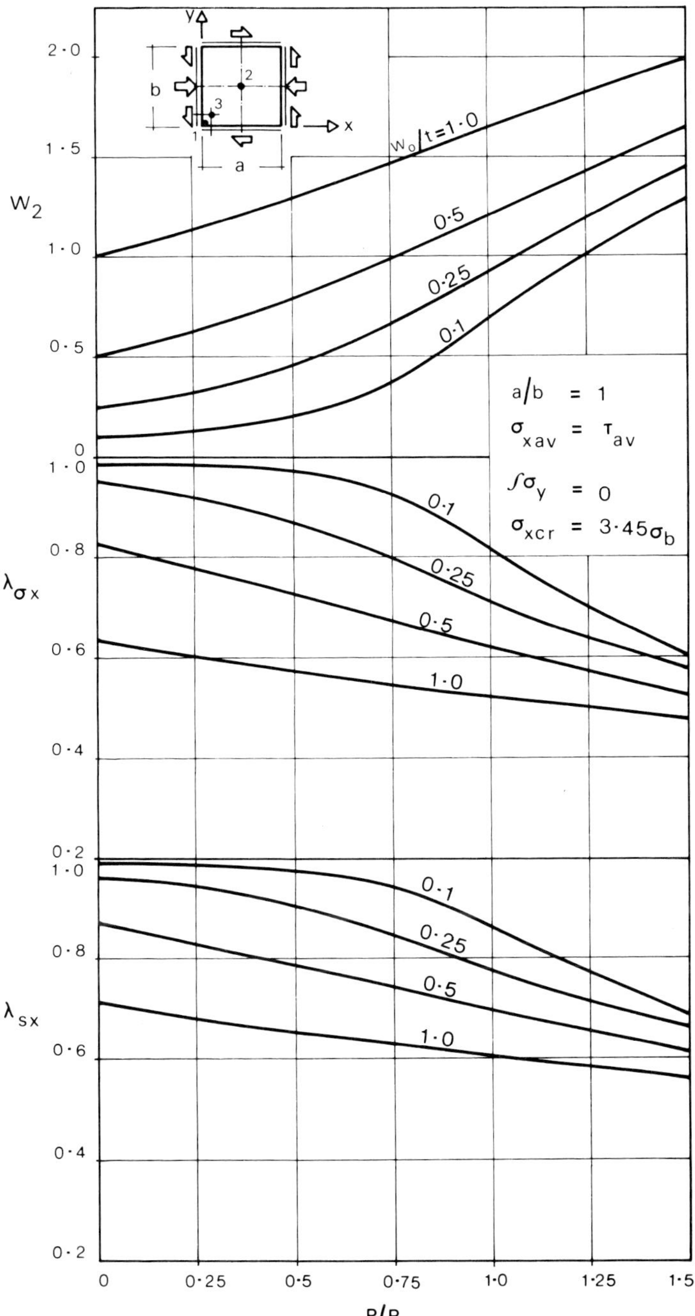


Fig. 4.46 Square ($a/b = 1$) plate subjected to combined uniform uniaxial compressive and shear displacement ($\sigma_{xav} = \tau_{av}$). Deflection and effectiveness

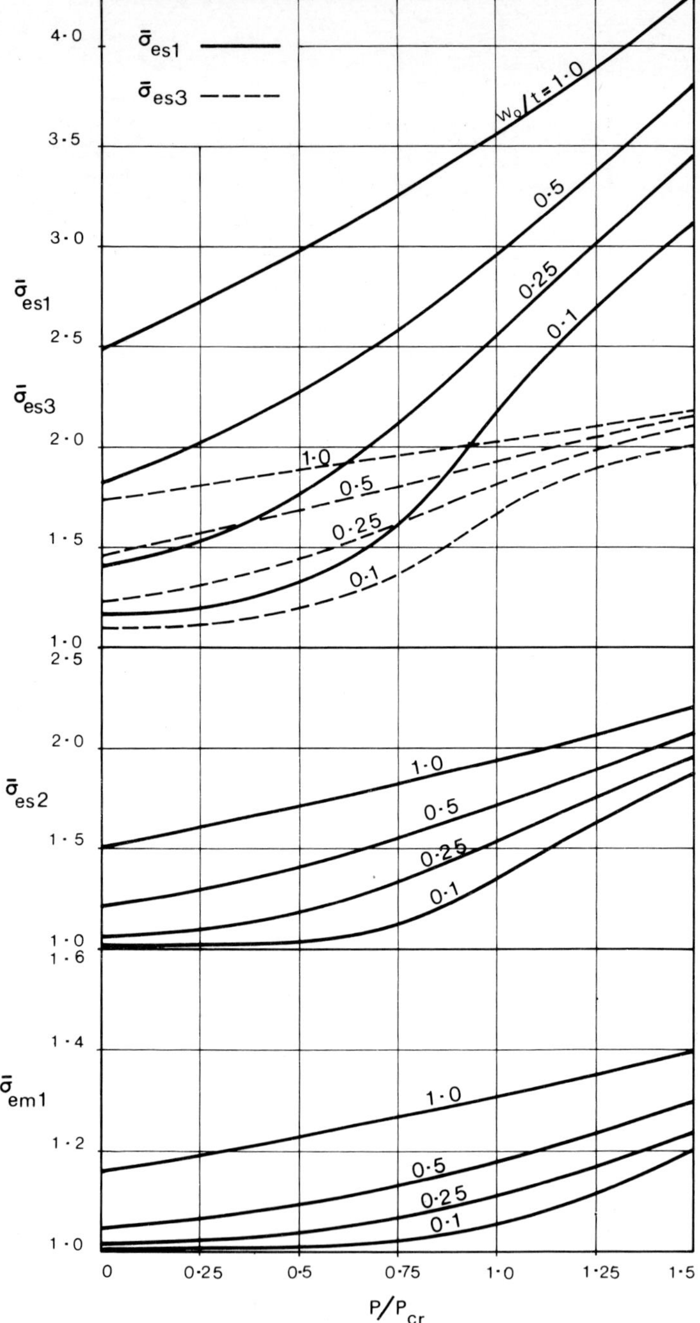


Fig. 4.47 Square ($a/b$ = 1) plate subjected to combined uniform uniaxial compressive and shear displacement ($\sigma_{xav} = \tau_{av}$). Equivalent membrane and surface stresses

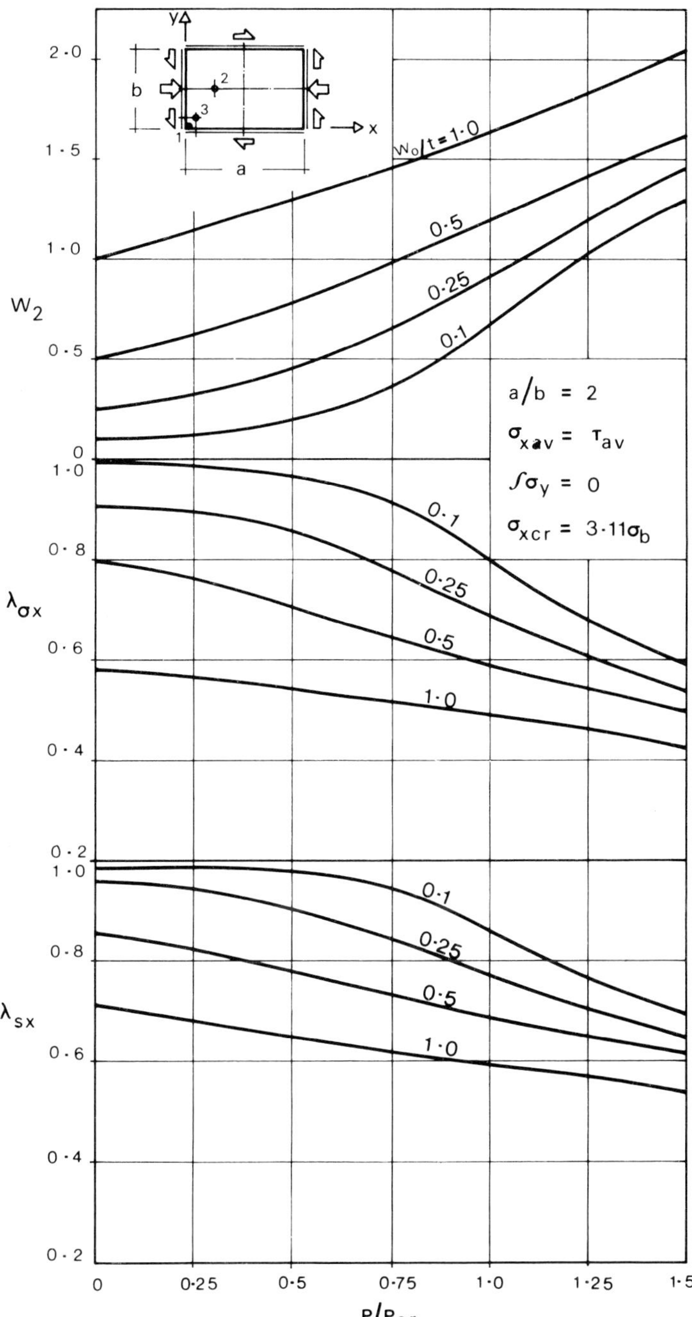


Fig. 4.48 Rectangular ($a/b = 2$) plate subjected to combined uniform uniaxial compressive and shear displacement ($\sigma_{xav} = \tau_{av}$). Deflection and effectiveness

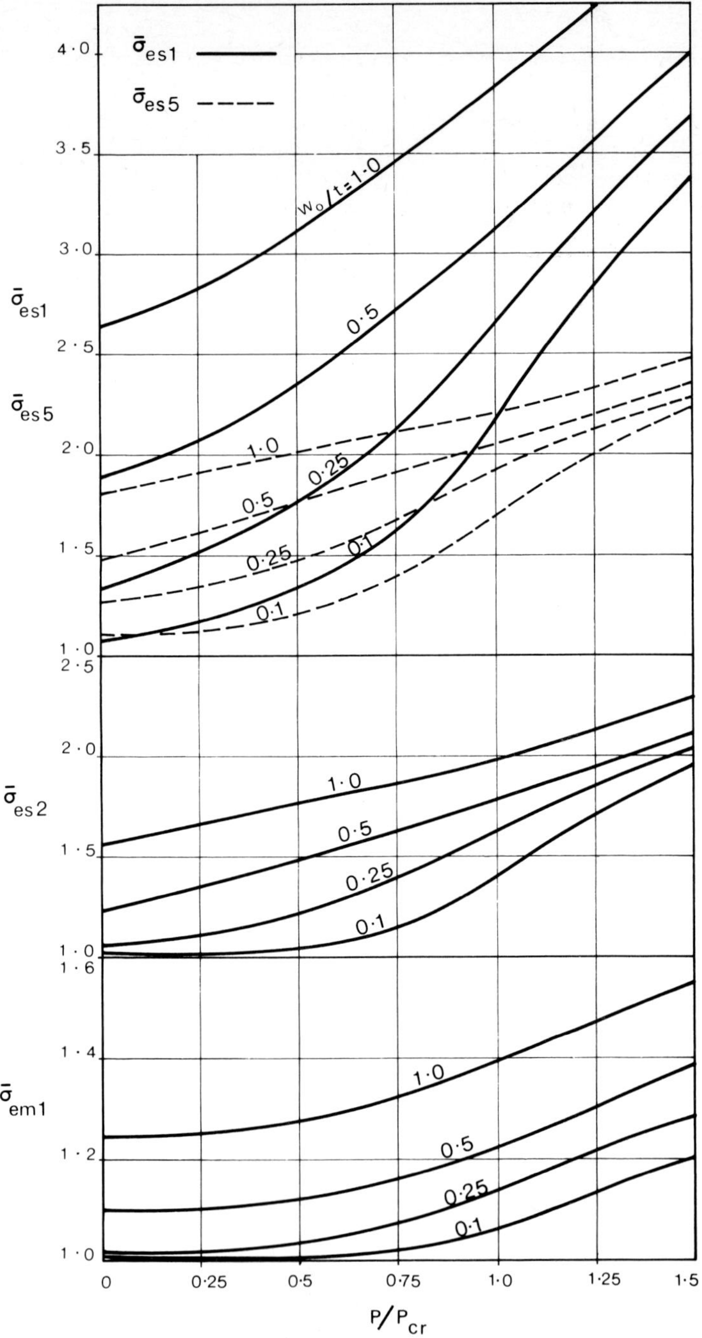


Fig. 4.49 Rectangular ($a/b = 2$) plate subjected to combined uniform uniaxial compressive and shear displacement ($\sigma_{xav} = \tau_{av}$). Equivalent membrane and surface stresses

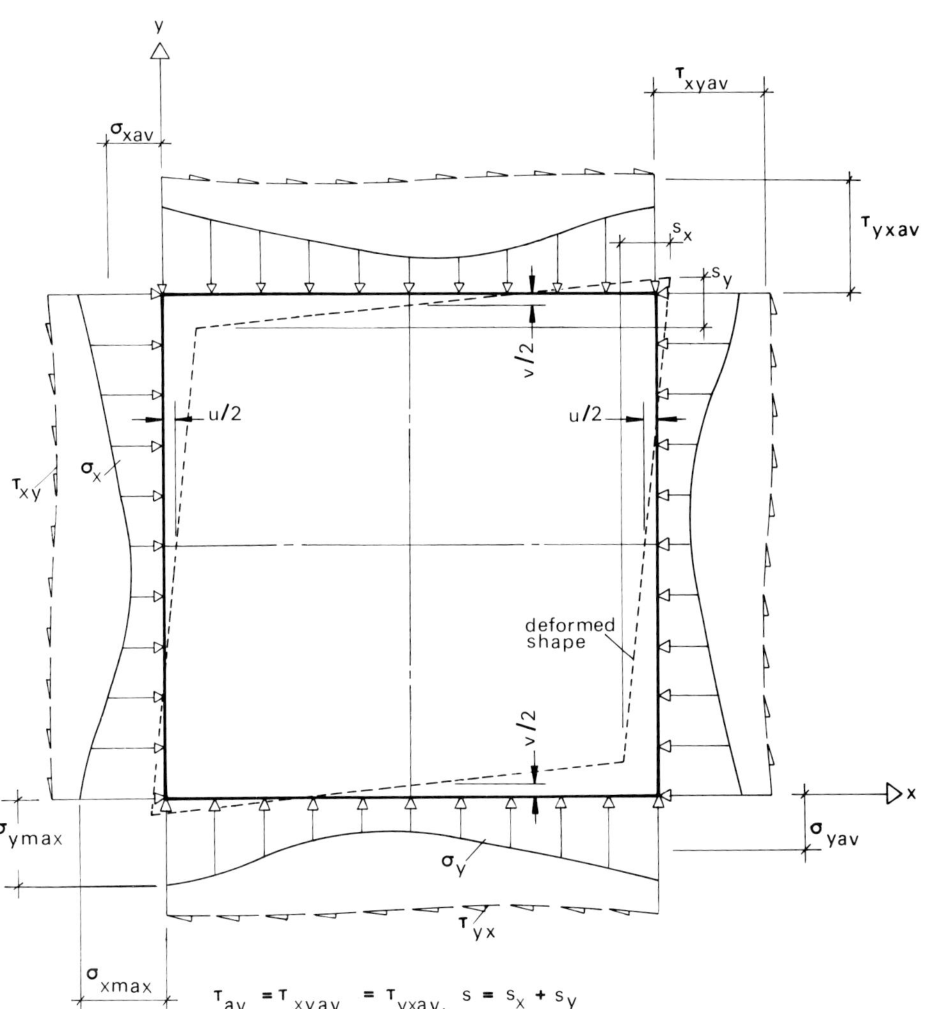


| Cases | | | | |
|---|---|---|---|---|
| a/b | I | J | Figures | Table |
| 1 | 10 | 10 | 4·52, 4·53 | 22 |
| 2 | 20 | 10 | 4·54, 4·55 | 23 |

Non dimensional factors

$\sigma_{eR} = (13)^{1/2}\sigma_{xav}$

$\sigma_{Rx} = \sigma_{xav}$

$\sigma_{Ry} = \sigma_{yav}$

$\tau_R = \tau_{av}$

$u_f = \{a(1-\nu)/E\}\sigma_{xav}$

$v_f = \{b(1-\nu)/E\}\sigma_{yav}$

$s_f = (a/G)\tau_{av}$

Fig. 4.50 Combined uniform biaxial compressive and shear displacement ($\sigma_{xav} = \sigma_{yav} = \tau_{av}/2$)

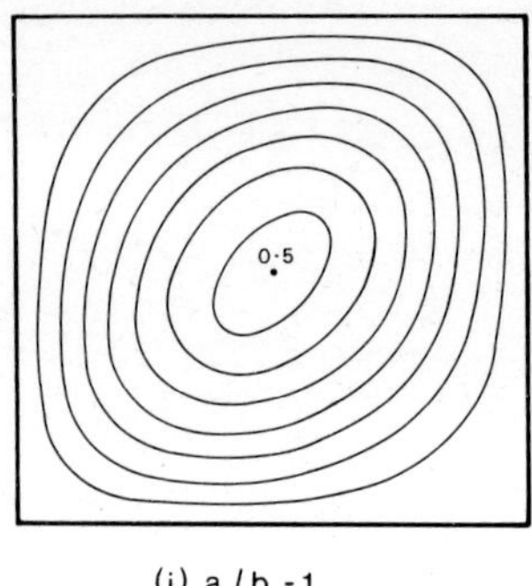


(i) a / b = 1

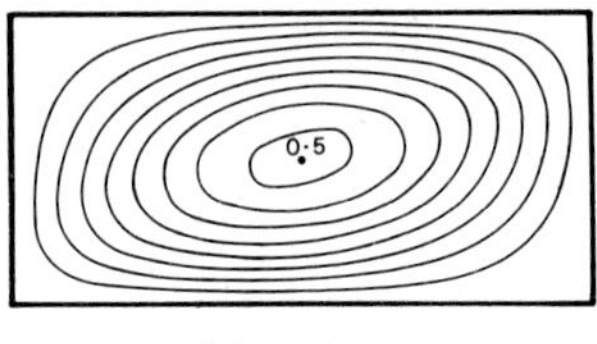


(ii) a / b = 2

Fig. 4.51 Initial imperfection contours ($w_0/t = 0{\cdot}5$). Combined biaxial compression and shear ($\sigma_{xav} = \sigma_{yav} = \tau_{av}/2$)

stress, even at higher loads. It is therefore recommended that corner surface yield be the serviceability criteria.

### 4.4.2 Unloaded sides stress-free

Typical boundary stresses and in-plane displacements: Fig. 4.62
Initial imperfections: Fig. 4.57
Design data: Figs. 4.63–66
Reference equivalent stress: $\sigma_{eR} = \sigma_{x1}$

The unloaded sides are free to 'follow' the buckle. As for restrained sides, $\sigma_{x1}$ is the value given by a global analysis assuming the plate is flat. The greater plate panel flexibility in this case, as compared to the case of restrained sides, results in maximum axial and equivalent membrane stresses, which occur partway along the unloaded side in line with maximum out-of-plane deflection (location 2). Maximum corner equivalent surface stresses are also higher, although the adjacent diagonal surface stress can be less than the applied value and very limited, if any, surface yield should be allowed.

## 4.5 Bending

Solutions are given for bending in the plane of the plate for side ratios of $a/b = 0{\cdot}667$, 1 and $1{\cdot}5$. Typical cases occur in vertically stiffened web panels of

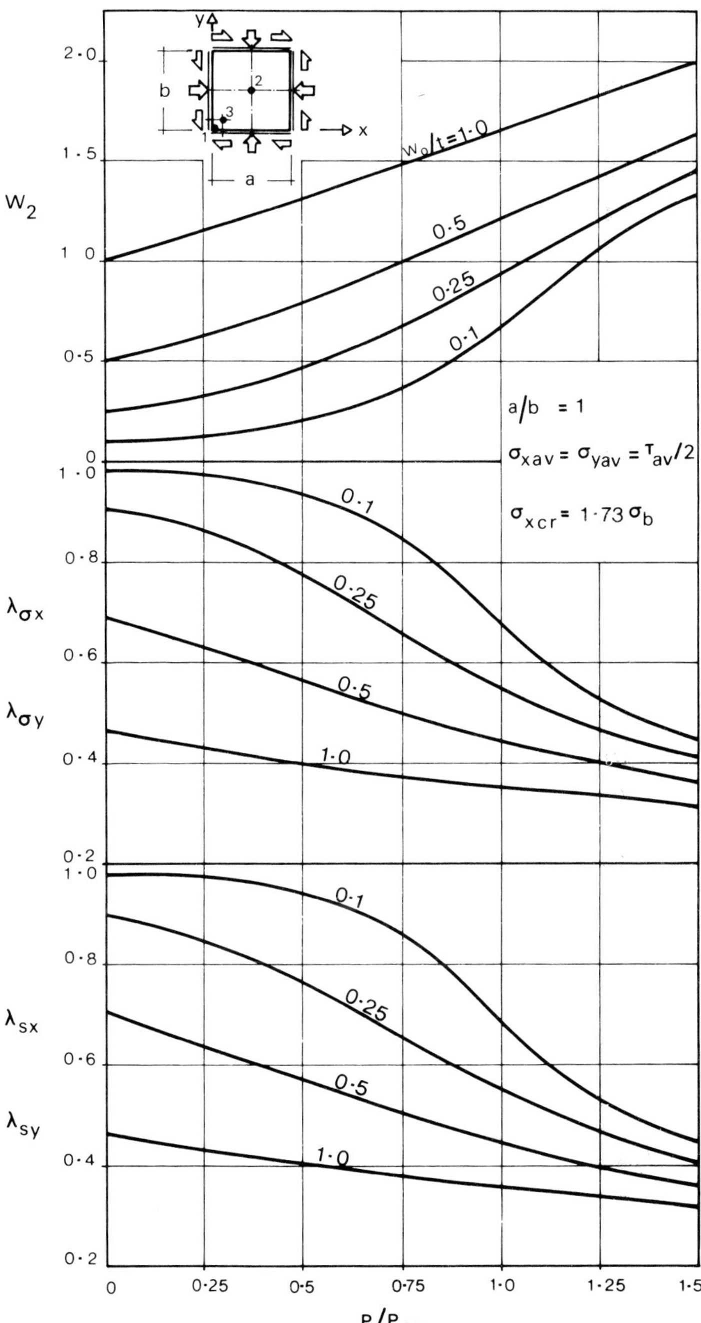


Fig. 4.52 Square ($a/b = 1$) plate subjected to combined uniform biaxial compressive and shear displacement ($\sigma_{xav} = \sigma_{yav} = \tau_{av}/2$). Deflection and effectiveness

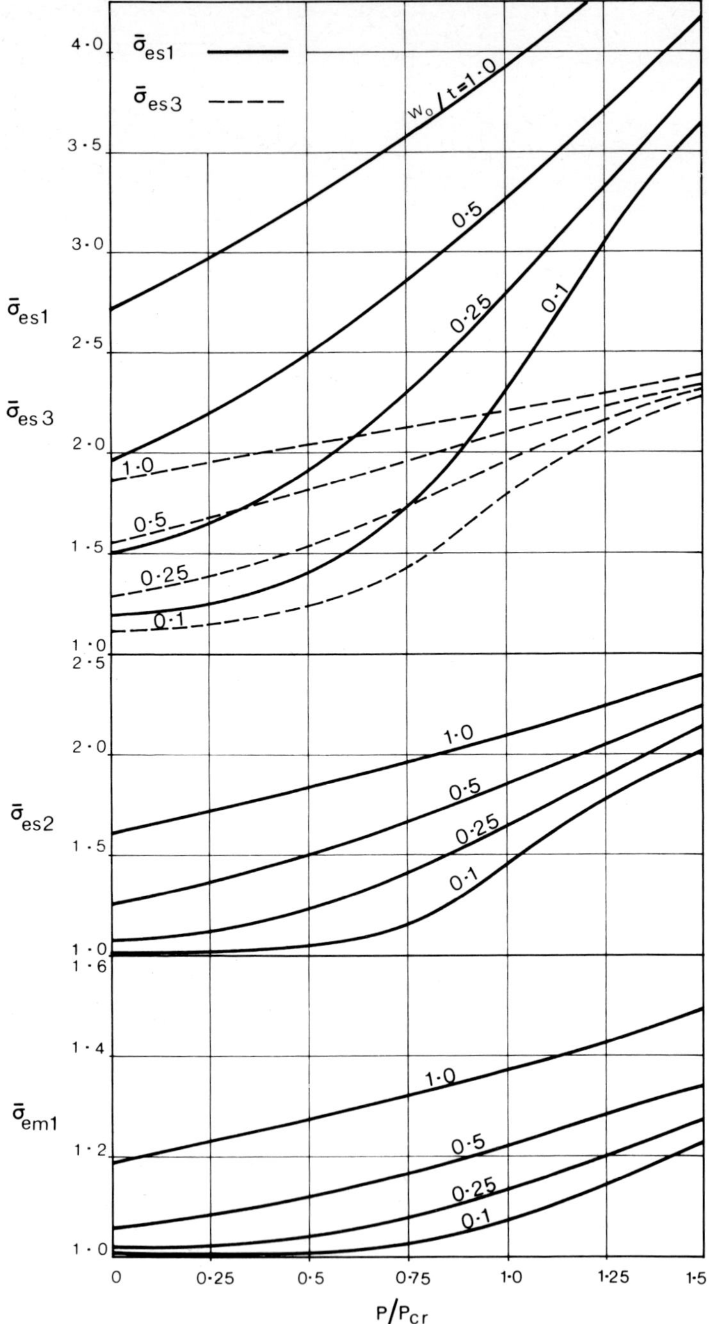


Fig. 4.53 Square ($a/b = 1$) plate subjected to combined uniform biaxial compressive and shear displacement ($\sigma_{xav} = \sigma_{yav} = \tau_{av}/2$). Equivalent membrane and surface stresses

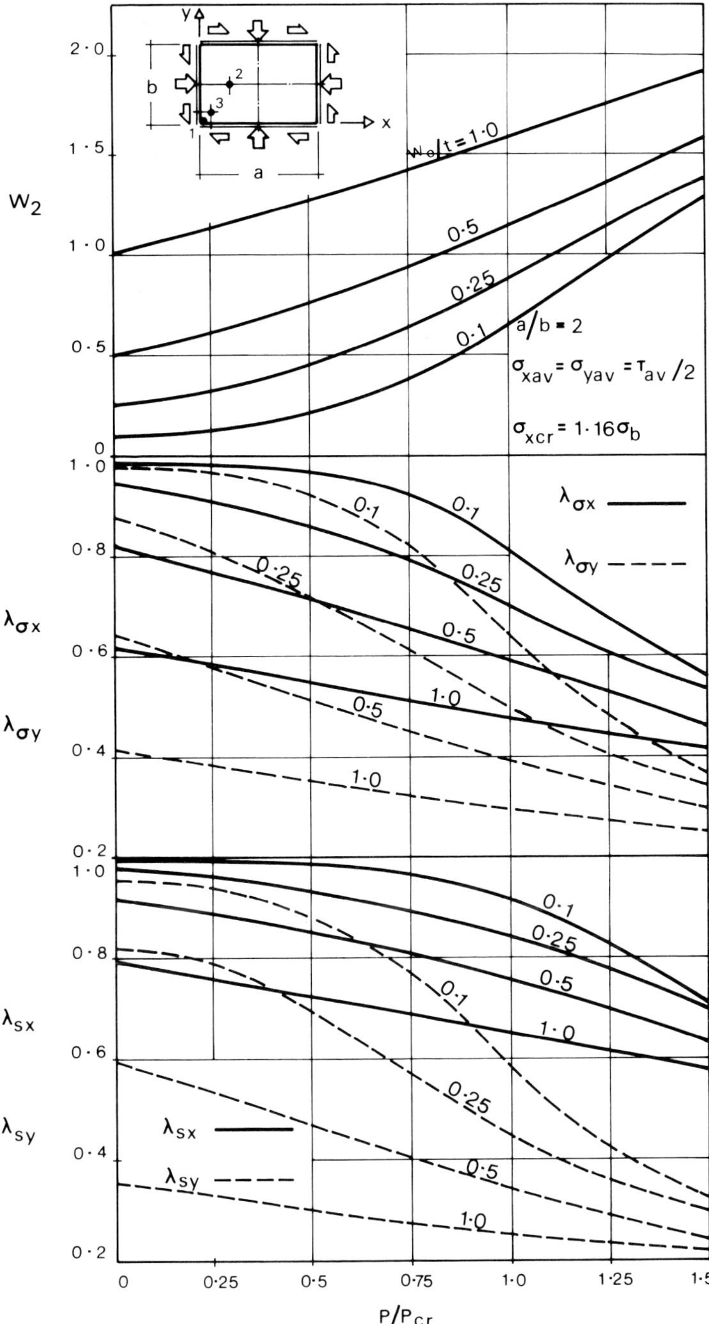


Fig. 4.54 Rectangular ($a/b = 2$) plate subjected to combined uniform biaxial compressive and shear displacement ($\sigma_{xav} = \sigma_{yav} = \tau_{av}/2$). Deflection and effectiveness

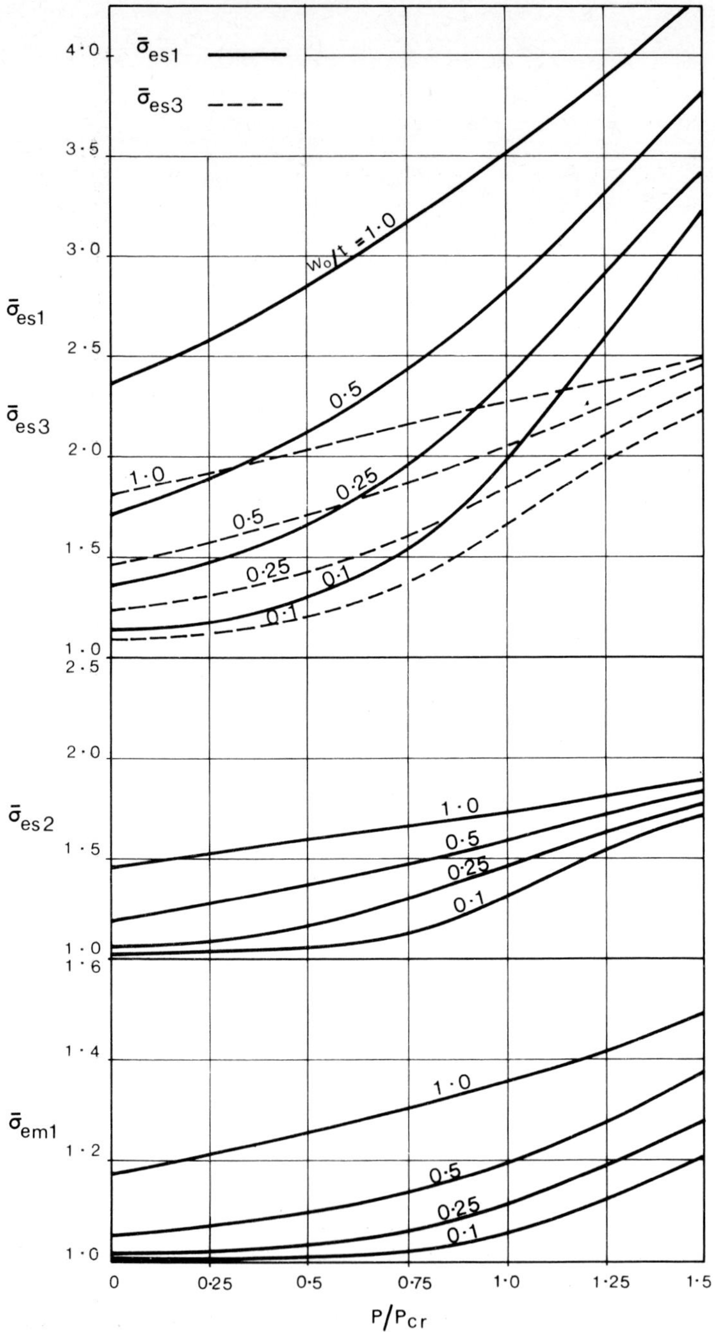


Fig. 4.55 Rectangular ($a/b = 2$) plate subjected to combined uniform biaxial compressive and shear displacement ($\sigma_{xav} = \sigma_{yav} = \tau_{av}/2$). Equivalent membrane and surface stresses

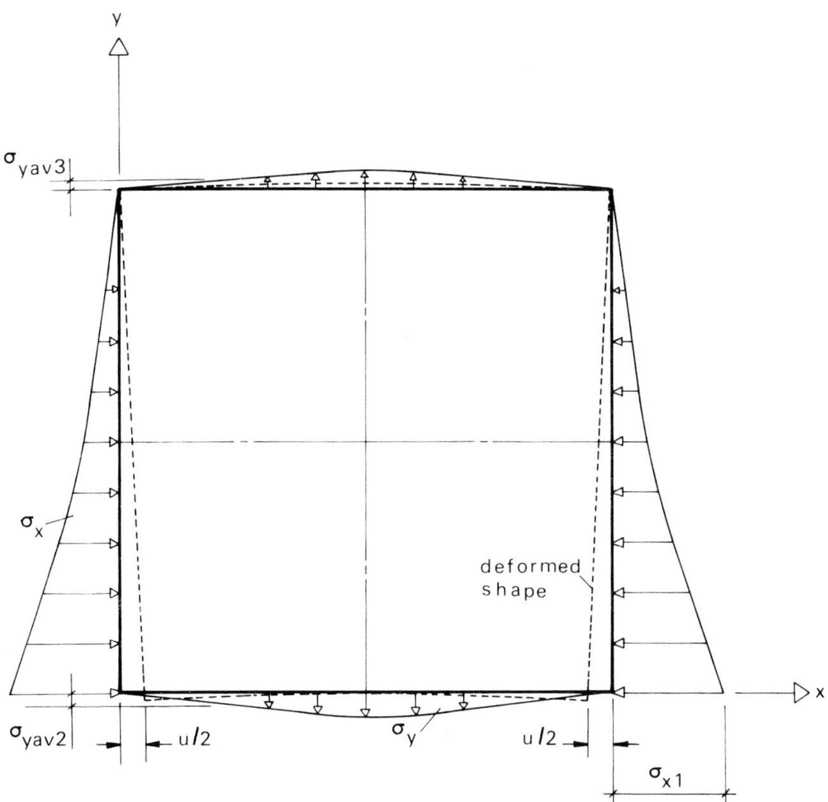


| Cases | | | | | Non dimensional factors |
|---|---|---|---|---|---|
| a/b | I | J | Figure | Table | $\sigma_{eR} = \sigma_{x1}$ |
| 0·667 | 10 | 14 | 4·58 | 24 | $\sigma_{Rx} = \sigma_{x1}$ |
| 1 | 10 | 10 | 4·59 | 25 | $\sigma_{Ry} = \sigma_{x1}$ |
| 1·5 | 14 | 10 | 4·60 | 26 | $u_f = (a/E)\sigma_{x1}$ |
| 2 | 20 | 10 | 4·61 | 27 | |

Fig. 4.56 Varying uniaxial compressive displacement; unloaded sides restrained

plate girders. Results cover applied displacement with and without the unloaded sides restrained and also applied stress with all sides unrestrained. All three conditions are identical for a flat plate, but in the case of the two applied displacement conditions, the stresses redistribute as the buckle develops. The restrained edge condition will usually apply in practice except where panels are discontinuous or not bounded by a stiff framing member.

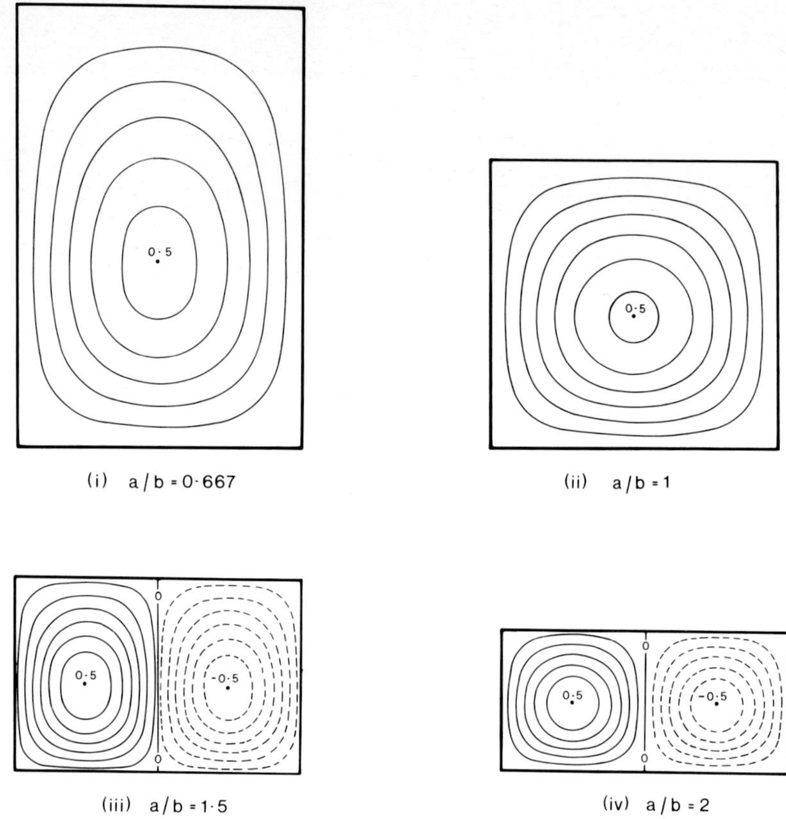


Fig. 4.57 Initial imperfection contours ($w_0/t = 0{\cdot}5$). Varying uniaxial compression

### 4.5.1 Bending displacement—unloaded sides restrained

Typical boundary stresses and in-plane displacements: Fig. 4.67
Initial imperfections: Fig. 4.68
Design data: Figs. 4.69–4.71
Reference equivalent stress: $\sigma_{eR} = \sigma_{x1}$

See comments for varying uniaxial compressive displacement, unloaded sides restrained in Section 4.4.1.

### 4.5.2 Bending displacement—unloaded sides stress-free

Typical boundary stresses and in-plane displacements: Fig. 4.72
Initial imperfections: Fig. 4.68
Design data: Figs. 4.73–75
Reference equivalent stress: $\sigma_{eR} = \sigma_{x1}$

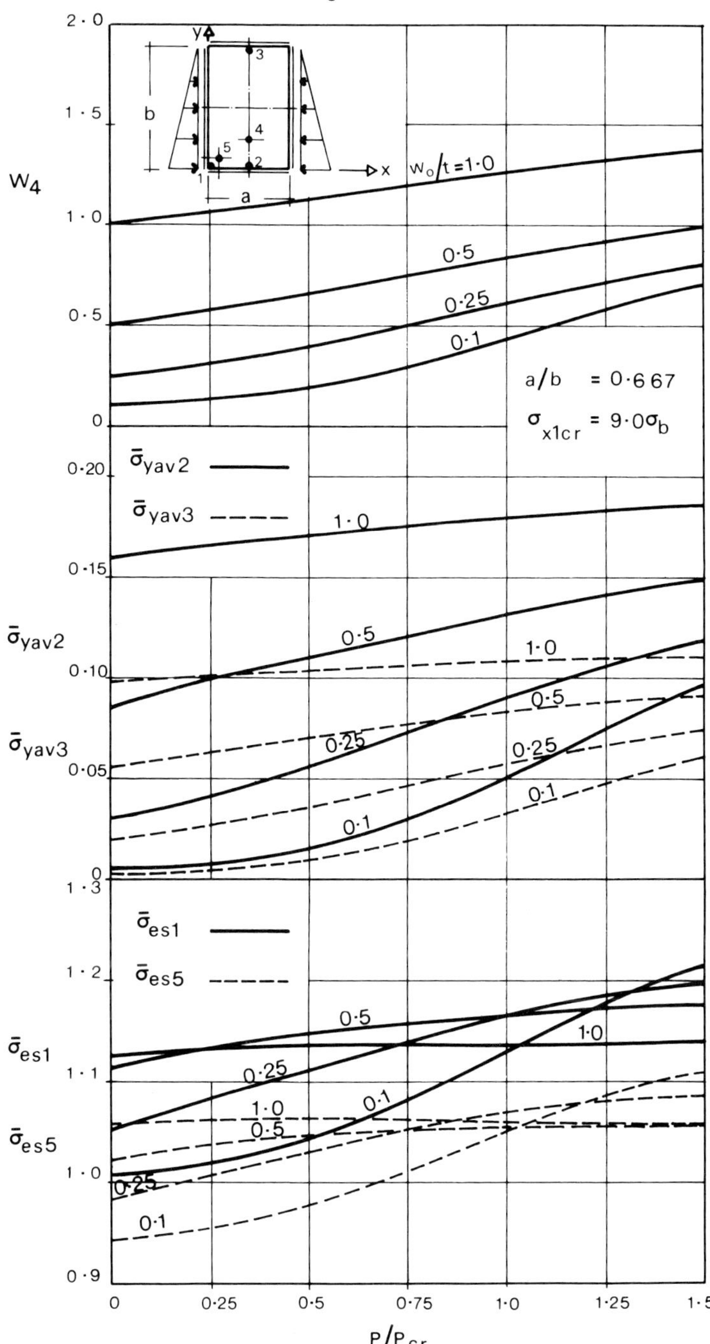


Fig. 4.58 Rectangular ($a/b = 0{\cdot}667$) plate subjected to varying uniaxial compressive displacement; unloaded sides restrained. Deflection, average edge restraining stresses and equivalent surface stresses

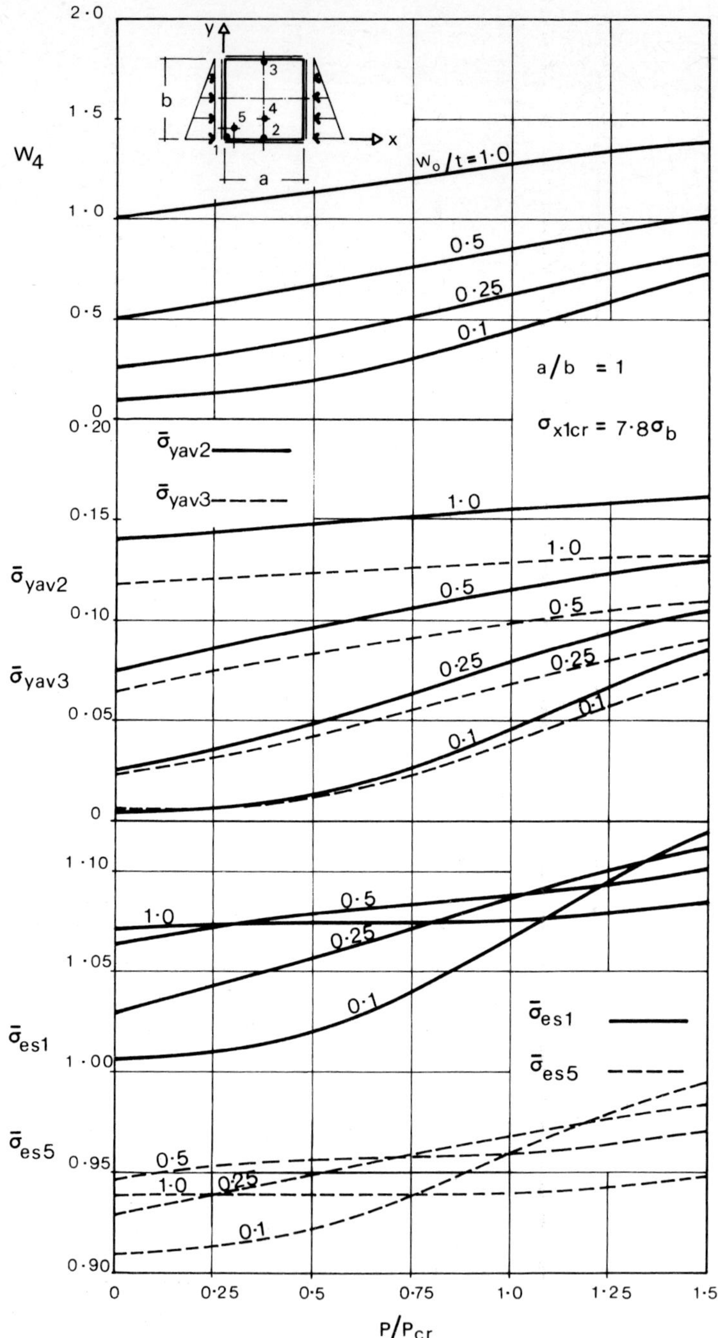


Fig. 4.59 Square ($a/b = 1$) plate subjected to varying uniaxial compressive displacement; unloaded sides restrained. Deflection, average edge restraining stresses and equivalent surface stresses

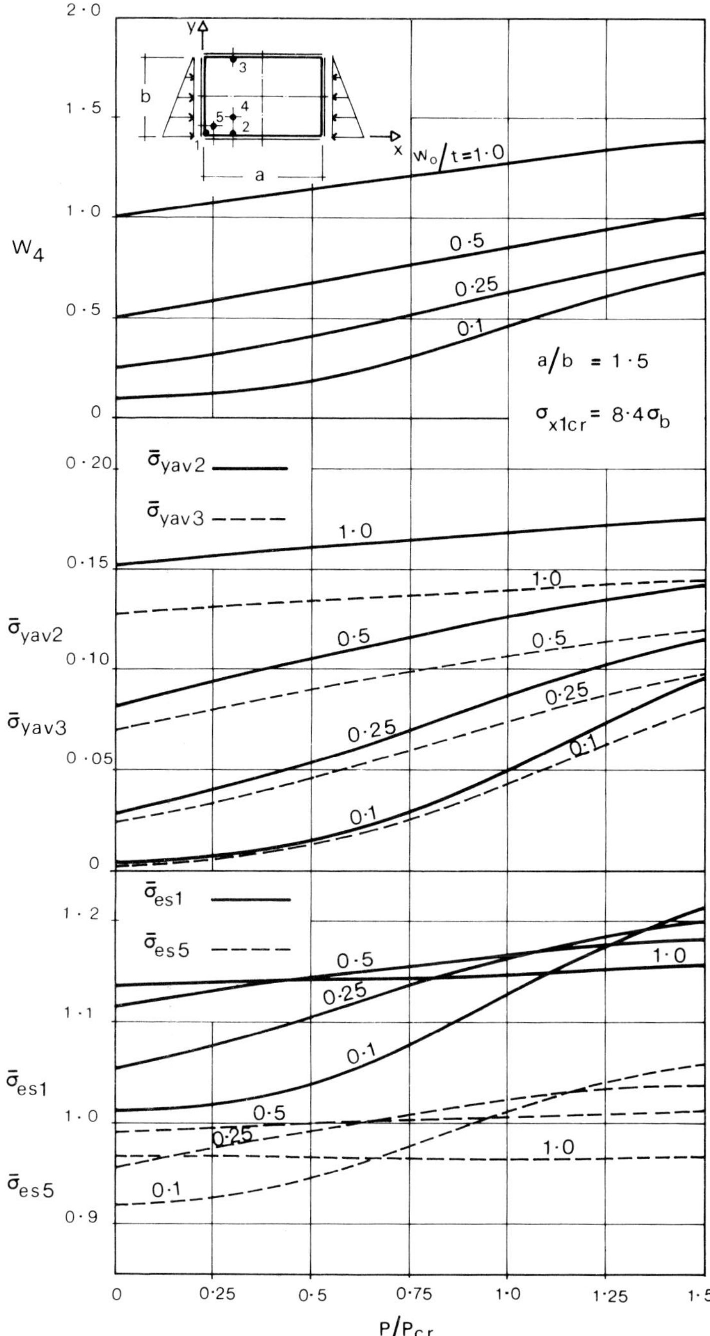


Fig. 4.60 Rectangular ($a/b = 1 \cdot 5$) plate subjected to varying uniaxial compressive displacement; unloaded sides restrained. Deflection, average edge restraining stresses and equivalent surface stresses

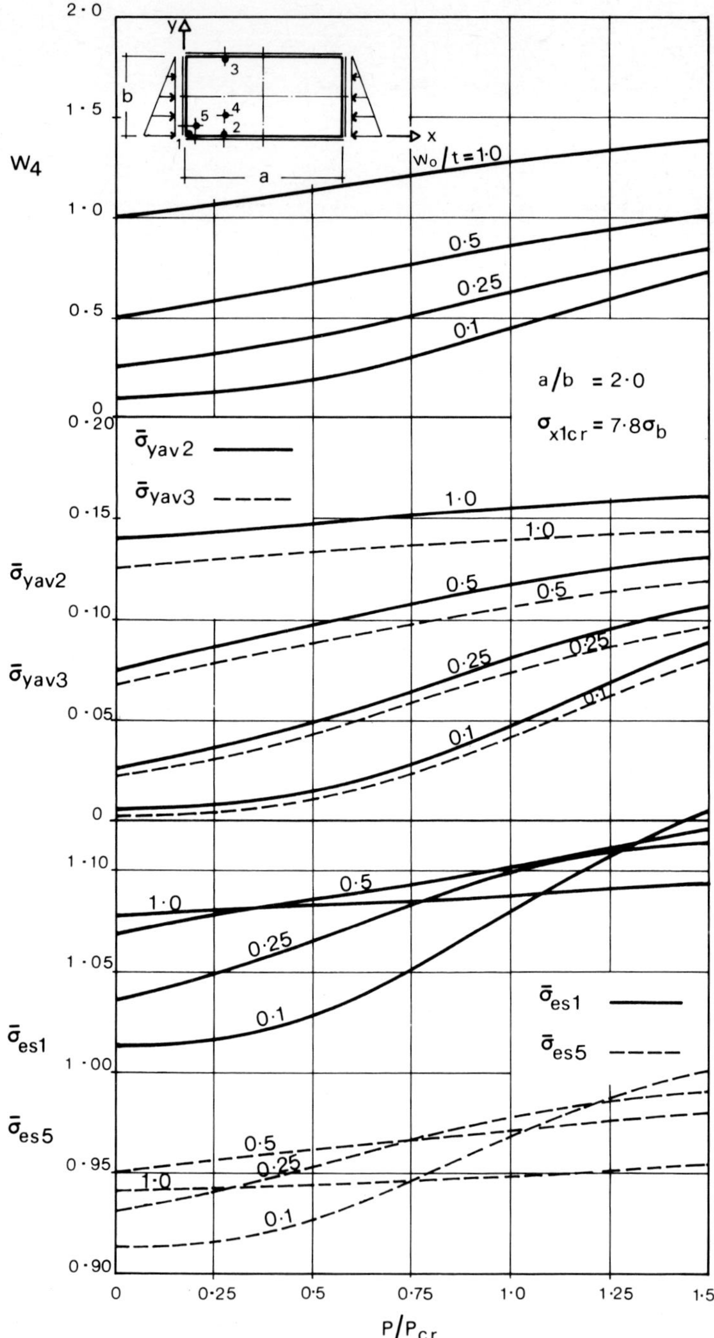


Fig. 4.61 Rectangular ($a/b = 2$) plate subjected to varying uniaxial compressive displacement; unloaded sides restrained. Deflection, average edge restraining stresses and equivalent surface stresses

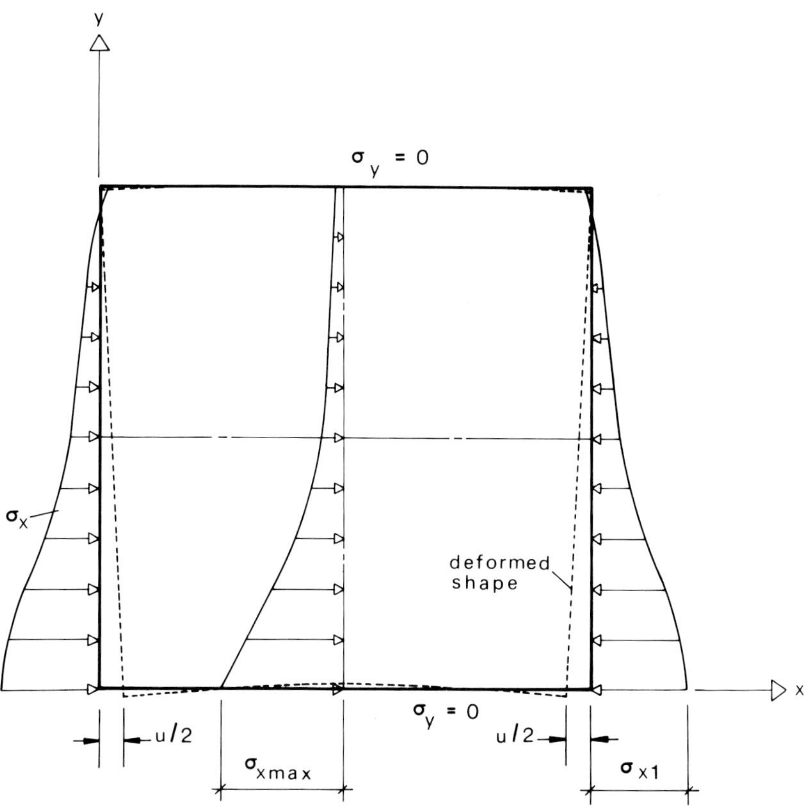


Cases

| a/b | I | J | Figure | Table |
|---|---|---|---|---|
| 0·667 | 10 | 14 | 4·63 | 28 |
| 1 | 10 | 10 | 4·64 | 29 |
| 1·5 | 14 | 10 | 4·65 | 30 |
| 2 | 20 | 10 | 4·66 | 31 |

Non dimensional factors

$\sigma_{eR} = \sigma_{x1}$

$\sigma_{Rx} = \sigma_{x1}$

$\sigma_{Ry} = \sigma_{x1}$

$u_f = (a/E)\sigma_{x1}$

Fig. 4.62 Varying uniaxial compressive displacement; unloaded sides stress free

The comments given in Section 4.4.2 for varying uniaxial compressive displacement, unloaded sides unrestrained apply, except that in this case the maximum equivalent surface stress occurs partway along the unloaded side (location 2) in certain cases, for which it is recommended that surface yield be the serviceability criteria.

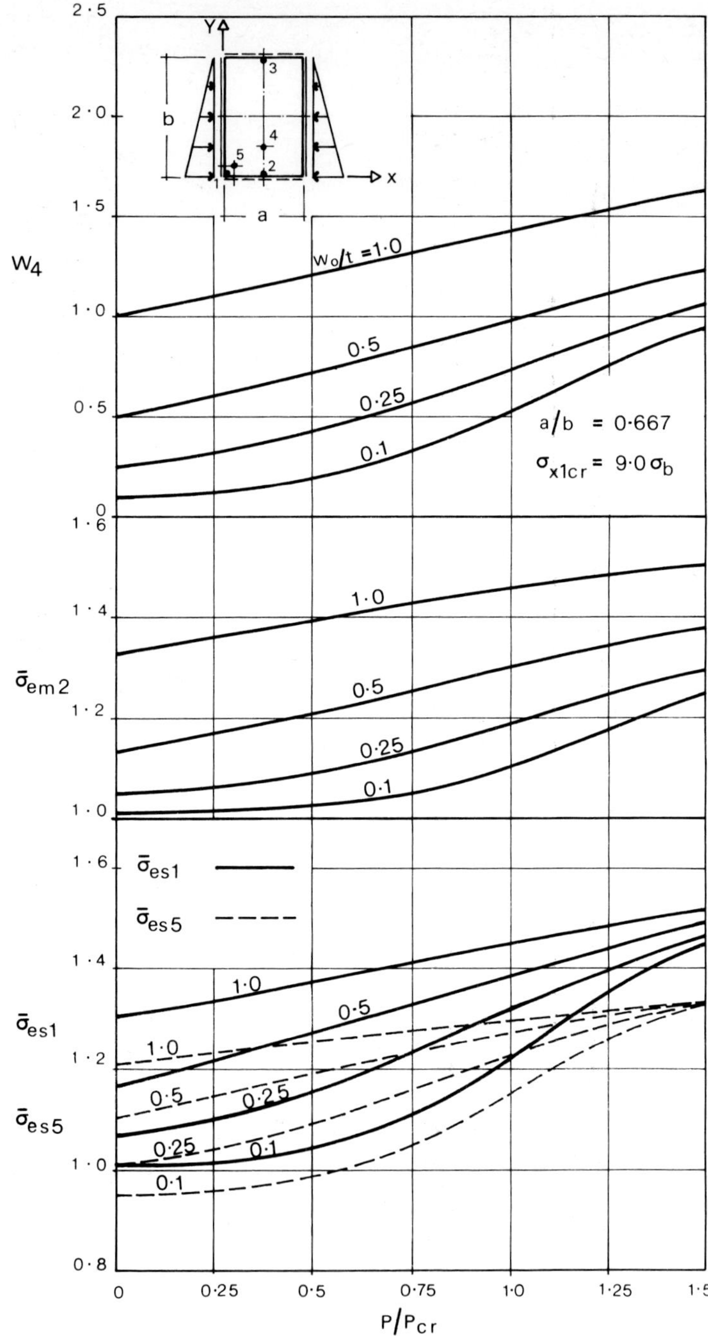


Fig. 4.63 Rectangular ($a/b = 0{\cdot}667$) plate subjected to varying uniaxial compressive displacement; unloaded sides stress-free. Deflection, equivalent membrane and surface stresses

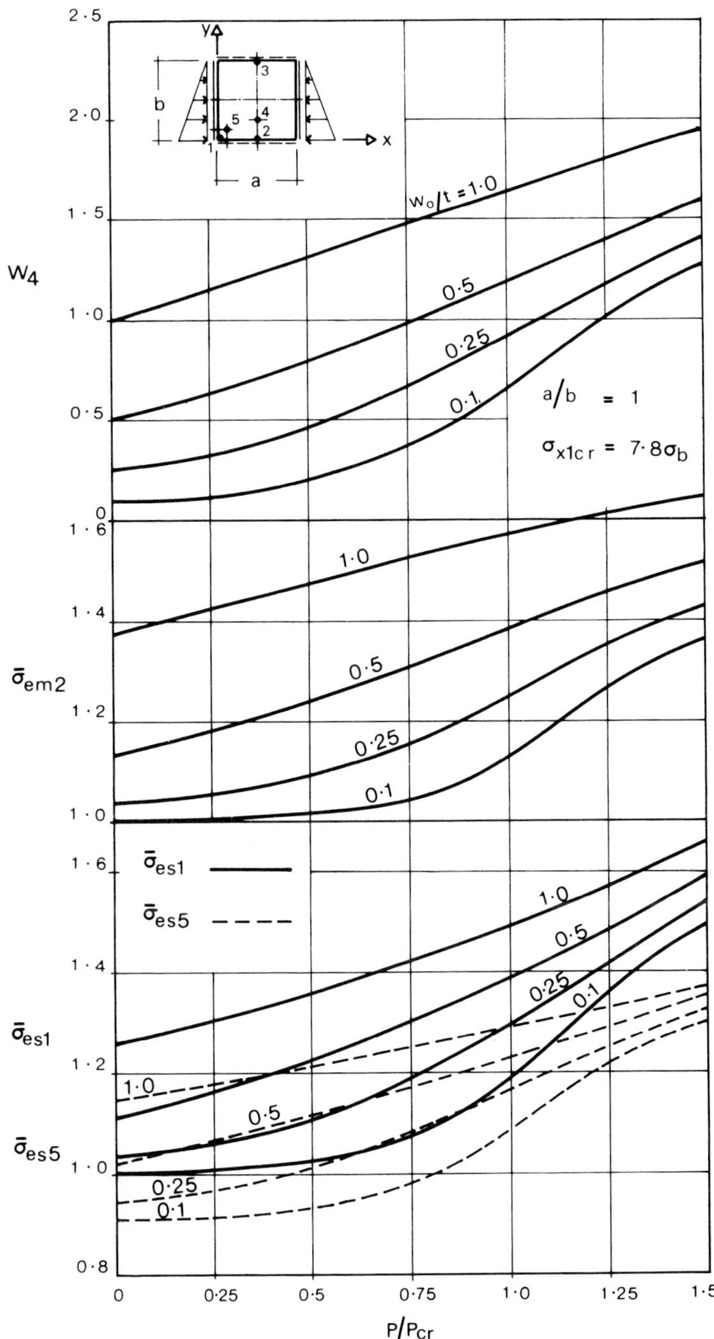


Fig. 4.64 Square ($a/b = 1$) plate subjected to varying uniaxial compressive displacement; unloaded sides stress-free. Deflection, equivalent membrane and surface stresses

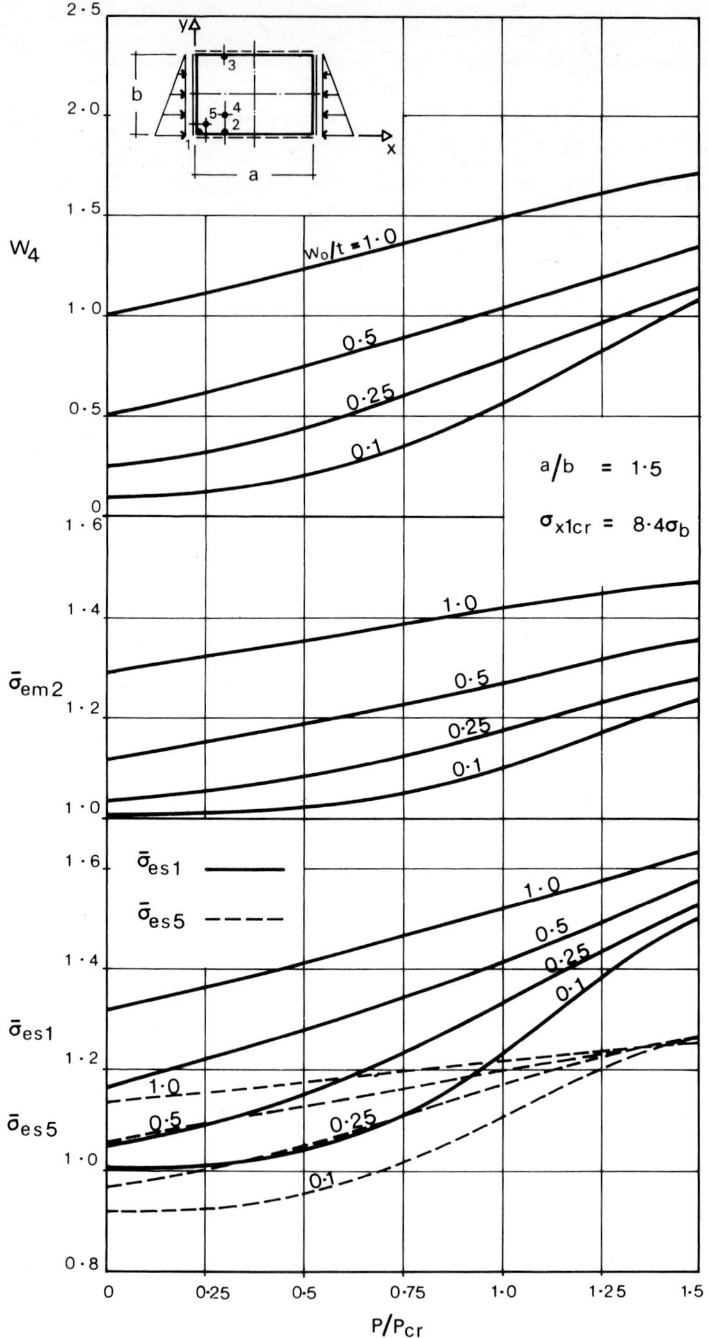


Fig. 4.65 Rectangular ($a/b = 1\cdot5$) plate subjected to varying uniaxial compressive displacement; unloaded sides stress-free. Deflection equivalent membrane and surface stresses

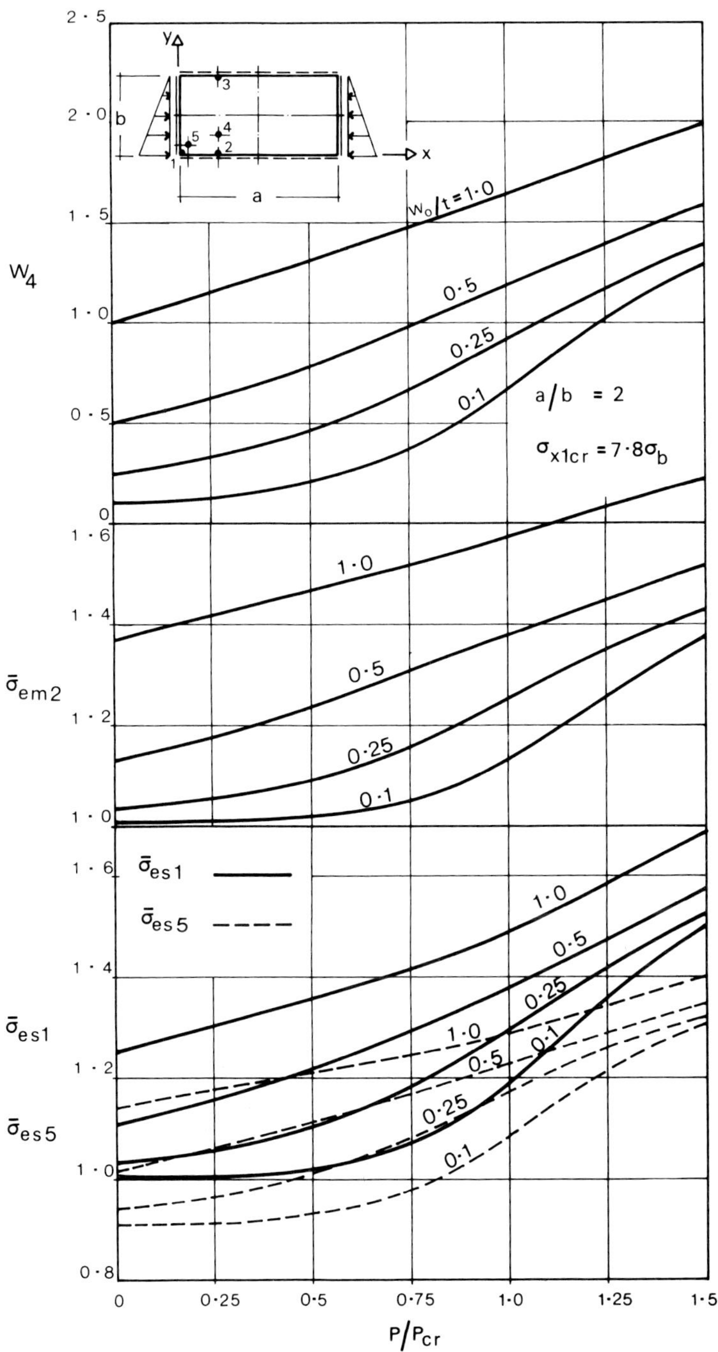


Fig. 4.66 Rectangular ($a/b = 2$) plate subjected to varying uniaxial compressive displacement; unloaded sides stress-free. Deflection, equivalent membrane and surface stresses

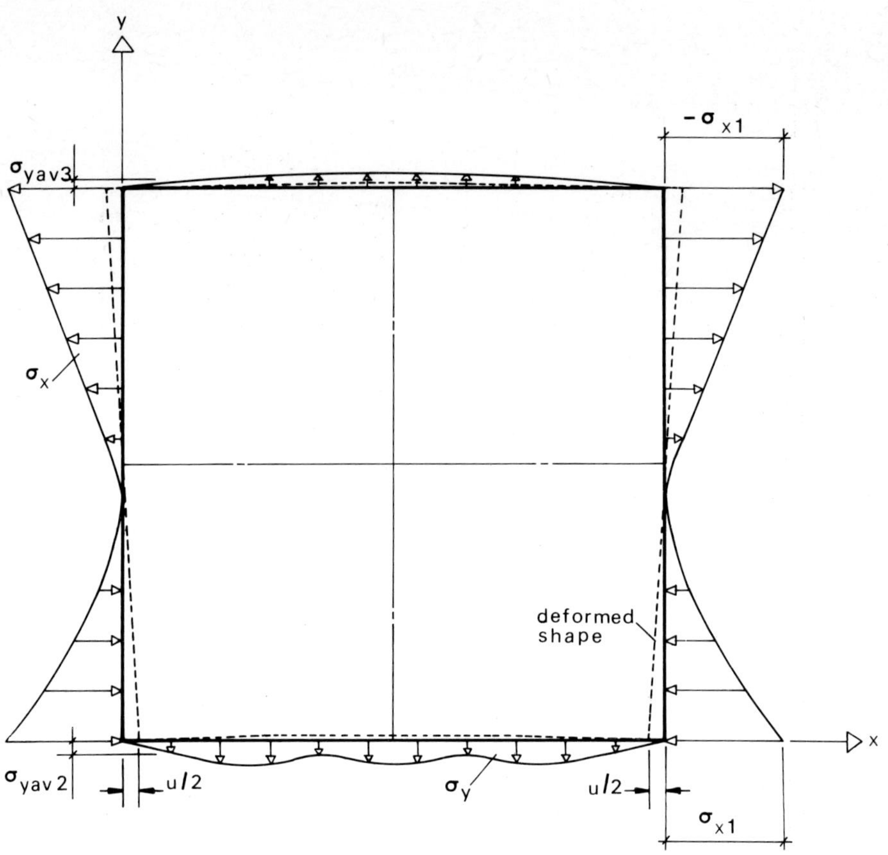


**Cases**

| a/b | I | J | Figure | Table |
|---|---|---|---|---|
| 0·667 | 11 | 15 | 4·69 | 32 |
| 1 | 11 | 11 | 4·70 | 33 |
| 1·5 | 15 | 11 | 4·71 | 34 |

**Non dimensional factors**

$\sigma_{eR} = \sigma_{x1}$

$\sigma_{Rx} = \sigma_{x1}$

$\sigma_{Ry} = \sigma_{x1}$

$u_f = (a/E)\sigma_{x1}$

Fig. 4.67 Bending displacement; unloaded sides restrained

### 4.5.3 Bending stress—unloaded sides stress-free

Typical boundary stresses and in-plane displacements: Fig. 4.76
Initial imperfections: Fig. 4.68
Design data: Figs. 4.77–79
Reference equivalent stress: $\sigma_{eR} = \sigma_{x1}$

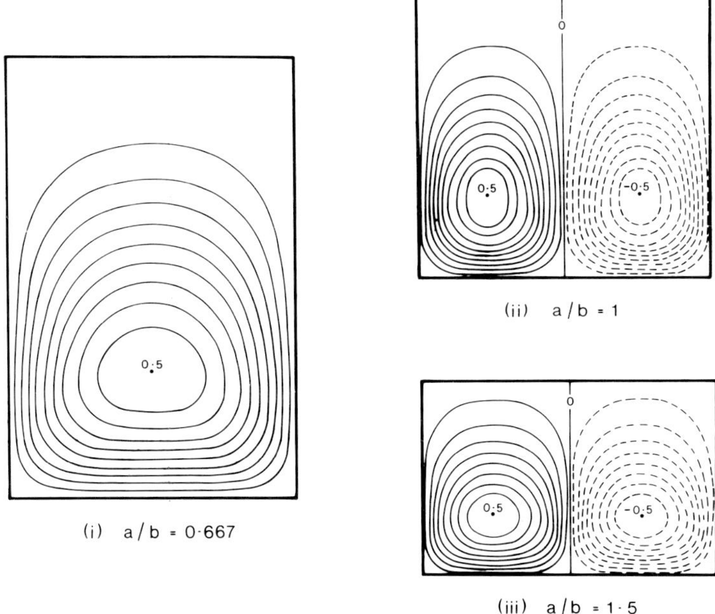


Fig. 4.68 Initial imperfection contours ($w_0/t = 0{\cdot}5$). Bending

The applied stress distribution is linear and remains so and the sides are free to 'follow' the buckle. Redistribution due to load shedding in the interior of the plate results in maximum axial and equivalent membrane stress occurring part-way along the unloaded sides (location 2). Maximum equivalent surface stress occurs at the corners (location 1) and the stress gradient at higher loads becomes steep enough to justify designing for restricted yield.

## 4.6 Combined bending and shear displacement

Typical boundary stresses and in-plane displacements: Fig. 4.80
Initial imperfections: Fig. 4.81
Design data: Figs. 4.82–84
Reference equivalent stress: $\sigma_{\mathrm{eR}} = (\sigma_{x1}^2 + 3\tau_{\mathrm{av}}^2)^{1/2}$

Solutions are given for bending combined with shear for side ratios of $a/b = 0{\cdot}667$, 1 and 1·5 for bending ($\sigma_{x1}$) and shear ($\tau_{\mathrm{av}}$) stress in the ratio of their respective independent critical buckling loads. All sides are restrained and there is therefore some redistribution of side loading, although not enough to significantly affect the latter bending to shear stress ratio. This condition applies to interior panel sides or sides restrained by stiff framing members. There will be cases of free edges for which the results are slightly unconservative.

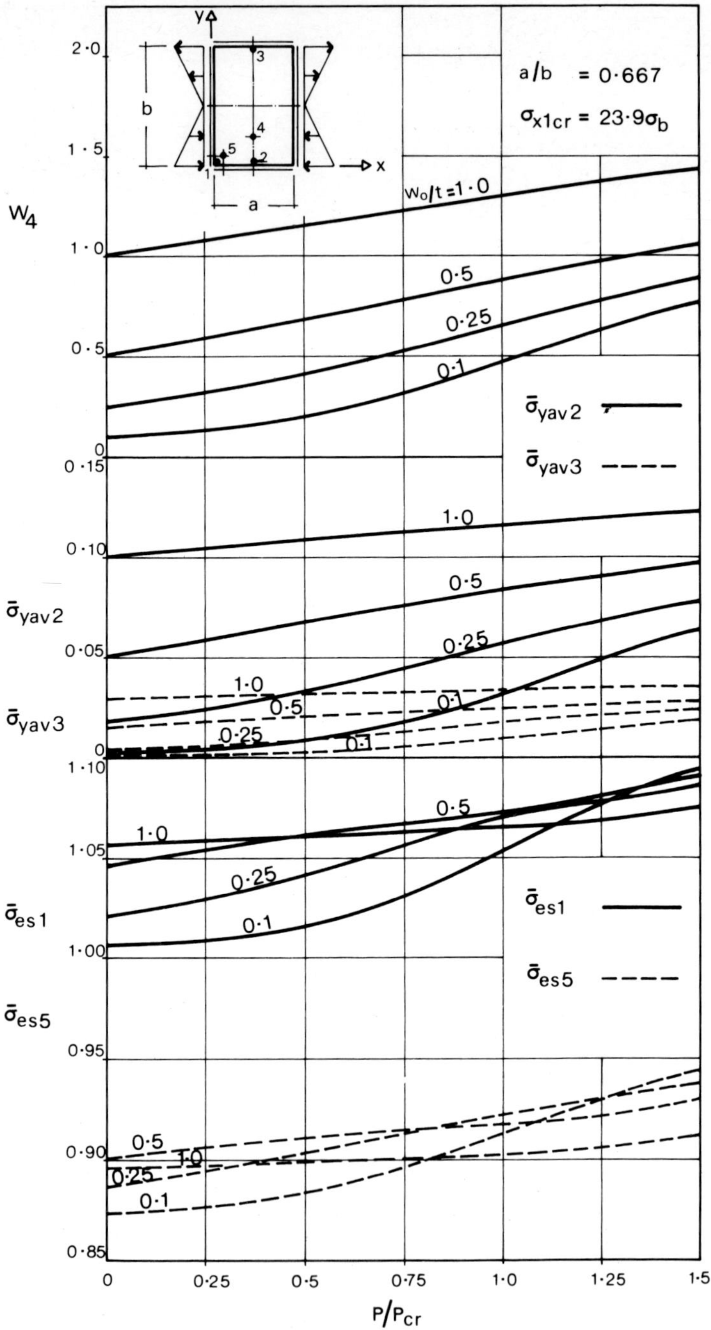


Fig. 4.69 Rectangular ($a/b = 0{\cdot}667$) plate subjected to bending displacement; unloaded sides restrained. Deflection, average edge restraining stresses, equivalent surface stresses

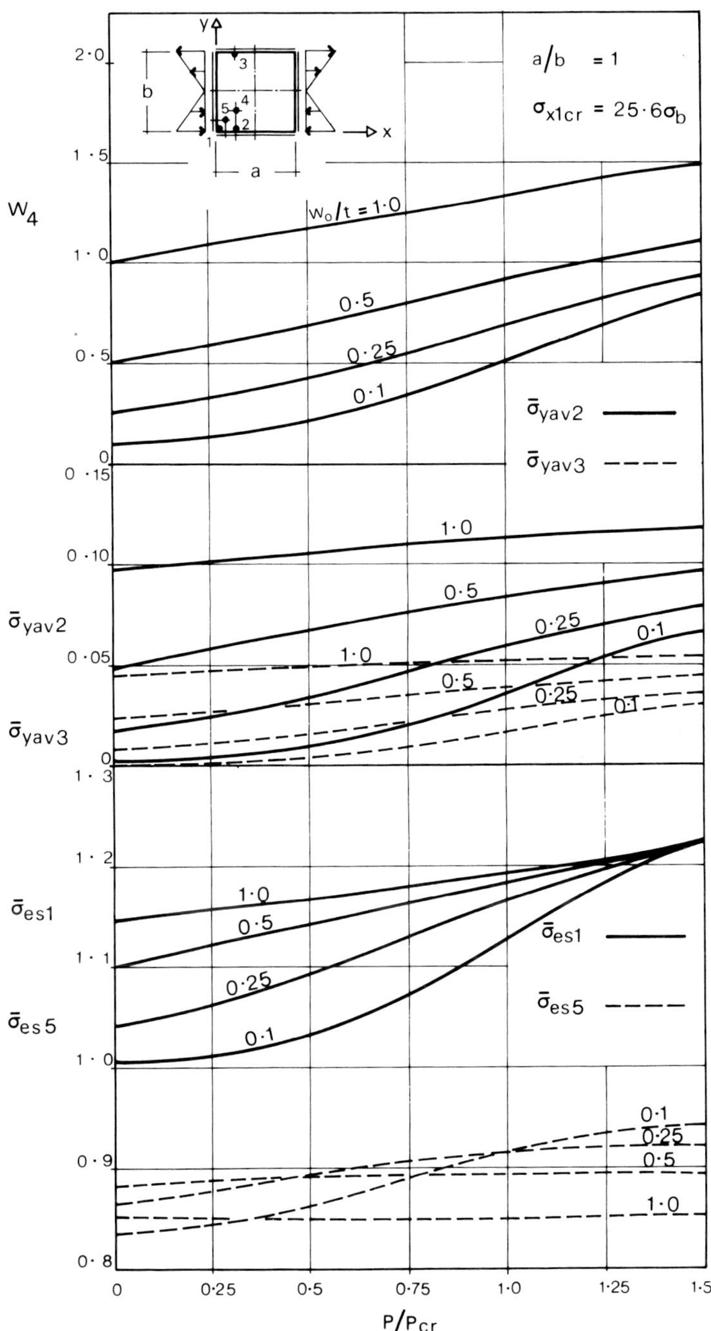


Fig. 4.70 Square ($a/b = 1$) plate subjected to bending displacement; unloaded sides restrained. Deflection, average edge restraining stresses, equivalent surface stresses

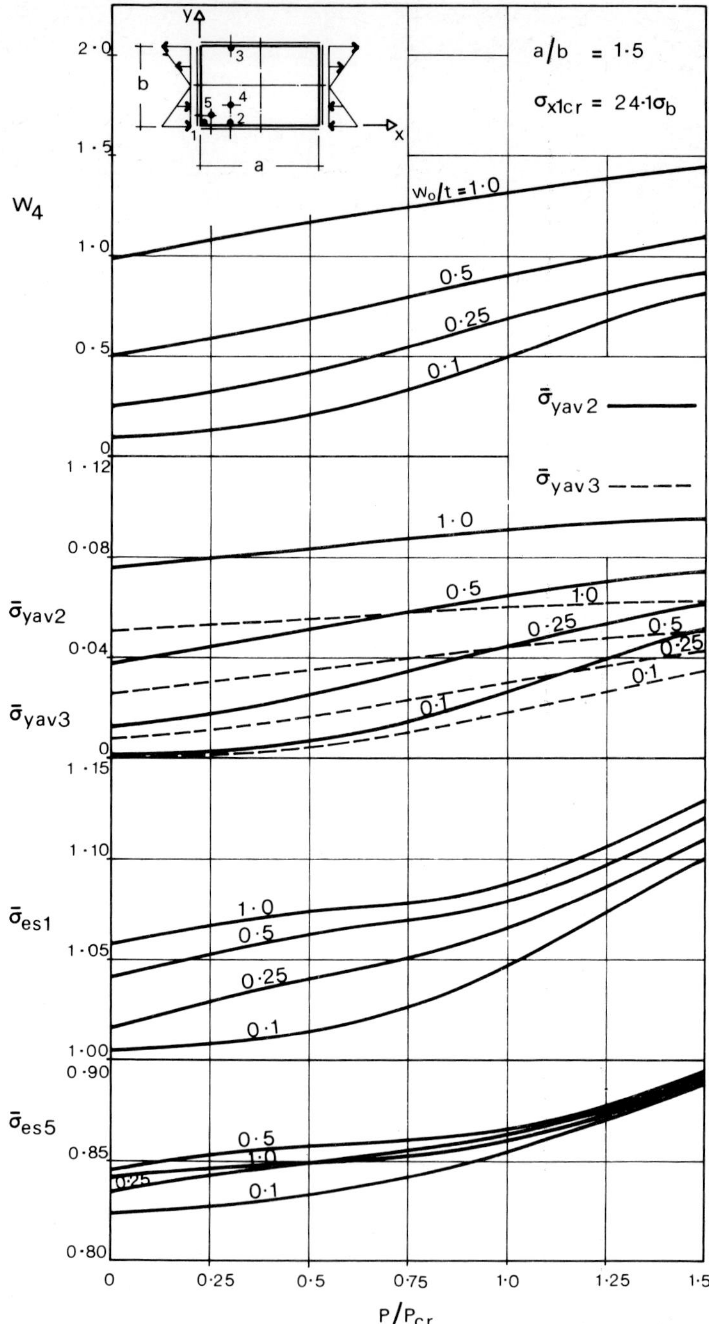


Fig. 4.71 Rectangular ($a/b$ = 1·5) plate subjected to bending displacement; unloaded sides restrained. Deflection, average edge restraining stresses, equivalent surface stresses

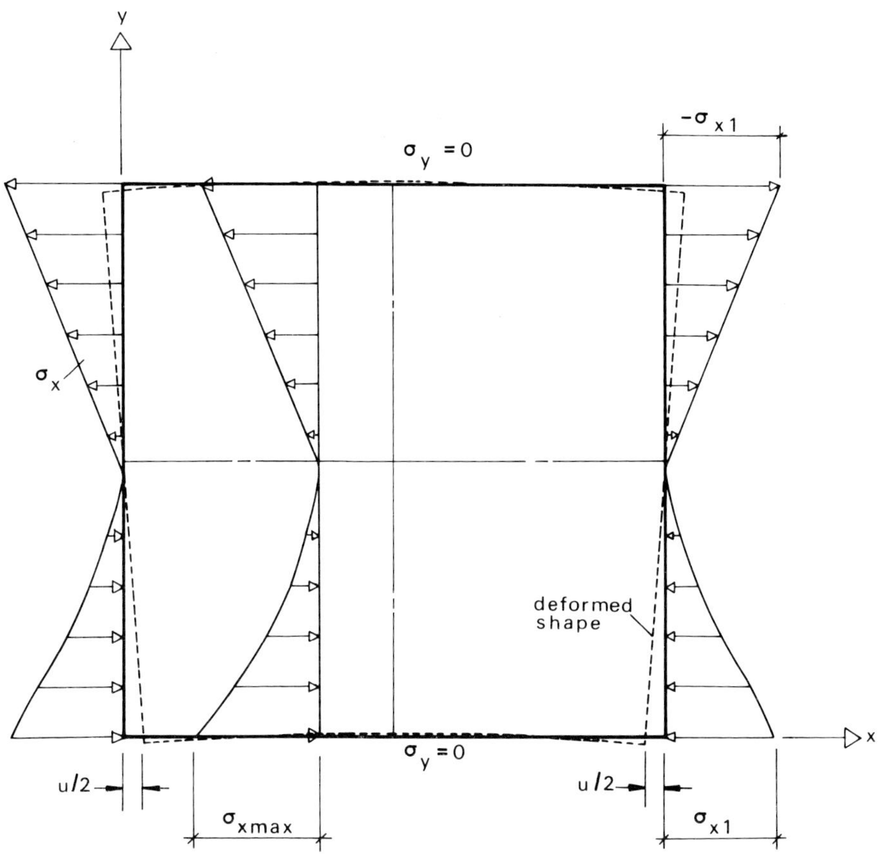


Cases

| a/b | I | J | Figure | Table |
|---|---|---|---|---|
| 0·667 | 11 | 15 | 4·73 | 35 |
| 1 | 11 | 11 | 4·74 | 36 |
| 1·5 | 15 | 11 | 4·75 | 37 |

Non dimensional factors

$\sigma_{eR} = \sigma_{x1}$

$\sigma_{Rx} = \sigma_{x1}$

$\sigma_{Ry} = \sigma_{x1}$

$u_f = (a/E)\sigma_{x1}$

Fig. 4.72 Bending displacement; unloaded sides stress-free

As for the restrained side load cases described in Sections 4.4.1 and 4.5.1 the applied stresses $\sigma_{x1}$ and $\tau_{av}$ can be taken as the values given from a global analysis assuming the plate is flat and the loss of effectiveness can be ignored in calculating global deflections. Average side restraining stresses are plotted to be used in appraising side framing with maximum values provided in the tables in Appendix 1 if required.

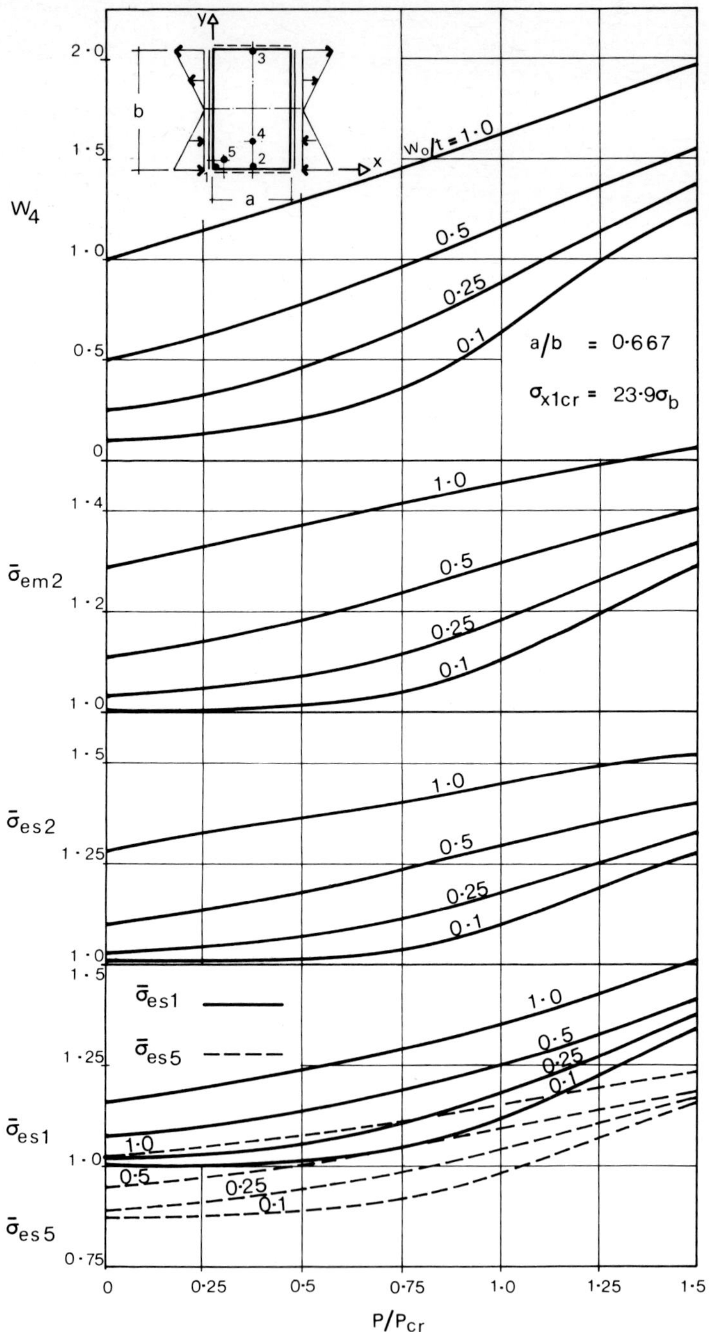


Fig. 4.73 Rectangular ($a/b = 0{\cdot}667$) plate subjected to bending displacement; unloaded sides stress-free. Deflection, equivalent membrane and surface stresses

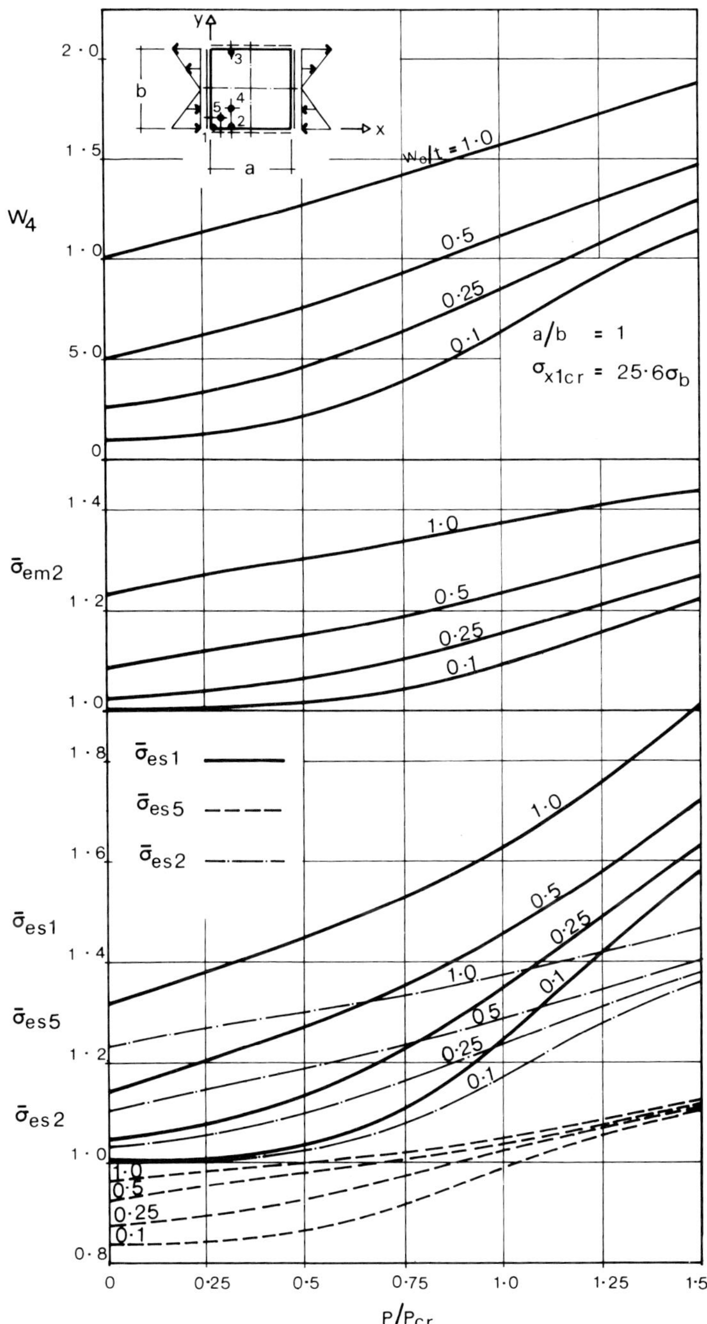


Fig. 4.74 Square ($a/b = 1$) plate subjected to bending displacement; unloaded sides stress-free. Deflection, equivalent membrane and surface stresses

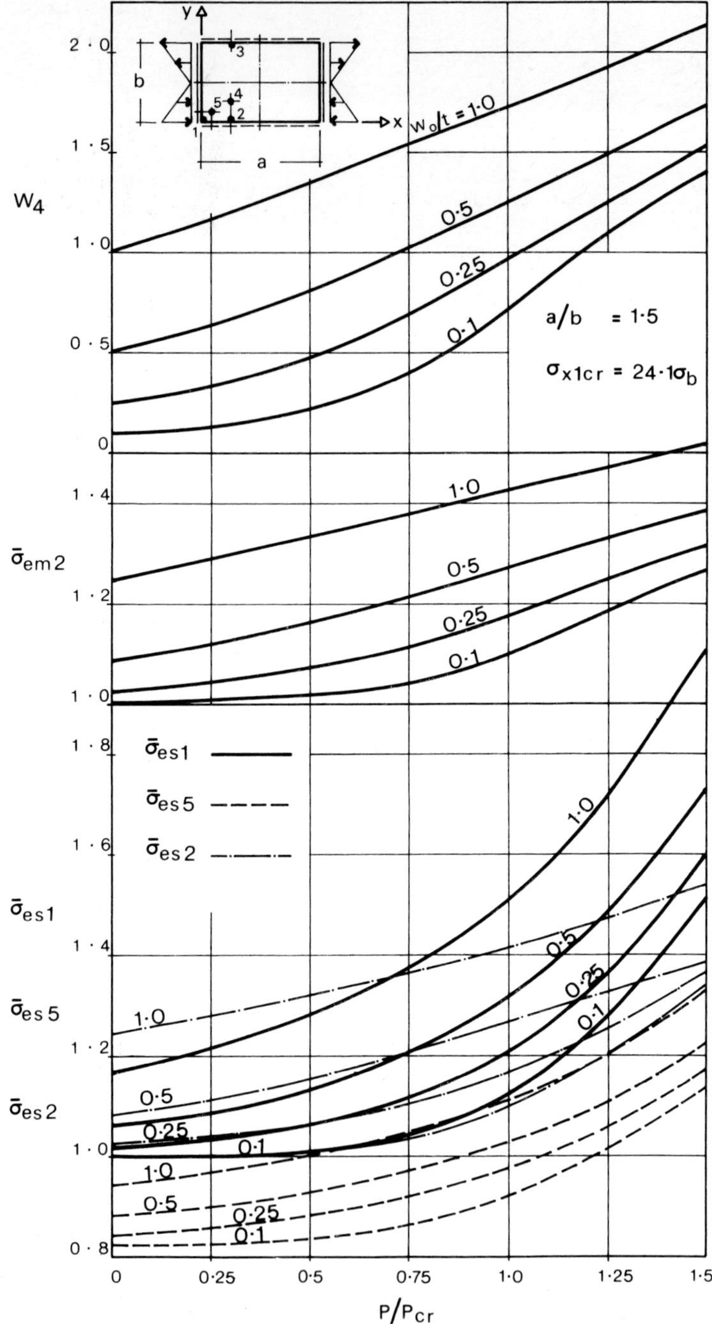


Fig. 4.75 Rectangular ($a/b = 1{\cdot}5$) plate subjected to bending displacement; unloaded sides stress-free. Deflection, equivalent membrane and surface stresses

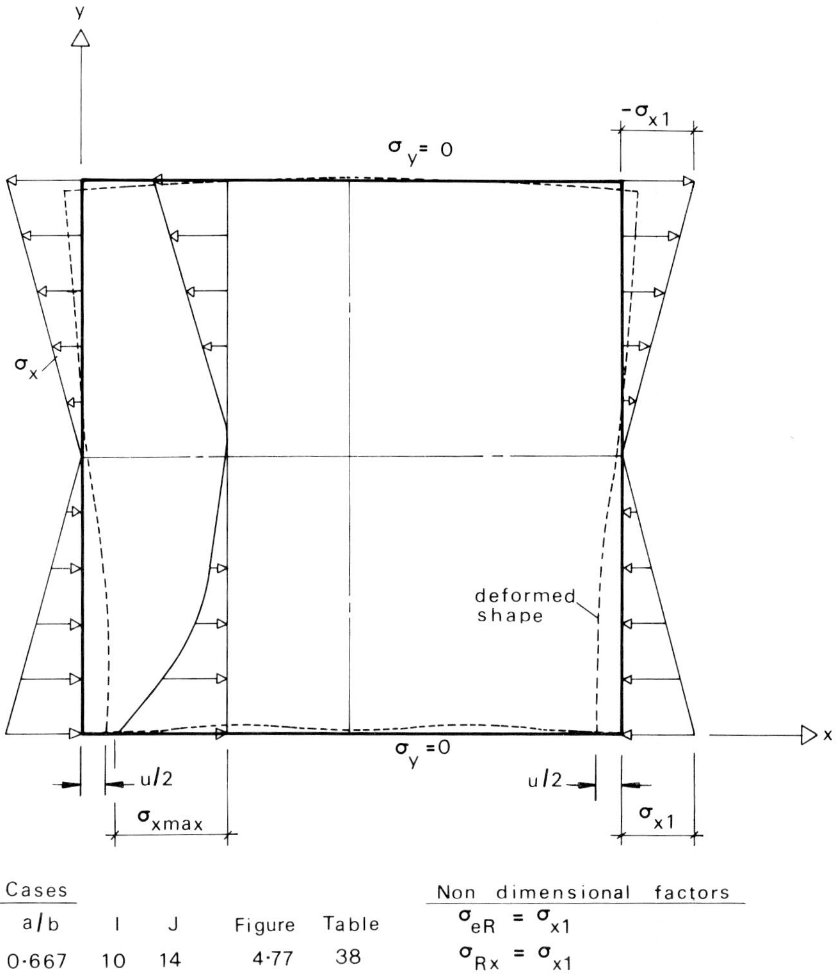


| Cases a/b | I | J | Figure | Table |
|---|---|---|---|---|
| 0·667 | 10 | 14 | 4·77 | 38 |
| 1 | 10 | 10 | 4·78 | 39 |
| 1·5 | 14 | 10 | 4·79 | 40 |

Non dimensional factors

$\sigma_{eR} = \sigma_{x1}$

$\sigma_{Rx} = \sigma_{x1}$

$\sigma_{Ry} = \sigma_{x1}$

$u_f = (a/E)\sigma_{x1}$

Fig. 4.76 Bending stress; unloaded sides stress-free

Maximum equivalent membrane stress occurs partway along the unloaded side (location 2). It is not plotted but can be computed from the tables as described in Section 4.4.1. Maximum equivalent surface stress occurs at the corners (location 1) and although the stresses at the adjacent diagonal are relatively low the stress gradient is very steep. Allowance should therefore be made for restricted local yield.

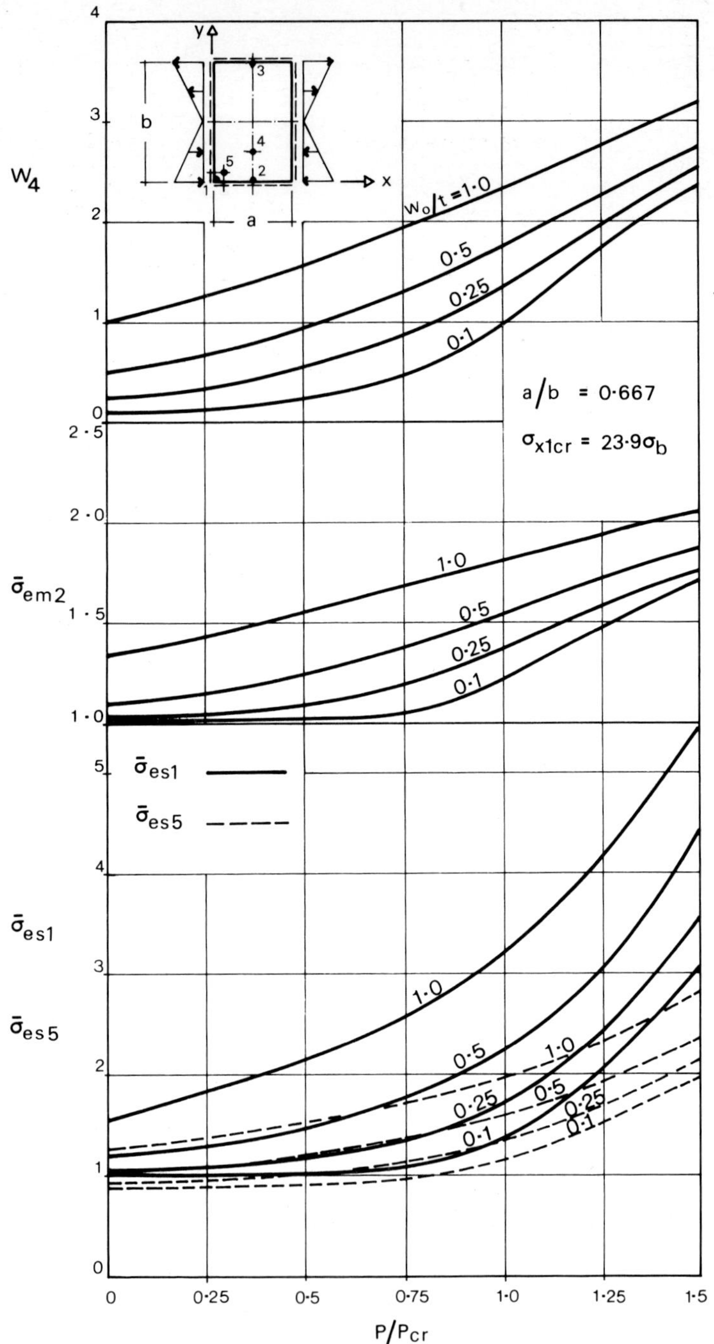


Fig. 4.77 Rectangular ($a/b$ = 0·667) plate subjected to bending stress; unloaded sides stress-free. Deflection, equivalent membrane and surface stresses

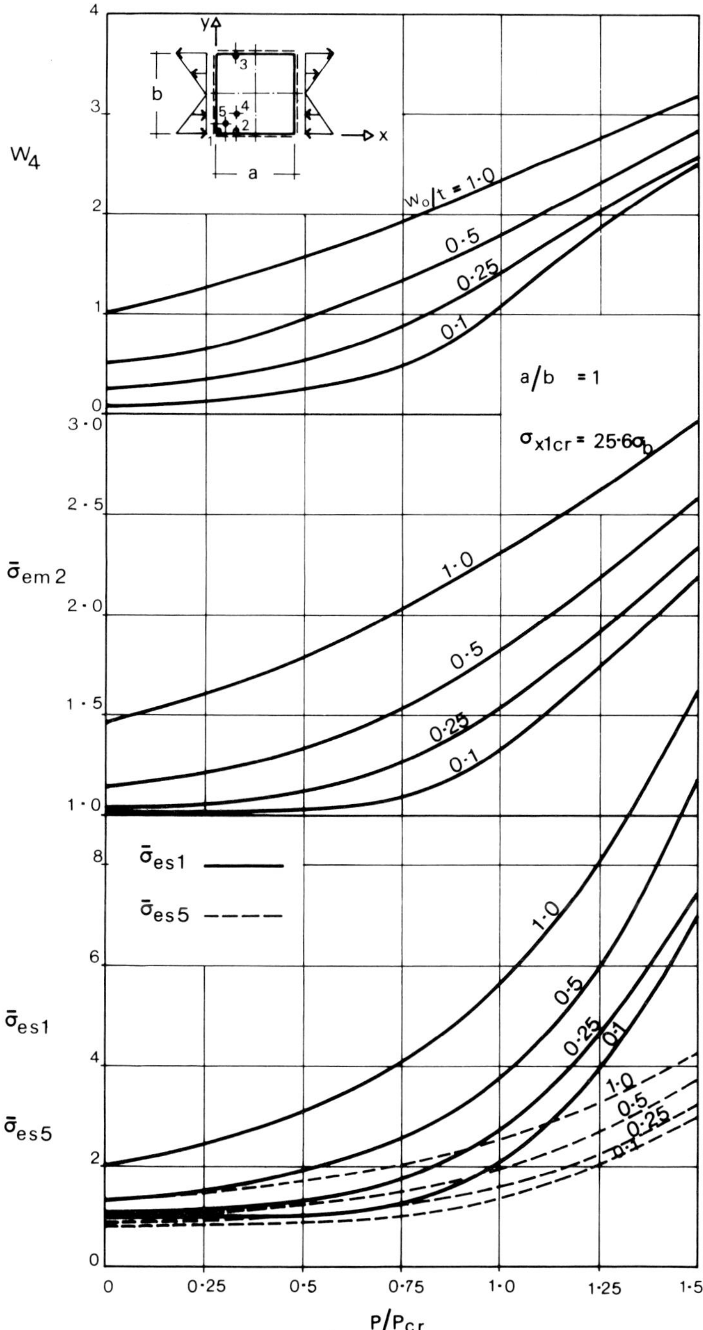


Fig. 4.78 Square ($a/b = 1$) plate subjected to bending stress; unloaded sides stress-free. Deflection, equivalent membrane and surface stresses

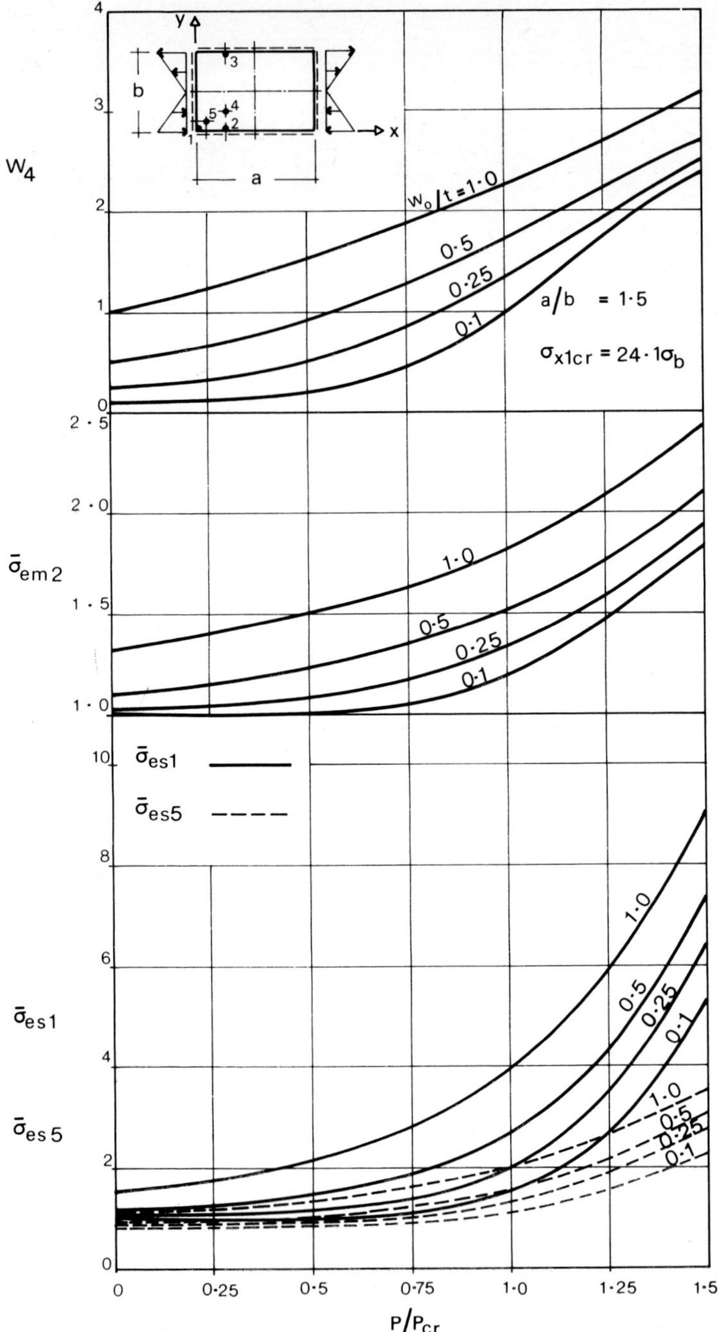


Fig. 4.79 Rectangular ($a/b$ = 1·5) plate subjected to bending stress; unloaded sides stress-free. Deflection, equivalent membrane and surface stresses

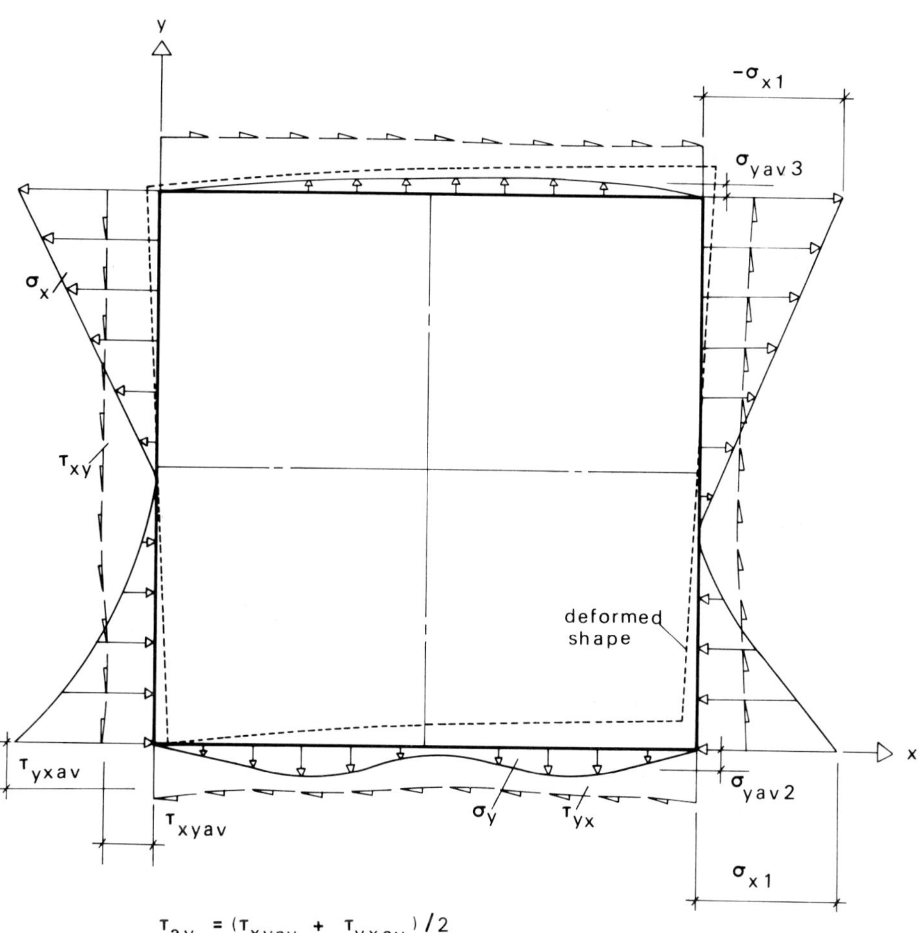


$\tau_{av} = (\tau_{xyav} + \tau_{yxav})/2$

$\sigma_{x1} = \alpha\tau_{av}$ ($\alpha$ = ratio of independent critical loads)

Cases

| a/b | I | J | α | Figure | Table |
|---|---|---|---|---|---|
| 0·667 | 11 | 15 | 1·5 | 4·82 | 41 |
| 1 | 11 | 11 | 2·75 | 4·83 | 42 |
| 1·5 | 15 | 11 | 4·0 | 4·84 | 43 |

Non Dimensional factors

$\sigma_{eR} = (\sigma_{x1}^2 + 3\tau_{av}^2)^{1/2}$

$\sigma_{Rx} = \sigma_{x1}$

$\sigma_{Ry} = \sigma_{x1}$

$\tau_R = \tau_{av}$

Fig. 4.80 Combined bending and shear displacement

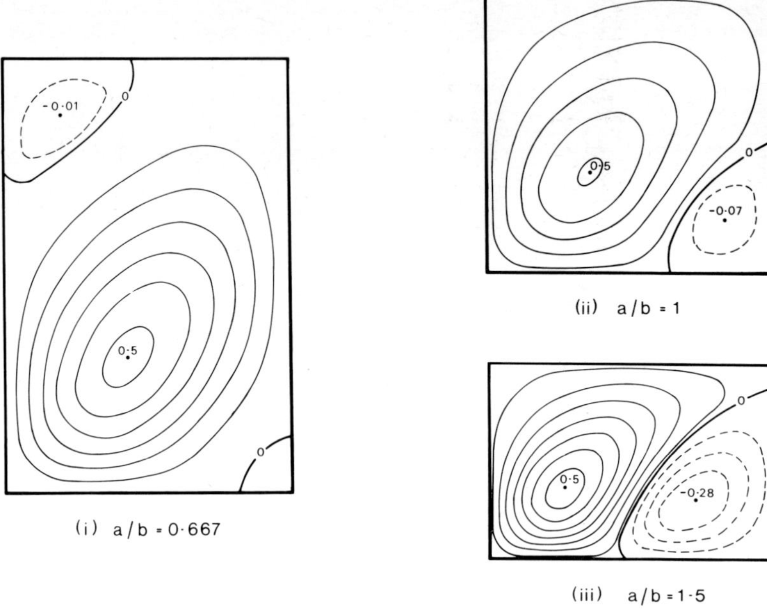


Fig. 4.81 Initial imperfection contours ($w_0/t = 0{\cdot}5$). Combined bending and shear

## 4.7 Bearing stress

Solutions are given for uniform and non-uniform bearing stress distributions for side ratios of $a/b = 0{\cdot}667$, 1 and 1·5. All sides are unrestrained, which will be conservative for most practical applications, which include cases such as the centre girder or web of a ship's double bottom during launching, where the timber slipway constitutes a semi-rigid loading medium.

### 4.7.1 Uniform bearing

Typical boundary stresses and in-plane displacements: Fig. 4.85
Initial imperfections: Fig. 4.86
Design data: Figs. 4.87–89
Reference equivalent stress:

$$\sigma_{eR} = (\sigma_{y1}^2 + 3\tau_1^2)^{1/2}$$
$$= (1 + 0{\cdot}75(a/b)^2)^{1/2}\sigma_{y1}$$

Maximum equivalent membrane stress is equal to the applied corner value (location 1) for low loads and hence is not plotted, but as load and imperfection increase, the membrane stresses generated by the buckle become significant and the maximum moves to the centre of the loaded side (location 2). Maximum

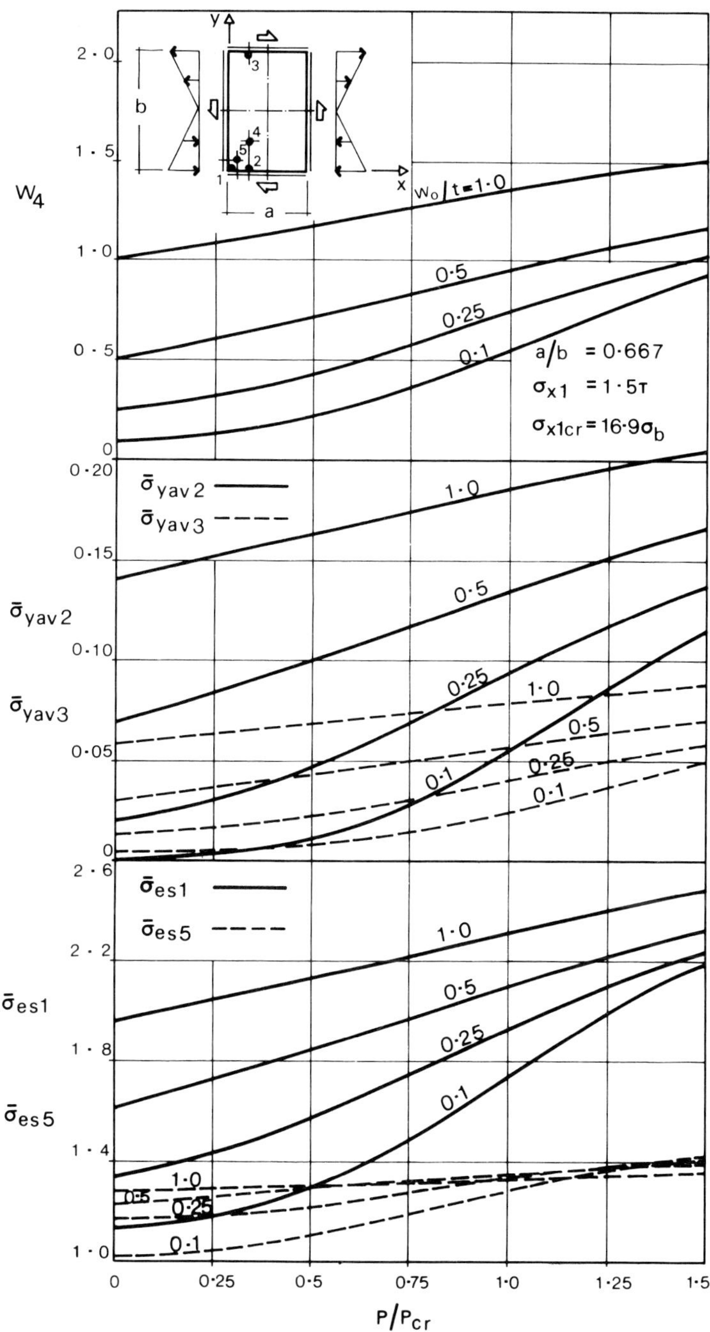


Fig. 4.82 Rectangular ($a/b = 0·667$) plate subjected to combined bending and shear displacement. Deflection, average edge restraining stresses, equivalent surface stresses

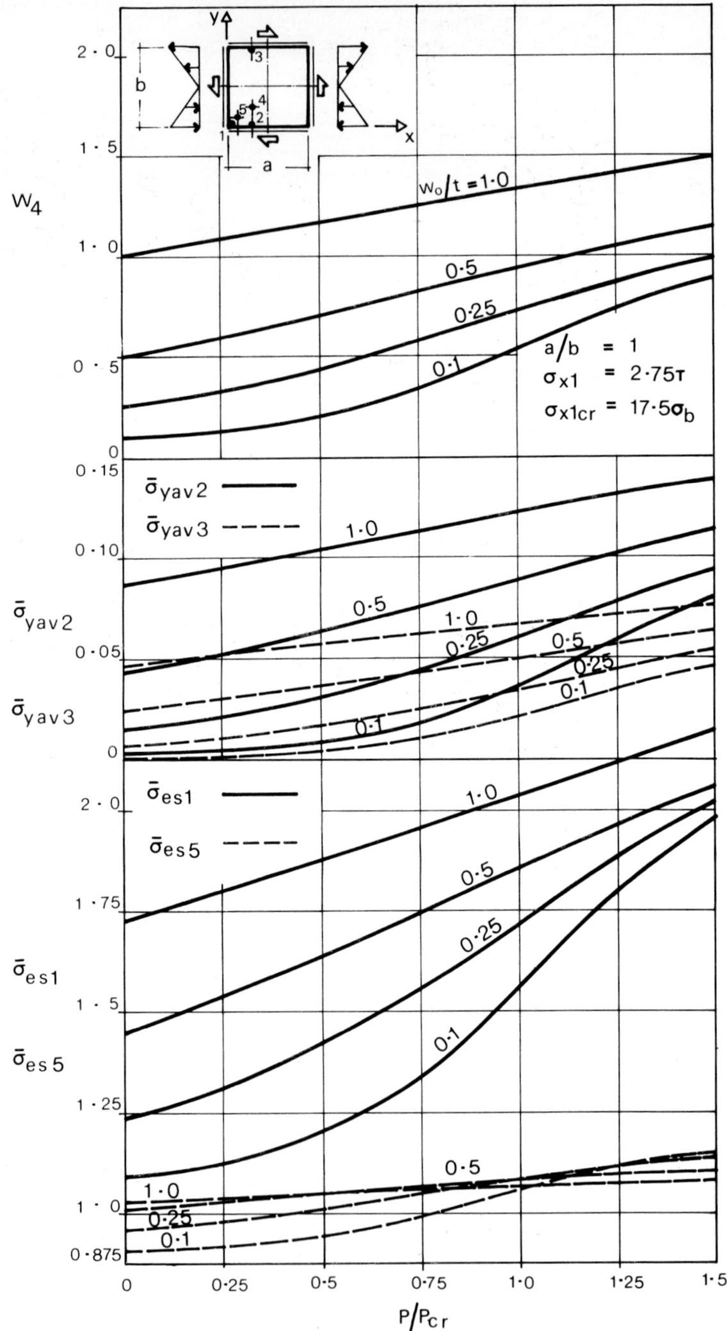


Fig. 4.83 Square ($a/b = 1$) plate subjected to combined bending and shear displacement. Deflection, average edge restraining stresses, equivalent surface stresses

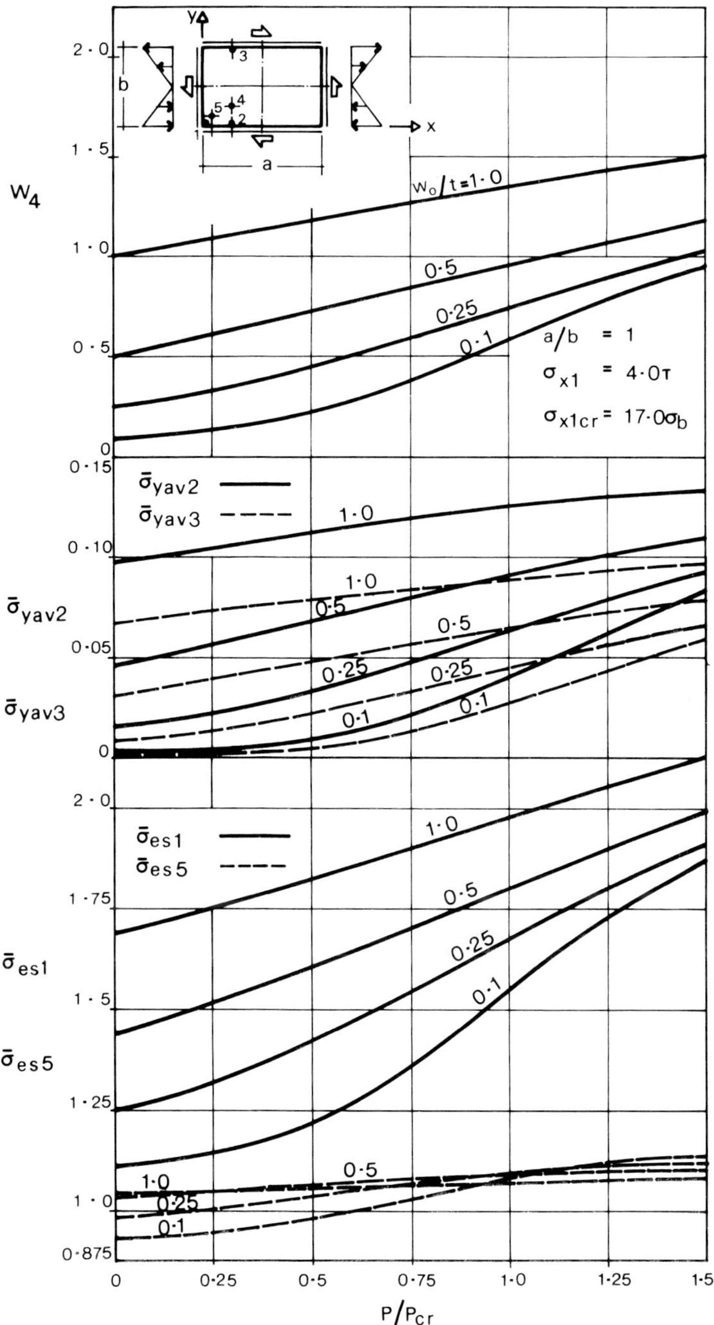


Fig. 4.84 Rectangular ($a/b = 1{\cdot}5$) plate subjected to combined bending and shear displacement. Deflection, average edge restraining stresses, equivalent surface stresses

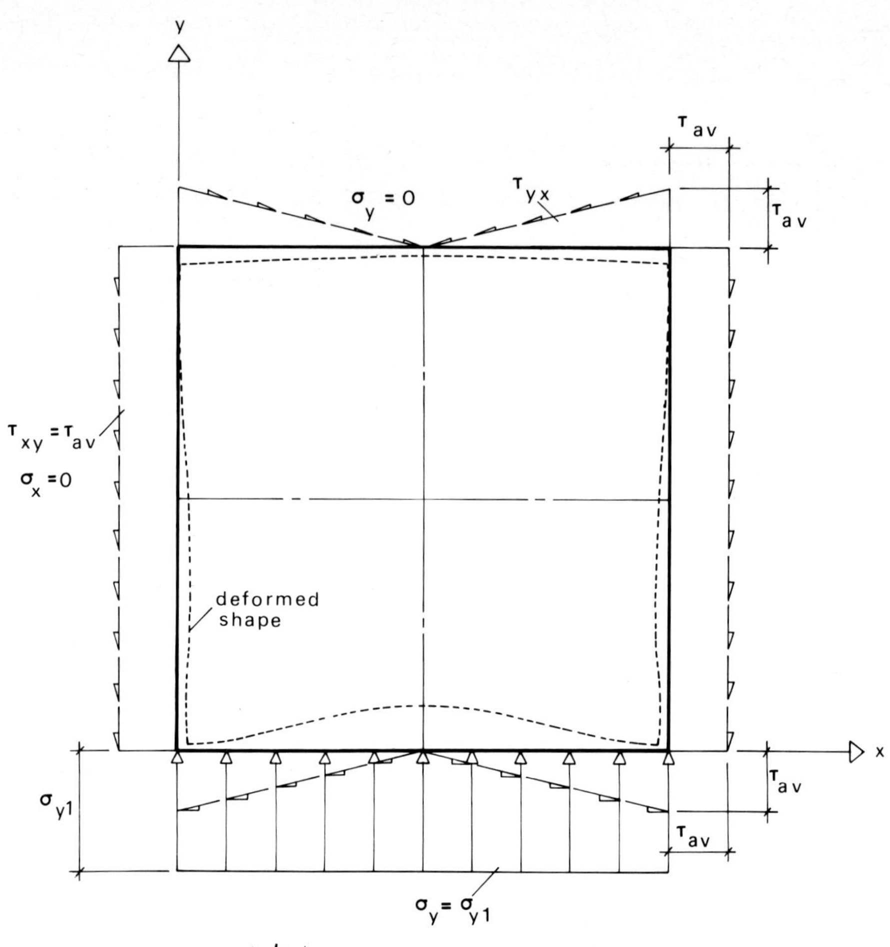


Cases

| a/b | I | J | Figure | Table |
|---|---|---|---|---|
| 0·667 | 10 | 14 | 4·87 | 44 |
| 1 | 10 | 10 | 4·88 | 45 |
| 1·5 | 10 | 8 | 4·89 | 46 |

Non dimensional factors

$$\sigma_{eR} = (1 + 0{\cdot}75\,(a/b)^2)^{1/2}\sigma_{y1}$$

$$\sigma_{Rx} = \sigma_{y1}$$

$$\sigma_{Ry} = \sigma_{y1}$$

Fig. 4.85 Uniform bearing stress

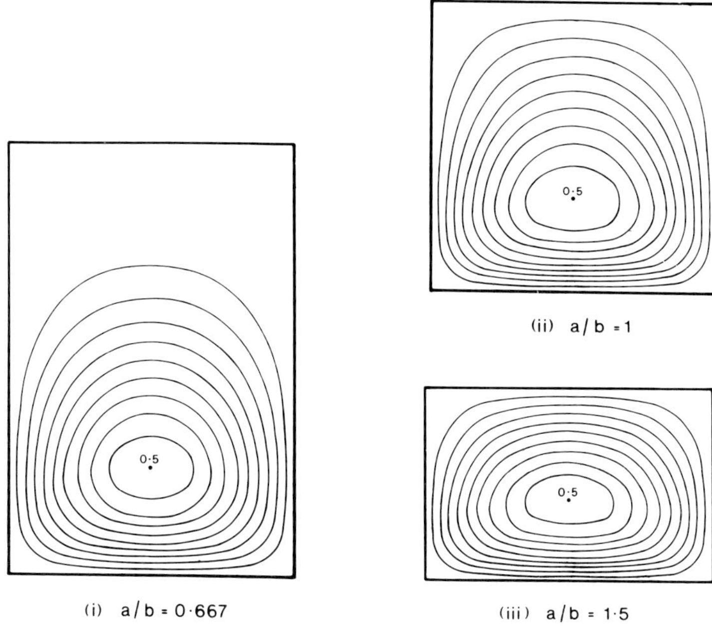


Fig. 4.86 Initial imperfection contours ($w_0/t = 0{\cdot}5$). Uniform bearing

equivalent surface stress occurs at the corners (location 1) and is relatively localised, justifying design for restricted local yielding.

### 4.7.2 Non-uniform bearing

Typical boundary stresses and in-plane displacements: Fig. 4.90
Initial imperfections: Fig. 4.91
Design data: Figs. 4.92–94
Reference equivalent stress: $\sigma_{eR} = \sigma_{y2}$

The non-uniform bearing distribution allows for a more flexible or localised bearing medium. The above comments for uniform bearing apply except that maximum equivalent membrane stress originates at location 2 from the onset of loading. For lower values of imperfection $\sigma_{em2}$ is less and $\sigma_{eR}$ because of the effect of $\sigma_{x2}$ generated by buckling. It is recommended that when $\sigma_{em2} < 1$, $\sigma_{eR}$ be taken as the limiting collapse stress.

## 4.8 Combined bearing and bending stress

Solutions are given for uniform and non-uniform bearing combined with bending in the plane of the plate for side ratios of $a/b = 0{\cdot}667$, 1 and $1{\cdot}5$. The maximum bending stress is equal to twice the bearing stress and all sides are unrestrained,

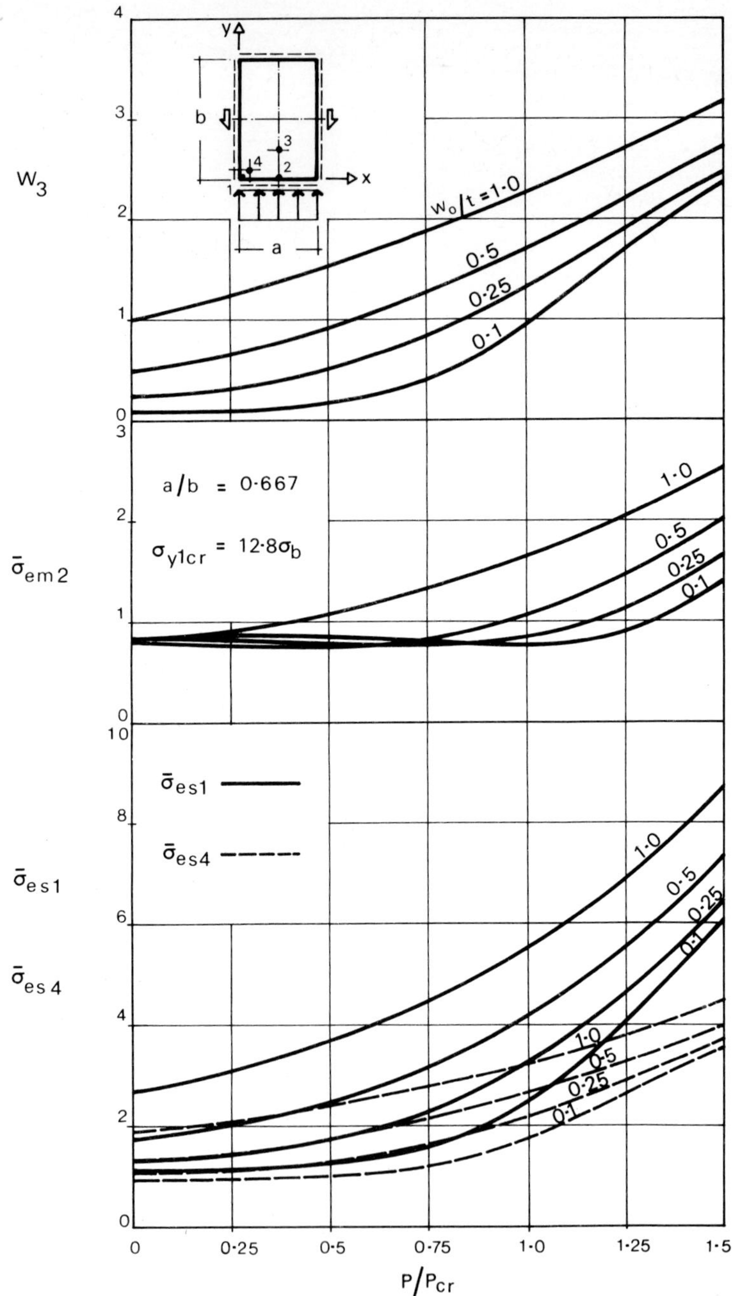


Fig. 4.87 Rectangular ($a/b = 0{\cdot}667$) plate subjected to uniform bearing stress. Deflection, equivalent membrane and surface stresses

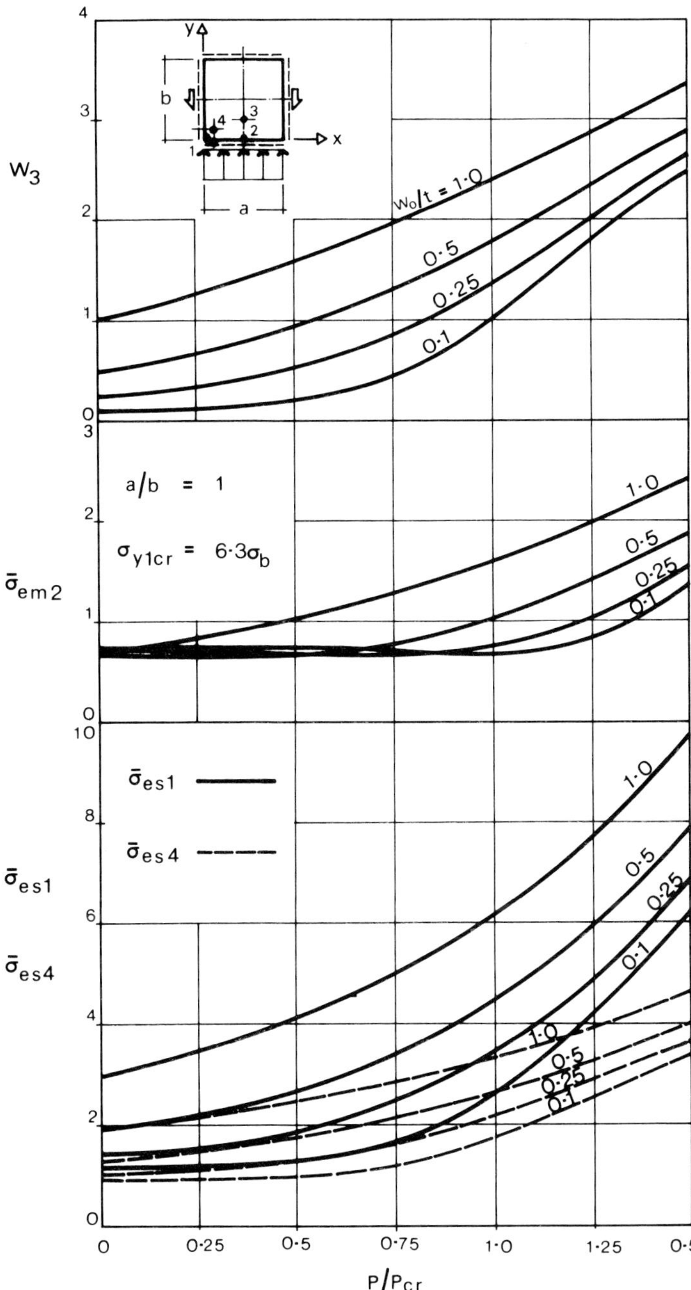


Fig. 4.88 Square ($a/b$ = 1) plate subjected to uniform bearing stress. Deflection, equivalent membrane and surface stresses

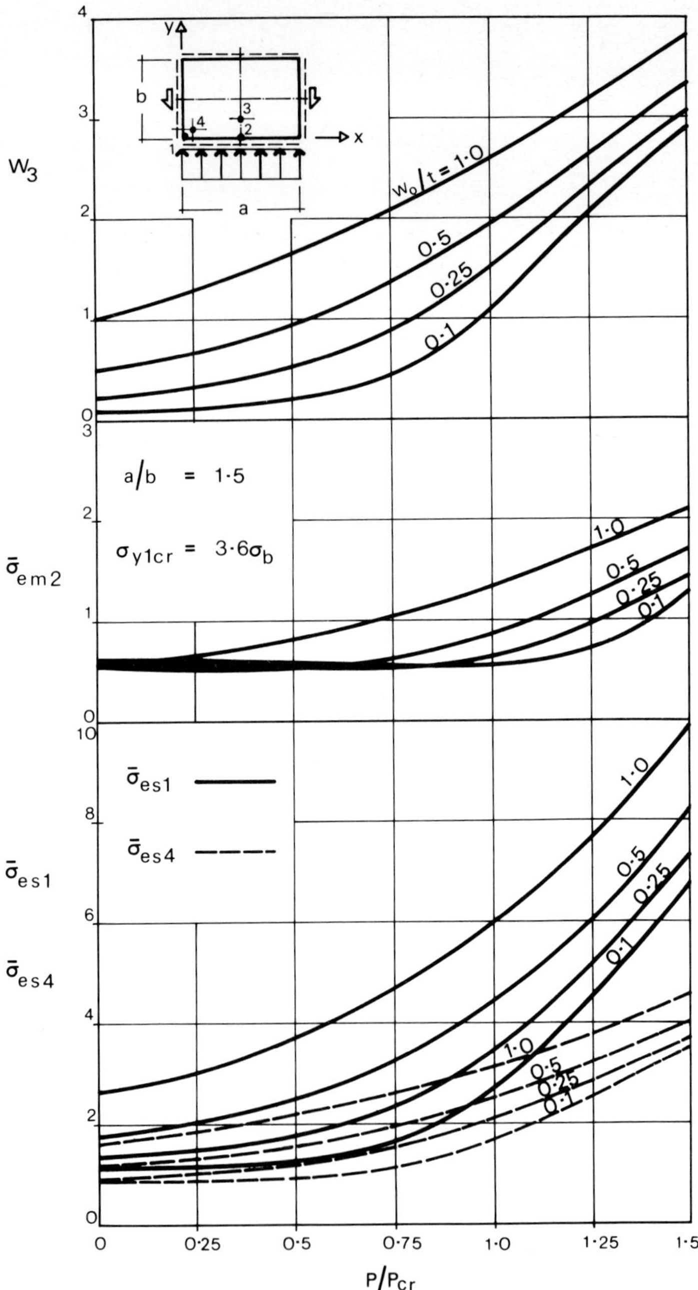


Fig. 4.89 Rectangular ($a/b = 1{\cdot}5$) plate subjected to uniform bearing stress. Deflection, equivalent membrane and surface stresses

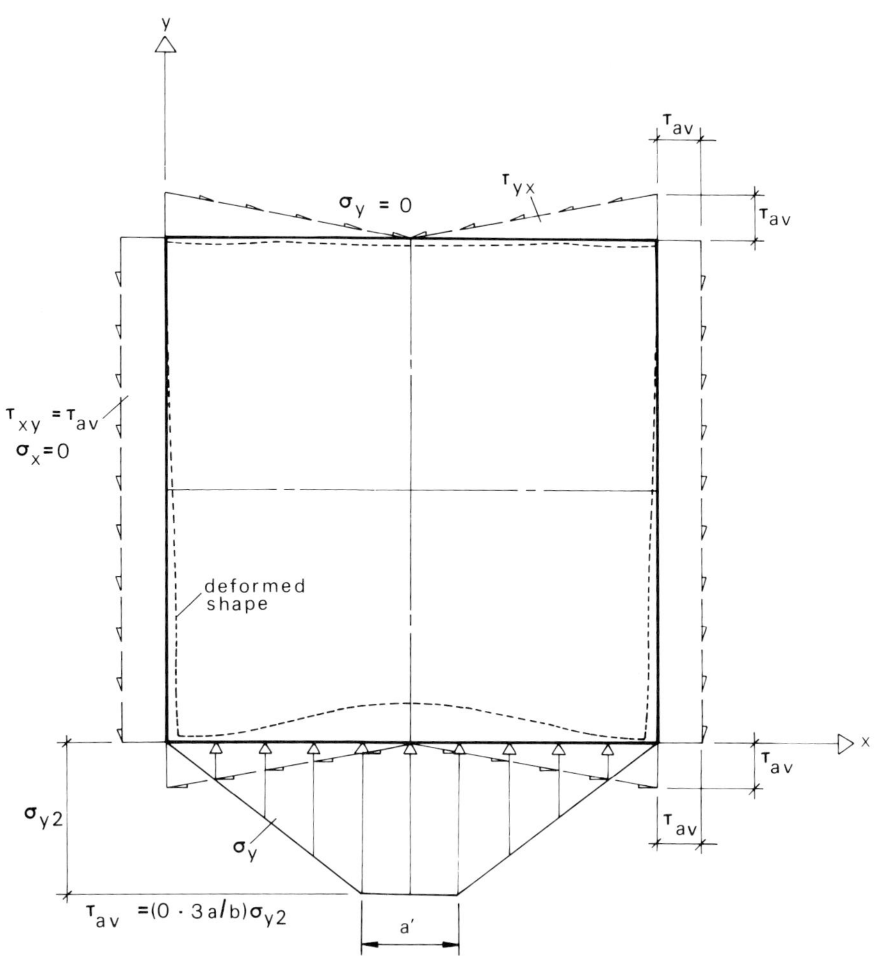


| Cases | | | | | |
|---|---|---|---|---|---|
| a/b | a'/a | I | J | Figure | Table |
| 0·667 | 0·2 | 10 | 14 | 4·92 | 47 |
| 1 | 0·2 | 10 | 10 | 4·93 | 48 |
| 1·5 | 0·2 | 10 | 8 | 4·94 | 49 |

Non dimensional factors

$\sigma_{eR} = \sigma_{y2}$

$\sigma_{Rx} = \sigma_{y2}$

$\sigma_{Ry} = \sigma_{y2}$

Fig. 4.90 Non-uniform bearing stress

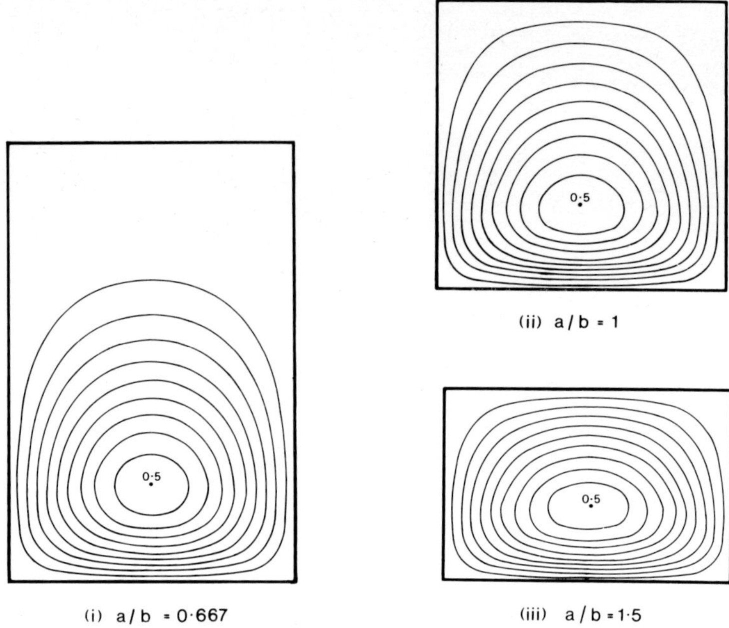


Fig. 4.91 Initial imperfection contours ($w_0/t = 0{\cdot}5$). Non-uniform bearing

which will be conservative for many practical applications. A typical example of combined bearing and bending occurs in the web panels of plate girders during erection by launching over semi-rigid bearings.

### 4.8.1 Uniform bearing and bending

Typical boundary stresses and in-plane displacements: Fig. 4.95
Initial imperfections: Fig. 4.96
Design data: Figs. 4.97–99
Reference equivalent stress:

$$\sigma_{eR} = (\sigma_{x1}^2 + \sigma_{y1}^2 - \sigma_{x1}\sigma_{y1} + 3\tau_1^2)^{1/2}$$
$$= (3 + 0{\cdot}75(a/b)^2)^{1/2}\sigma_{y1}$$

See comments for uniform bearing in Section 4.7.1

### 4.8.2 Non-uniform bearing and bending

Typical boundary stresses and in-plane displacements: Fig. 4.100
Initial imperfections: Fig. 4.101

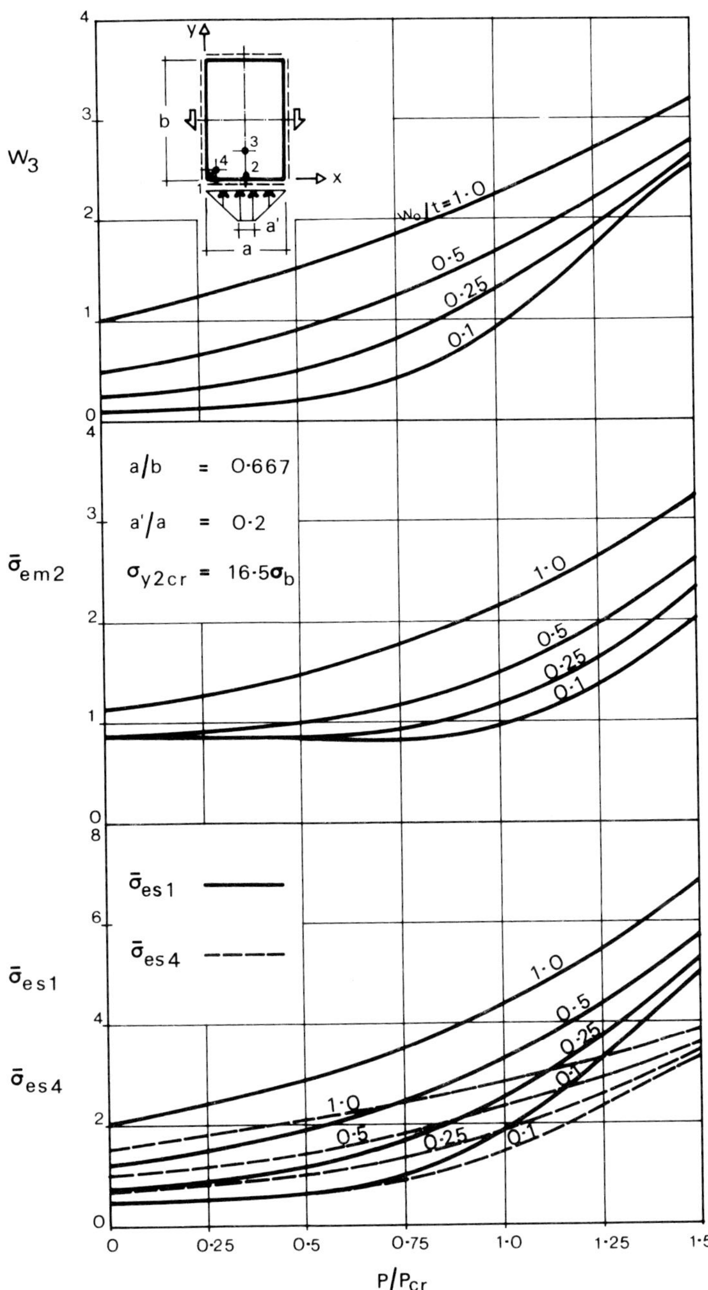


Fig. 4.92 Rectangular ($a/b = 0{\cdot}667$) plate subjected to non-uniform bearing stress. Deflection, equivalent membrane and surface stresses

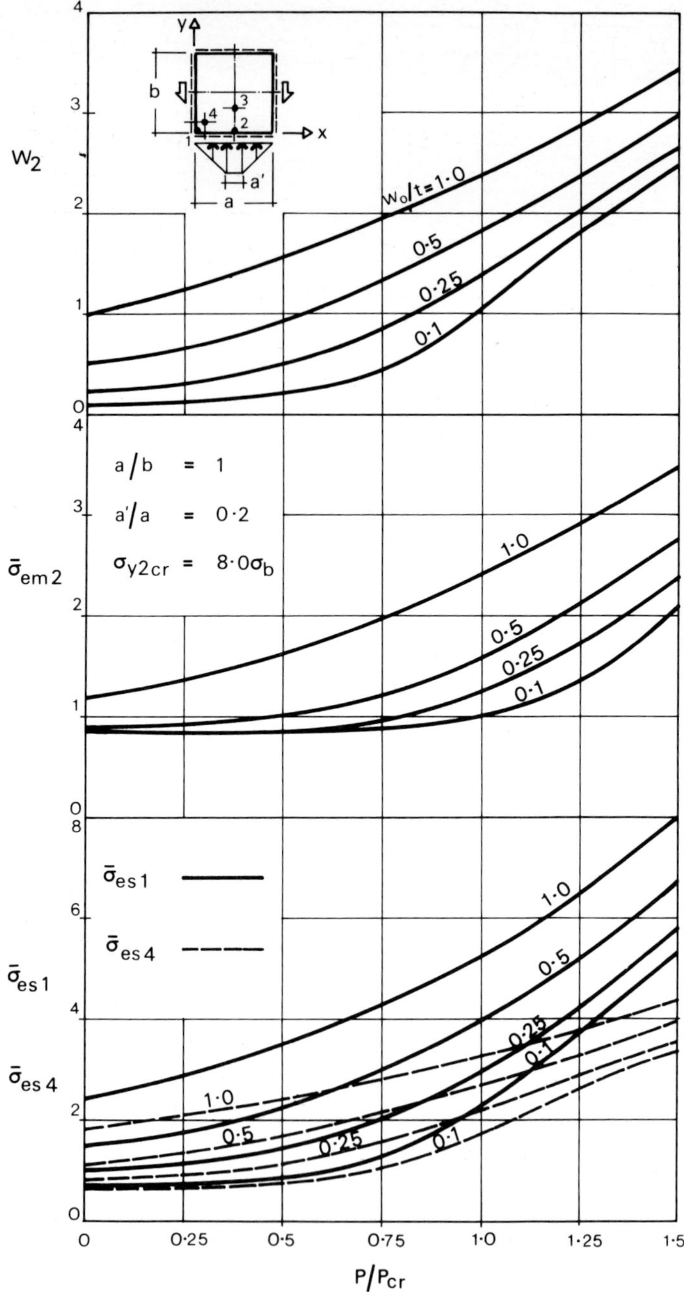


Fig. 4.93 Square ($a/b = 1$) plate subjected to non-uniform bearing stress. Deflection, equivalent membrane and surface stresses

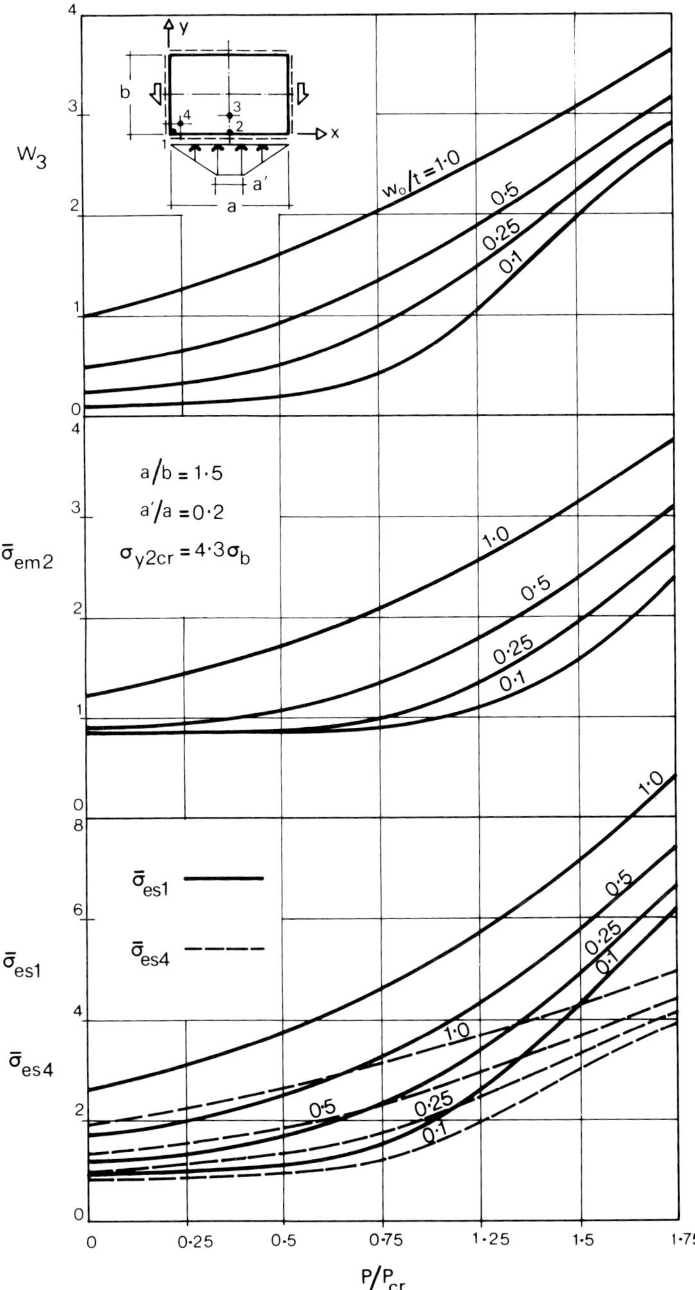


Fig. 4.94 Rectangular ($a/b = 1{\cdot}5$) plate subjected to non-uniform bearing stress. Deflection, equivalent membrane and surface stresses

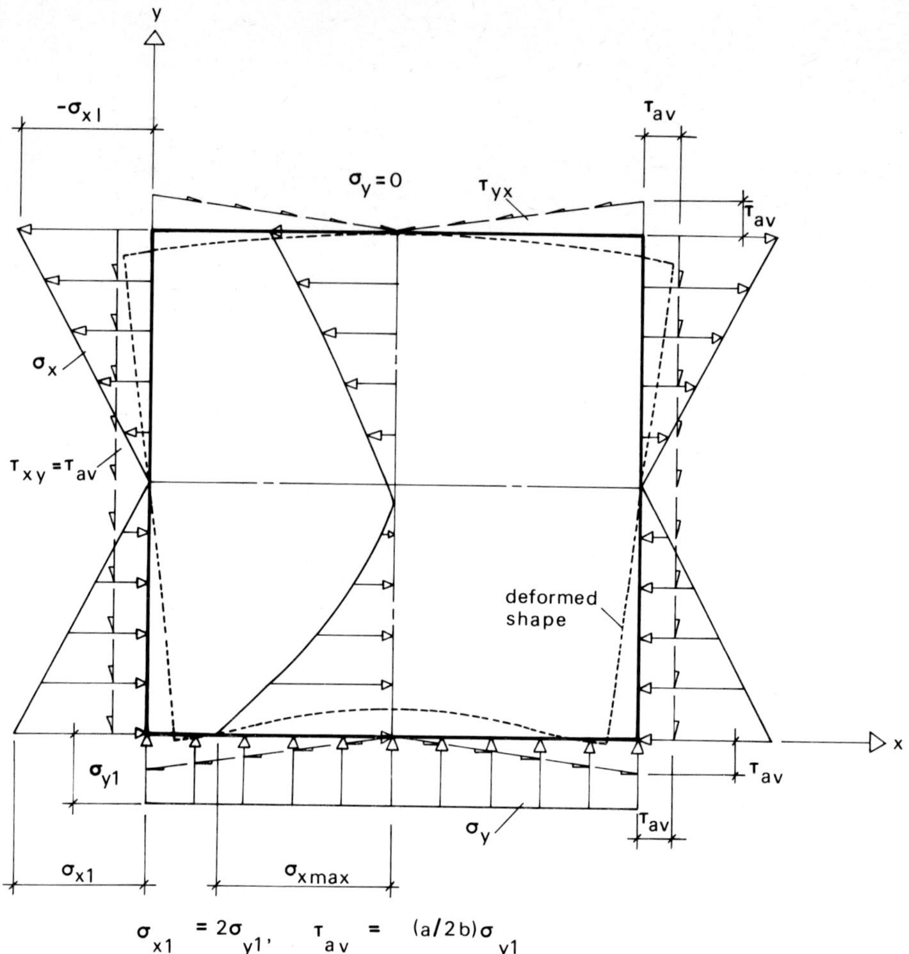


$$\sigma_{x1} = 2\sigma_{y1}, \quad \tau_{av} = (a/2b)\sigma_{y1}$$

| Cases | | | | |
|---|---|---|---|---|
| a/b | I | J | Figure | Table |
| 0·667 | 10 | 14 | 4·97 | 50 |
| 1 | 10 | 10 | 4·98 | 51 |
| 1·5 | 10 | 8 | 4·99 | 52 |

Non dimensional factors

$$\sigma_{eR} = (3 + 0{\cdot}75(a/b)^2)^{1/2}\sigma_{y1}$$

$$\sigma_{Rx} = \sigma_{y1}$$

$$\sigma_{Ry} = \sigma_{y1}$$

Fig. 4.95 Combined uniform bearing and bending stress

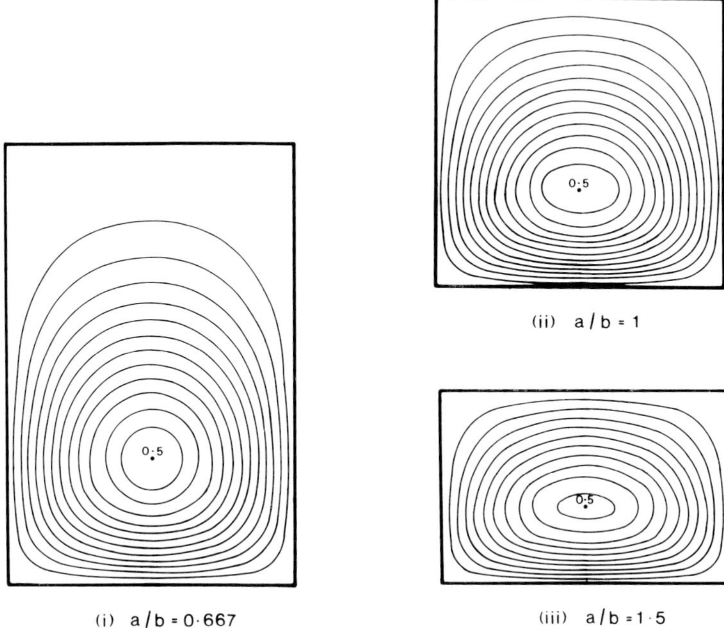


Fig. 4.96 Initial imperfection contours ($w_0/t = 0{\cdot}5$). Uniform bearing and bending

Design data: Figs. 4.102–104
Reference equivalent stress:

$$\sigma_{eR} = (\sigma_{x1}^2 + \sigma_{y1}^2 - \sigma_{x1}\sigma_{y1} + 3\tau_1^2)^{1/2}$$
$$= (4 + 0{\cdot}27(a/b)^2)^{1/2}\sigma_{y2}$$

Comments as for non-uniform bearing in Section 4.7.2, except that maximum equivalent membrane stress exceeds the applied value from the onset of loading.

## 4.9 Interaction

A rudimentary interaction procedure can be used to obtain solutions for combined-load cases not covered by the graphical results. This is applicable to cases for which solutions are given for the components of load applied independently and for one load combination. The method is applicable for identical boundary conditions for a particular side ratio and is illustrated in Fig. 4.105 for the case of combined uniaxial compression and shear, where

$\sigma_{y2}, \tau_2$ = load combination for which appraisal required

$\sigma_{yy0}, \tau_{y0}$ = loads applied independently to cause surface or membrane yield (obtain from graphs by trial)

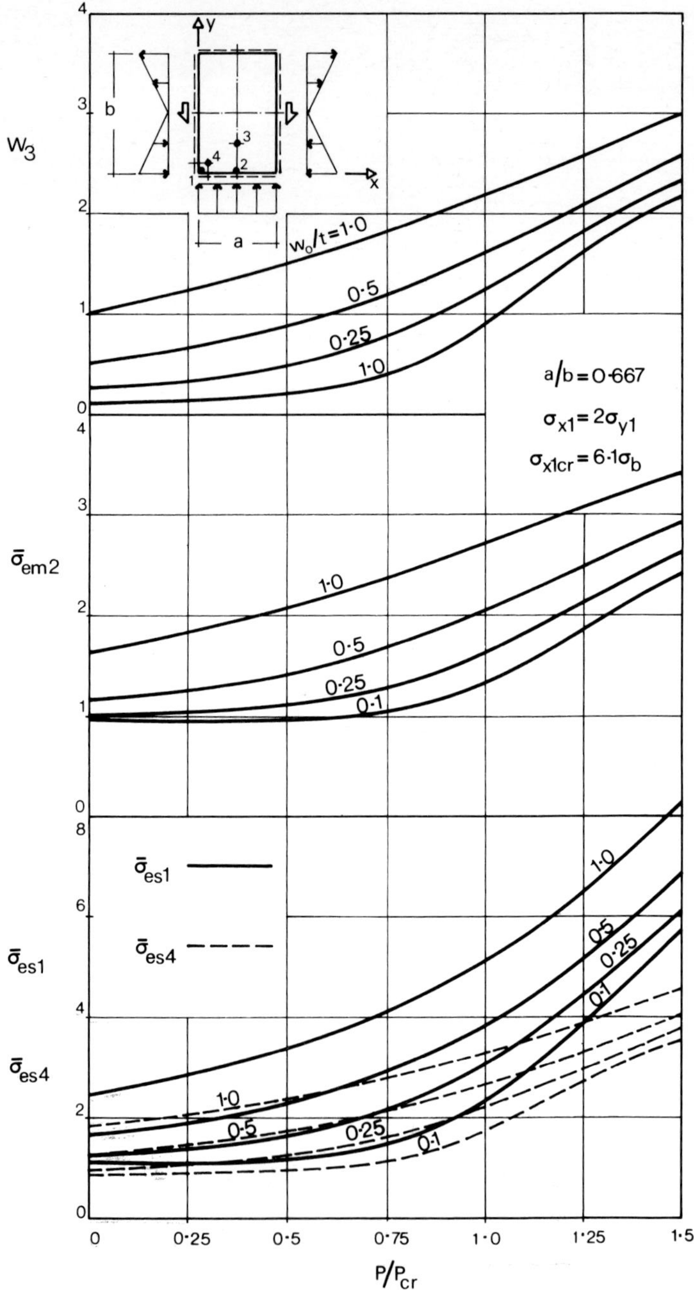


Fig. 4.97 Rectangular ($a/b$ = 0·667) plate subjected to combined uniform bearing and bending stress. Deflection, equivalent membrane and surface stresses

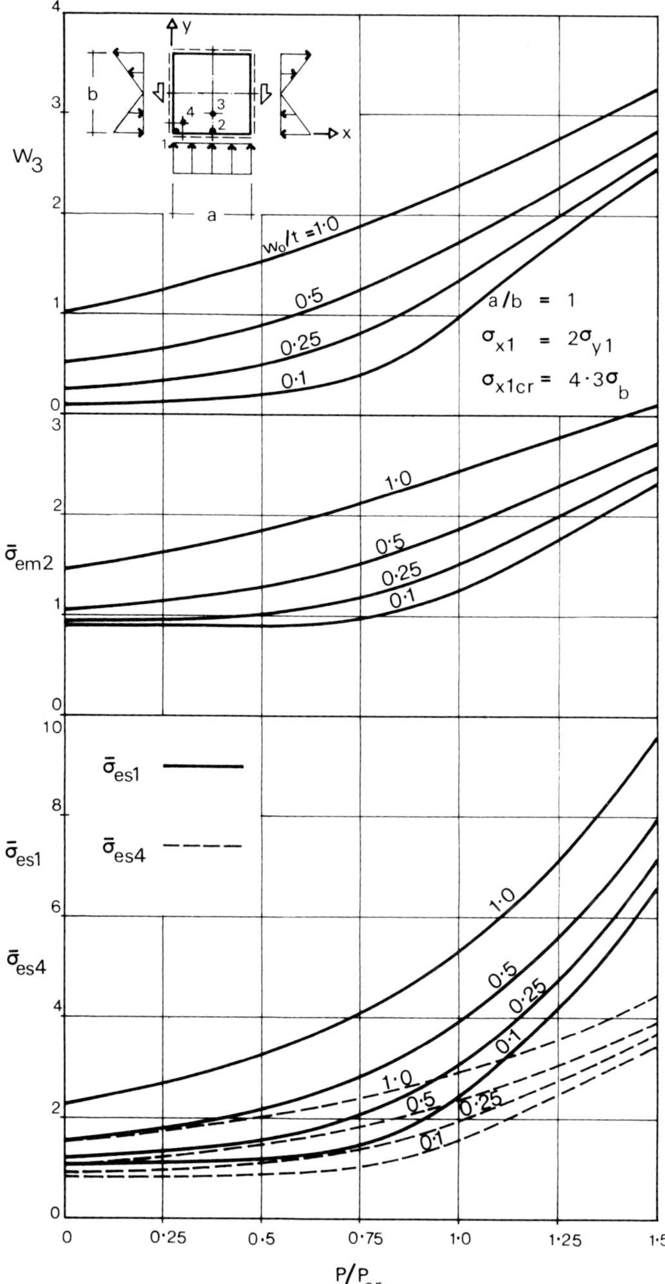


Fig. 4.98 Square ($a/b = 1$) plate subjected to combined uniform bearing and bending stress. Deflection, equivalent membrane and surface stresses

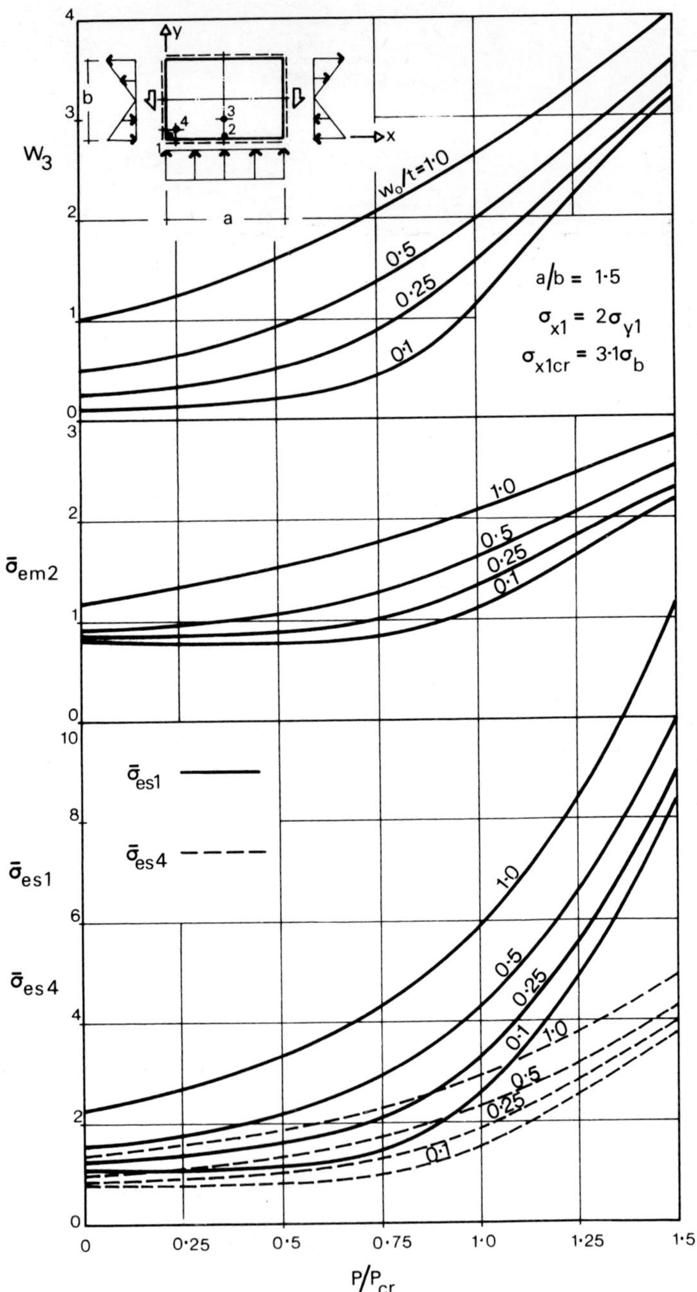


Fig. 4.99 Rectangular ($a/b = 1{\cdot}5$) plate subjected to combined uniform bearing and bending stress. Deflection, equivalent membrane and surface stresses

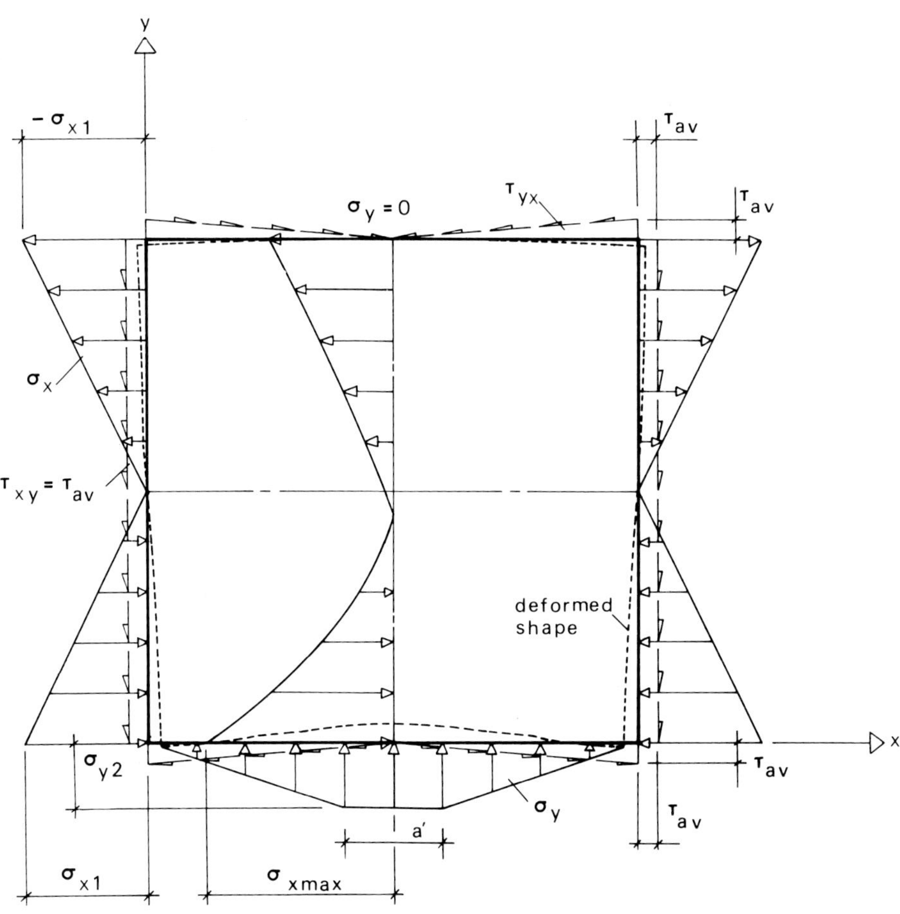


$$\sigma_{x1} = 2\sigma_{y2}, \quad \tau_{av} = (0\cdot3\,a/b)\,\sigma_{y2}$$

Cases

| a/b | a′/a | I | J | Figure | Table |
|---|---|---|---|---|---|
| 0·667 | 0·2 | 10 | 14 | 4·102 | 53 |
| 1 | 0·2 | 10 | 10 | 4·103 | 54 |
| 1·5 | 0·2 | 10 | 8 | 4·104 | 55 |

Non dimensional factors

$$\sigma_{eR} = (4 + 0\cdot27(a/b)^2)^{1/2}\sigma_{y2}$$

$$\sigma_{Rx} = \sigma_{y2}$$

$$\sigma_{Ry} = \sigma_{y2}$$

Fig. 4.100 Combined non-uniform bearing and bending stress

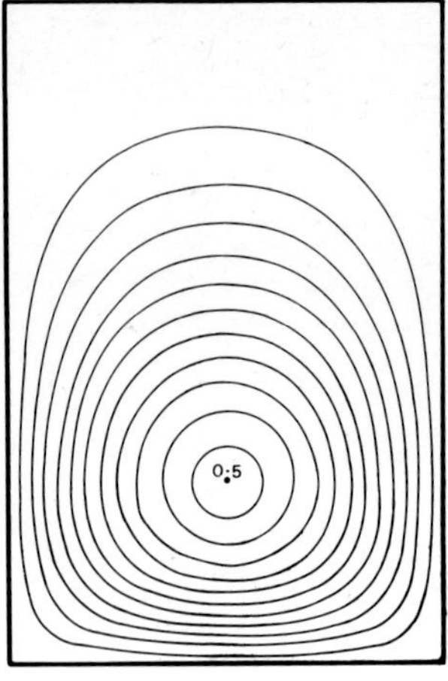


(i) a / b = 0·667

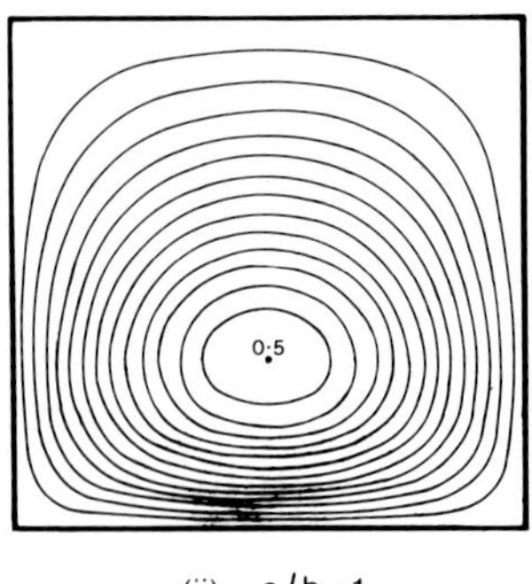


(ii) a / b = 1

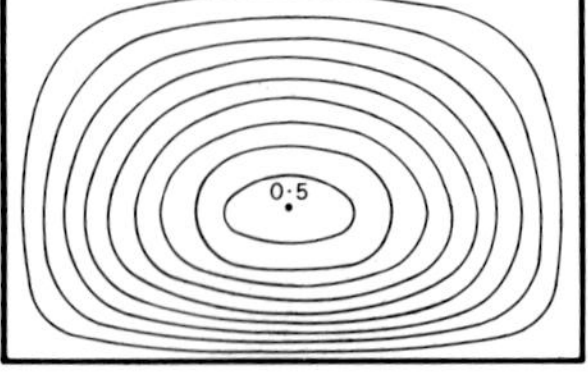


(iii) a / b = 1·5

Fig. 4.101 Initial imperfection contours ($w_0/t = 0{\cdot}5$). Non-uniform bearing and bending

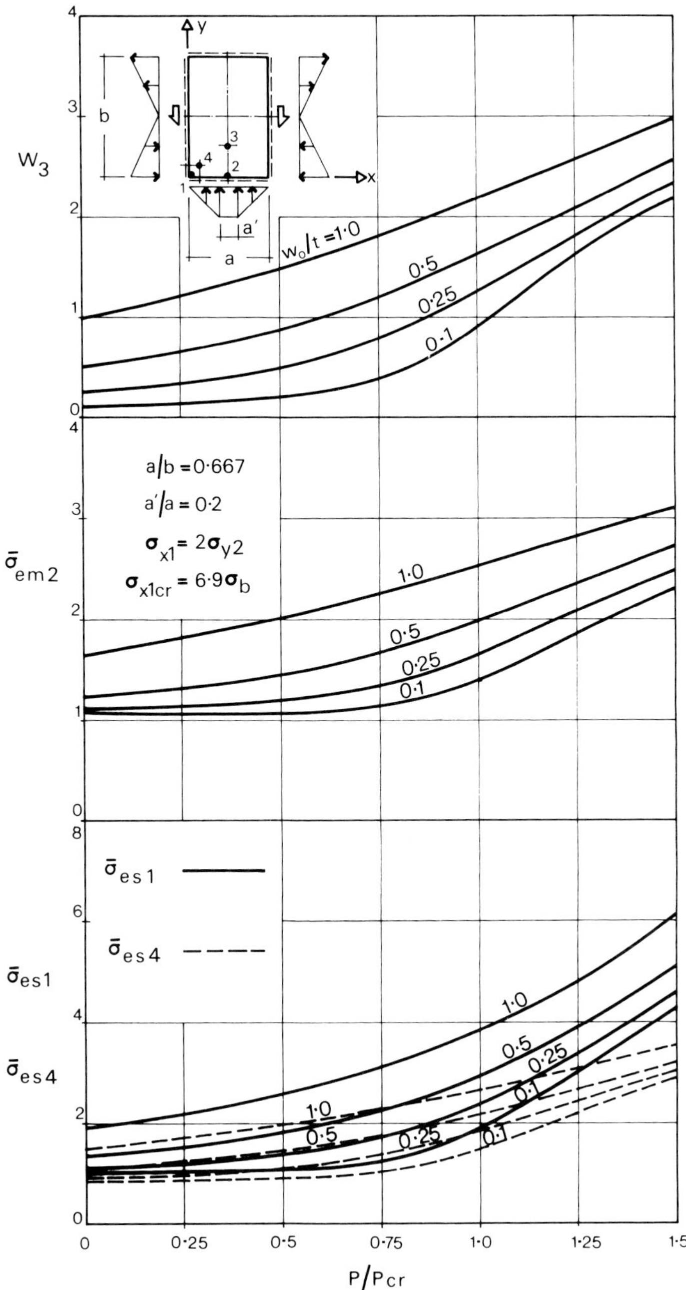


Fig. 4.102 Rectangular ($a/b = 0{\cdot}667$) plate subjected to combined non-uniform bearing and bending stress. Deflection, equivalent membrane and surface stresses

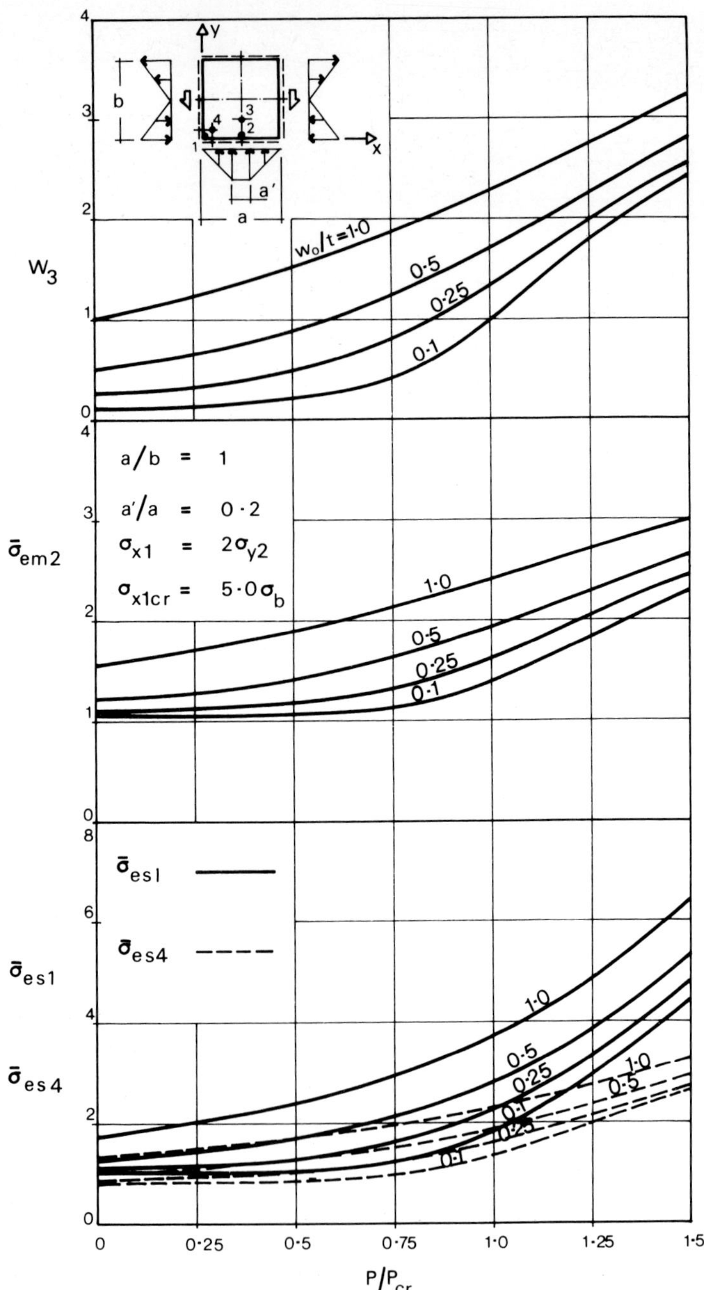


Fig. 4.103 Square (*a*/*b* = 1) plate subjected to combined non-uniform bearing and bending stress. Deflection, equivalent membrane and surface stresses

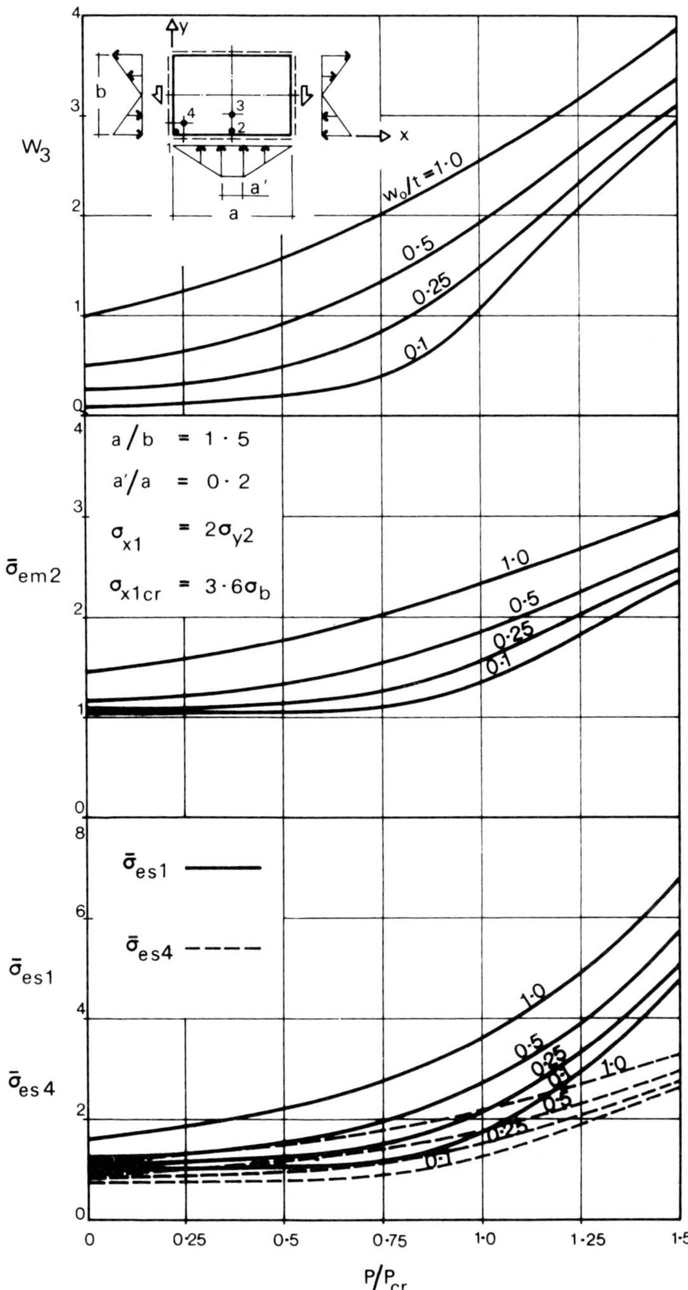


Fig. 4.104 Rectangular ($a/b = 1{\cdot}5$) plate subjected to combined non-uniform bearing and bending stress. Deflection, equivalent membrane and surface stresses

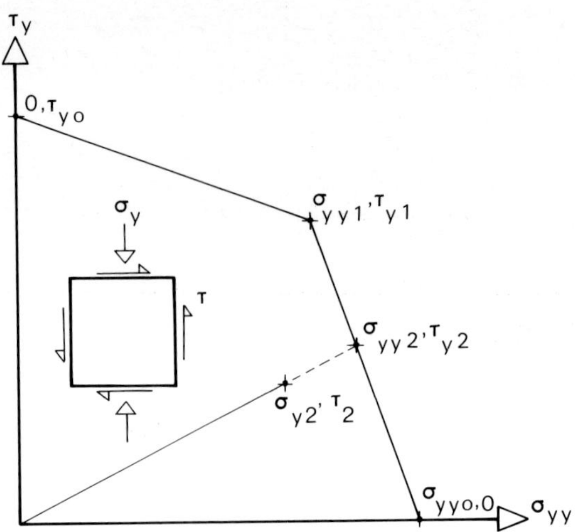


Fig. 4.105 Interaction diagram for combined uniaxial compression and shear

$\sigma_{yy1}, \tau_{y1}$ = particular combined load case causing surface or membrane yield (obtain from graphs by trial)

The solution is $\sigma_{yy2}, \tau_{y2}$, the level to which load combination $\sigma_{y2}, \tau_2$ must be increased (or decreased) to give surface or membrane yield. The serviceability or collapse safety factor is then given by $\sigma_{yy2}/\sigma_{y2}$. Examples are given in Chapter 5. The same procedure can be applied for other load combinations for which solutions are given for the two independent and one combined-load condition.

A straight-line interaction diagram can be used where solutions are given for the independent load cases only, but it will give conservative results in most cases particularly for membrane yield.

CHAPTER FIVE

# Numerical Examples

The application of the design data given in this book is illustrated in the following examples. In applying serviceability and strength design criteria, load factors are assumed which may differ significantly from those applying in practice, since they depend on type of structure and load combinations.

### Example 5.1 Welded steel box column subjected to pure axial compression

Appraise deformations and stresses in the plating of the box column shown in Fig. 5.1(i) for an axial load of 6750 kN (1517 x $10^3$ lbf). Figure 5.1(ii) illustrates the weld detail and Fig. 5.1(iii) shows the discrete plate-panel idealisation where the long panels are assumed to buckle in square waves, the antinodes defining the idealised plate-panel boundaries in the loaded direction.

Plate parameters

Geometry

$$a = 570 \text{ mm } (22{\cdot}4 \text{ in}),\ b = 570 \text{ mm } (22{\cdot}4 \text{ in}),\ t = 15 \text{ mm } (0{\cdot}591 \text{ in})$$

$$a/b = 570/570 = 1,\ b/t = 570/15 = 38{\cdot}0$$

Material properties—high tensile steel

$$E = 210 \text{ kN/mm}^2 \ (30\,450 \text{ ksi}),\ \nu = 0{\cdot}3,\ \sigma_y = 350 \text{ N/mm}^2 \ (50{\cdot}8 \text{ ksi})$$

Critical buckling load

$$\sigma_b = \pi^2 E/[12(1-\nu^2)(b/t)^2] = 131{\cdot}4 \text{ N/mm}^2 \ (19{\cdot}1 \text{ ksi}) \qquad \text{(eq. 3.2)}$$

$$k = 4 \qquad \text{(Fig. 4.9)}$$

$$\sigma_{ycr} = k\sigma_b = 526 \text{ N/mm}^2 \quad (76{\cdot}3 \text{ ksi}) \qquad \text{(eq. 3.1)}$$

Unfactored applied load

$$\sigma_{yav} = (6750 \times 10^3)/[4 \times 15 \times (600-15)] = 192 \text{ N/mm}^2 \quad (27{\cdot}8 \text{ ksi})$$

$$P/P_{cr} = \sigma_{yav}/\sigma_{ycr} = 0{\cdot}365$$

Initial geometric imperfection ($a > 3b, t < 25$ mm (0·984 in)) (Fig. 2.1)
(for purposes of calculation of $G, a$ is the long side dimension)

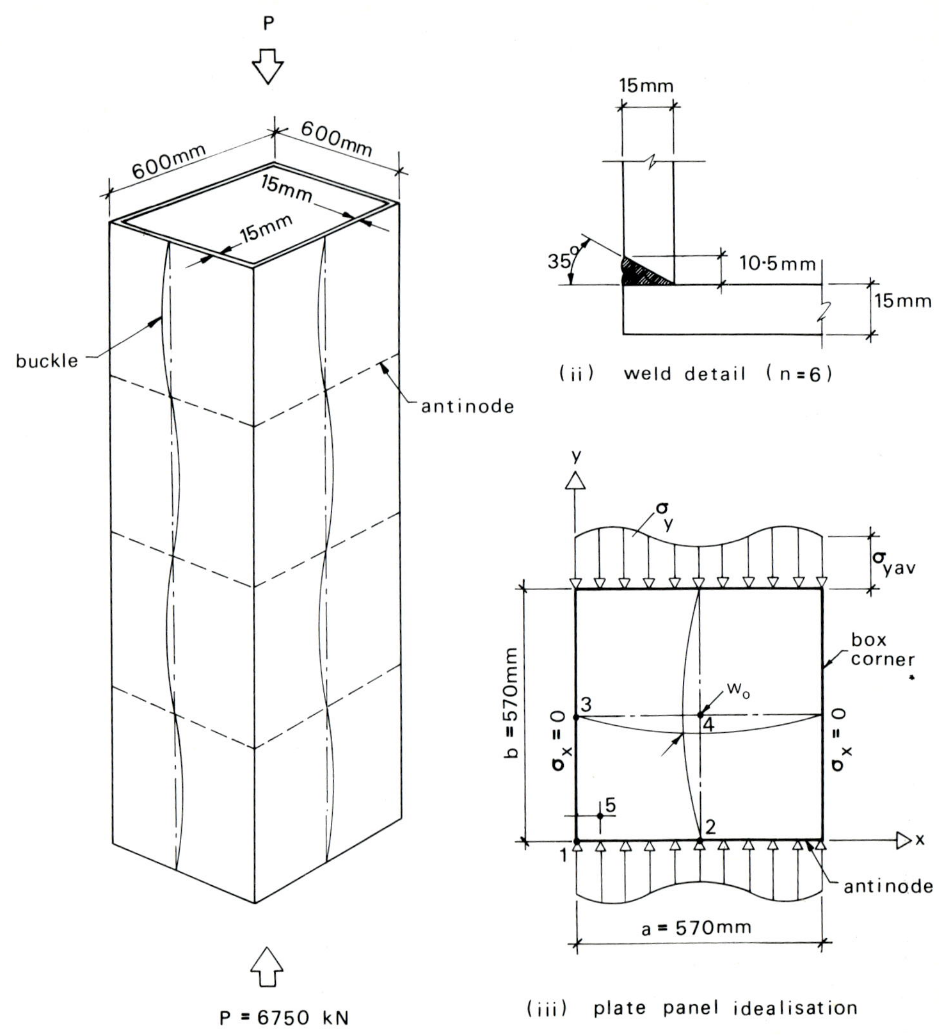


Fig. 5.1 Box column example

$$G = 2b = 1140 \text{ mm } (44{\cdot}9 \text{ in})$$

$$\Delta_x = G(1 + b/5000)/(30t) = 2{\cdot}82 \text{ mm } (0{\cdot}111 \text{ in})$$

$$w_g = 1{\cdot}2\Delta_x = 3{\cdot}39 \text{ mm } (0{\cdot}133 \text{ in})$$

Residual welding stress
single weld joint–multi-run (Fig. 2.10c)

$$A = (15 \times 10{\cdot}5)/2 = 78{\cdot}75 \text{ mm}^2 \ (0{\cdot}122 \text{ in}^2), n = 6$$

$$F_c = 10A/(\sqrt[3]{n})^2 = 239 \text{ kN } (53{\cdot}7 \times 10^3 \text{ lbf}) \quad \text{(eq. 2.8)}$$

stress area = 4 x 15(600–15) = 35 100 mm² (54·4 in²)

$$\sigma_{Rs} = 4F_c/35\ 100 = 27{\cdot}2 \text{ N/mm}^2 \ (3{\cdot}94 \text{ ksi})$$

$$\sigma_{Reff\ y} = \sigma_{Rs}(b/t - 20)/20 = 24{\cdot}5 \text{ N/mm}^2 \ (3{\cdot}55 \text{ ksi}) \quad \text{(eq. 2.16)}$$

Total imperfection (incorporating residual stress)

$$L_y = 570 \text{ mm } (22{\cdot}4 \text{ in})$$

$$w_0 = (w_g^2 + 8L_y^2\sigma_{Reff\ y}/\pi^2 E)^{1/2} = 6{\cdot}50 \text{ mm } (0{\cdot}256 \text{ in}) \quad \text{(eq. 2.18)}$$

$$w_0/t = 0{\cdot}43$$

Serviceability

serviceability load factor = 1·10 (say)
serviceability factored load = 1·10 x $P/P_{cr}$ = 0·40
deflection–Fig. 4.9

$$W_4 = 0{\cdot}65, w_4 = W_4 \times t = 9{\cdot}8 \text{ mm } (0{\cdot}386 \text{ in}) \quad \text{(eq. 3.14)}$$

stiffness effectiveness (to determine column compression)–Fig. 4.9

$$\lambda_{sy} = 0{\cdot}83$$

equivalent surface stresses–Fig. 4.10

$$\sigma_{eR} = 1{\cdot}10 \times \sigma_{yav} = 211 \text{ N/mm}^2 \ (30{\cdot}6 \text{ ksi}) \quad \text{(Fig. 4.8)}$$

$$\bar{\sigma}_{es1} = \bar{\sigma}_{es5} = 1{\cdot}53, \sigma_{es1} = \bar{\sigma}_{es1} \times \sigma_{eR} = 323 \text{ N/mm}^2 \ (46{\cdot}8 \text{ ksi}) \quad \text{(eq. 3.15)}$$

$$\bar{\sigma}_{es4} = 1{\cdot}55, \sigma_{es4} = \bar{\sigma}_{es4} \times \sigma_{eR} = 327 \text{ N/mm}^2 \ (47{\cdot}4 \text{ ksi}) < \sigma_y$$

Strength

collapse load factor = 1·26 (say)
collapse factored load = 1·26 x $P/P_{cr}$ = 0·46
equivalent membrane stress–Fig. 4.10

$$\sigma_{eR} = 1{\cdot}26 \times \sigma_{yav} = 242 \text{ N/mm}^2 \ (35{\cdot}1 \text{ ksi}) \quad \text{(Fig. 4.8)}$$

$$\bar{\sigma}_{em3} = 1{\cdot}41, \sigma_{em3} = \bar{\sigma}_{em3} \times \sigma_{eR} = 341 \text{ N/mm}^2 \ (49{\cdot}4 \text{ ksi}) < \sigma_y \quad \text{(eq. 3.16)}$$

### Alternative solution

Superposition of residual stress on applied load
unfactored applied load

$$P/P_{cr} = (\sigma_{yav} + \sigma_{Reff\ y})/\sigma_{cr} = 0{\cdot}41$$

initial imperfection

$$w_0/t = w_g/t = 0{\cdot}23$$

### Serviceability

serviceability load factor = 1·10 (say)
serviceability factored load = 1·10 x $P/P_{cr}$ = 0·45
deflection–Fig. 4.9

$$W_4 = 0{\cdot}42,\ w_4 = W_4 \times t = 6{\cdot}3 \text{ mm } (0{\cdot}248 \text{ in}) \qquad \text{(eq. 3.14)}$$

stiffness effectiveness (to determine column compression) – Fig. 4.9

$$\lambda_{sy} = 0{\cdot}92$$

equivalent surface stresses–Fig. 4.10

$$\sigma_{eR} = 1{\cdot}10(\sigma_{yav} + \sigma_{Reff\ y}) = 238 \text{ N/mm}^2 \ (34{\cdot}5 \text{ ksi}) \qquad \text{(Fig. 4.8)}$$

$$\bar{\sigma}_{es1} = \bar{\sigma}_{es5} = 1{\cdot}24,\ \sigma_{es1} = \bar{\sigma}_{es1} \times \sigma_{eR} = 295 \text{ N/mm}^2 \ (42{\cdot}8 \text{ ksi}) \qquad \text{(eq. 3.15)}$$

$$\bar{\sigma}_{es4} = 1{\cdot}42,\ \sigma_{es4} = \bar{\sigma}_{es4} \times \sigma_{eR} = 338 \text{ N/mm}^2 \ (49{\cdot}0 \text{ ksi}) < \sigma_y$$

### Strength

collapse load factor = 1·26 (say)
collapse factored load = 1·26 x $P/P_{cr}$ = 0·52
equivalent membrane stress – Fig. 4.10

$$\sigma_{eR} = 1{\cdot}26(\sigma_{yav} + \sigma_{Reff\ y}) = 273 \text{ N/mm}^2 \ (39{\cdot}6 \text{ ksi})$$

$$\bar{\sigma}_{em3} = 1{\cdot}19,\ \sigma_{em3} = \bar{\sigma}_{em3} \times \sigma_{eR} = 325 \text{ N/mm}^2 \ (47{\cdot}1 \text{ ksi}) < \sigma_y$$

Table 5.1
**Comparison of two methods of allowing for residual stress**

| Variable | (1) Inclusion in $w_0$ | (2) Added to applied load | (1)/(2) |
|---|---|---|---|
| $w_4 - w_0$ | 3·3 mm (0·130 in) | 2·9 mm (0·114 in) | 1·14 |
| $\lambda_{sy}$ | 0·83 | 0·92 | 0·90 |
| $\sigma_{es1}$ | 323 N/mm$^2$ (46·8 ksi) | 295 N/mm$^2$ (42·8 ksi) | 1·09 |
| $\sigma_{es4}$ | 327 N/mm$^2$ (47·4 ksi) | 338 N/mm$^2$ (49·0 ksi) | 0·97 |
| $\sigma_{em3}$ | 341 N/mm$^2$ (49·4 ksi) | 325 N/mm$^2$ (47·1 ksi) | 1·05 |

The differences in the above results are within the range of practical accuracy of the method.

**Example 5.2 Bottom flange panel adjacent to load-bearing diaphragm in a welded steel box-girder bridge**

Appraise the deformations and stresses in the box girder bottom flange panel shown in Fig. 5.2(i) for average applied stresses, given by a global analysis, of

$$\sigma_{xav} = 2\sigma_{yav} = 180 \text{ N/mm}^2 \text{ (26·1 ksi)}$$

Figure 5.2(ii) illustrates the weld details and Fig. 5.2(iii) shows the discrete plate panel idealisation.

Plate parameters

Geometry

$$a = 750 \text{ mm (29·5 in)}, b = 375 \text{ mm (14·8 in)}, t = 12·5 \text{ mm (0·492 in)}$$

$$a/b = 750/375 = 2, b/t = 375/12·5 = 30$$

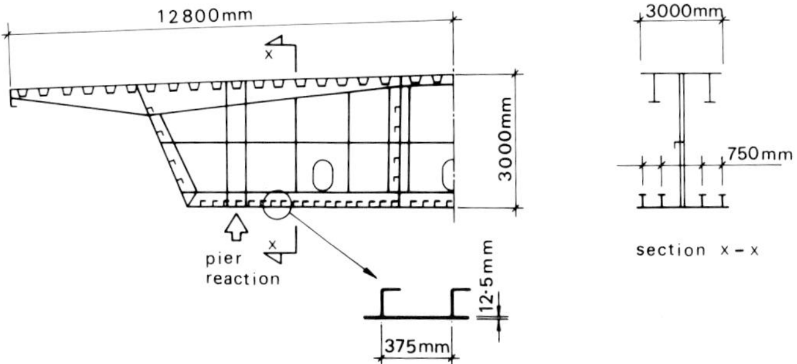


(i) section of box girder at load bearing diaphragm

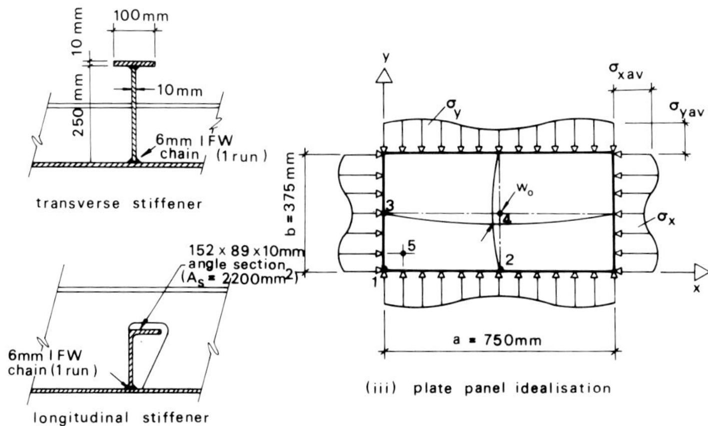


(iii) plate panel idealisation

(ii) stiffener and weld details

Fig. 5.2 Box-girder bridge example

Material properties—high tensile steel

$$E = 210\ \text{kN/mm}^2\ (30\,450\ \text{ksi}),\ \nu = 0{\cdot}3,\ \sigma_y = 350\ \text{N/mm}^2\ \ (50{\cdot}8\ \text{ksi})$$

Critical buckling load

$$\sigma_b = \pi^2 E/[12(1-\nu^2)(b/t)^2] = 210{\cdot}9\ \text{N/mm}^2\ (30{\cdot}6\ \text{ksi}) \qquad \text{(eq. 3.2)}$$

$$k = 25/12 \qquad \text{(Fig. 4.32)}$$

$$\sigma_{xcr} = k\sigma_b = 439\ \text{N/mm}^2\ (63{\cdot}7\ \text{ksi}) \qquad \text{(eq. 3.1)}$$

Unfactored applied load

$$\sigma_{x\,av} = 2\sigma_{y\,av} = 180\ \text{N/mm}^2\ (26{\cdot}1\ \text{ksi})$$

$$P/P_{cr} = \sigma_{x\,av}/\sigma_{xcr} = 0{\cdot}41$$

Initial geometric imperfection ($a < 3b$, $t < 25$ mm (0·984 in)) (Fig. 2.1)

$$G = a = 750\ \text{mm}\ (29{\cdot}5\ \text{in})$$

$$\Delta_x = G(1 + b/5000)/(30t) = 2{\cdot}15\ \text{mm}\ (0{\cdot}085\ \text{in})$$

$$N = 20$$

$$w_g = 1{\cdot}2b\Delta_x \sqrt[3]{[1/(N+1)]}/G = 0{\cdot}47\ \text{mm}\ (0{\cdot}019\ \text{in}) \qquad \text{(eq. 2.1)}$$

Residual welding stresses

(i) longitudinal intermittent weld—chain runs laid consecutively

$$A = 2 \times 6^2/2 = 36\ \text{mm}^2\ (0{\cdot}056\ \text{in}^2),\ n = 2 \qquad \text{(Figs. 2.11c, 2.13)}$$

$$w = 10\ \text{mm}\ (0{\cdot}394\ \text{in}),\ \Sigma t = 35\ \text{mm}\ (1{\cdot}38\ \text{in})$$

$$L_w = 100\ \text{mm}\ (3{\cdot}94\ \text{in}),\ L_m = 200\ \text{mm}\ (7{\cdot}87\ \text{in})$$

$$60A/[\Sigma t(\sqrt[3]{n})^2] = 38{\cdot}8\ \text{mm}\ (1{\cdot}53\ \text{in}) > w$$

(eqs. 2.9, 2.14)

$$F_c = L_w/(L_m + L_w)\,(10{\cdot}0A/(\sqrt[3]{n})^2 + 0{\cdot}1w\Sigma t)$$
$$= 87{\cdot}1\ \text{kN}\ (19{\cdot}6 \times 10^3\ \text{lbf})$$

stress area = 375 x 12·5 + 2200 = 6887·5 mm² (10·68 in²)

$$\sigma_{Rs} = F_c/6887{\cdot}5 = 12{\cdot}65\ \text{N/mm}^2\ (1{\cdot}83\ \text{ksi})$$

$$\sigma_{\text{Reff}\,x} = \sigma_{Rs}(b/t - 20)/20 = 6{\cdot}32\ \text{N/mm}^2\ (0{\cdot}92\ \text{ksi}) \qquad \text{(eq. 2.16)}$$

(ii) transverse intermittent weld—chain runs laid consecutively

$$F_c = 87{\cdot}1\ \text{kN}\ (19{\cdot}6 \times 10^3\ \text{lbf})\ \text{(as above)}$$

stress area = 750 x 12·5 + (250 + 100)10 = 12 875 mm² (19·96 in²)

$$\sigma_{Rs} = 6{\cdot}77\ \text{N/mm}^2\ (0{\cdot}98\ \text{ksi})$$

$$\sigma_{\text{Reff}\,y} = 3{\cdot}38\ \text{N/mm}^2\ (0{\cdot}49\ \text{ksi})$$

Total imperfection (incorporating residual stresses)

$L_x = 750$ mm (29·5 in), $L_y = 375$ mm (14·8 in)

$$w_0 = (w_g^2 + 8L_x^2\sigma_{\mathrm{Reff}\,x}/\pi^2 E + 8L_y^2\sigma_{\mathrm{Reff}\,y}/\pi^2 E)^{1/2} = 3{\cdot}97 \text{ mm } (0{\cdot}156 \text{ in}) \quad \text{(eq. 2.18)}$$

$w_0/t = 0{\cdot}32$

Serviceability

serviceability load factor = 1·17 (say)
serviceability factored load = 1·17 x $P/P_{\mathrm{cr}}$ = 0·48
deflection—Fig. 4.32

$$W_4 = 0{\cdot}51,\ w_4 = W_4 \times t = 6{\cdot}4 \text{ mm } (0{\cdot}252 \text{ in}) \quad \text{(eq. 3.14)}$$

stiffness effectiveness (required to determine global deflections)—Fig. 4.32

$$\lambda_{sx} = 0{\cdot}94, \qquad \lambda_{sy} = 0{\cdot}48$$

equivalent surface stresses—Fig. 4.33

$$\sigma_{\mathrm{eR}} = 1{\cdot}17(3/4)^{1/2}\sigma_{x\mathrm{av}} = 182 \text{ N/mm}^2 \text{ (26·4 ksi)} \quad \text{(Fig. 4.29)}$$

$$\bar{\sigma}_{\mathrm{es1}} = \bar{\sigma}_{\mathrm{es5}} = 1{\cdot}47,\ \sigma_{\mathrm{es1}} = \bar{\sigma}_{\mathrm{es1}} \times \sigma_{\mathrm{eR}} = 267 \text{ N/mm}^2 \text{ (38·7 ksi)} \quad \text{(eq. 3.15)}$$

$$\bar{\sigma}_{\mathrm{es4}} = 1{\cdot}81,\ \sigma_{\mathrm{es4}} = \bar{\sigma}_{\mathrm{es4}} \times \sigma_{\mathrm{eR}} = 329 \text{ N/mm}^2 \text{ (47·7 ksi)} < \sigma_y$$

Strength

collapse load factor = 1·65 (say)
collapse factored load = 1·65 x $P/P_{\mathrm{cr}}$ = 0·68
equivalent membrane stress—Fig. 4.33

$$\sigma_{\mathrm{eR}} = 1{\cdot}65(3/4)^{1/2}\sigma_{x\mathrm{av}} = 257 \text{ N/mm}^2 \text{ (Fig. 4.29) (37·3 ksi)} \quad \text{(eq. 3.16)}$$

$$\bar{\sigma}_{\mathrm{em1}} = 1{\cdot}16,\ \sigma_{\mathrm{em1}} = \bar{\sigma}_{\mathrm{em1}} \times \sigma_{\mathrm{eR}} = 298 \text{ N/mm}^2 \text{ (43·2 ksi)} < \sigma_y$$

**Example 5.3 Web panel in rivetted aluminium aircraft beam**

Appraise the deformations and stresses in a web panel of the rivetted aircraft cantilever beam section shown in Fig. 5.3(i) for a point load of 21 kN (4719 lbf). Figure 5.3(ii) shows the discrete panel idealisation. Load factors of 1·00 will be assumed for both serviceability and strength and hence the applied load is the factored load in both cases.

**Plate parameters**

Geometry

$$a = 350 \text{ mm (13·8 in)},\ b = 350 \text{ mm (13·8 in)},\ t = 2 \text{ mm (0·079 in)}$$

$$a/b = 350/350 = 1,\ b/t = 350/2 = 175$$

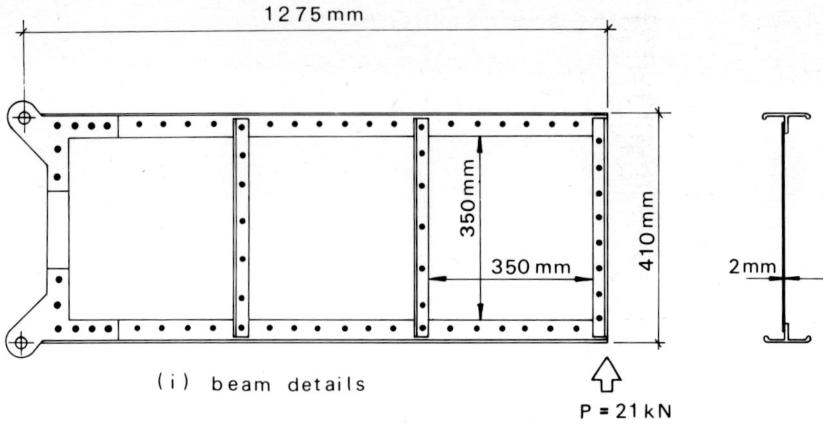


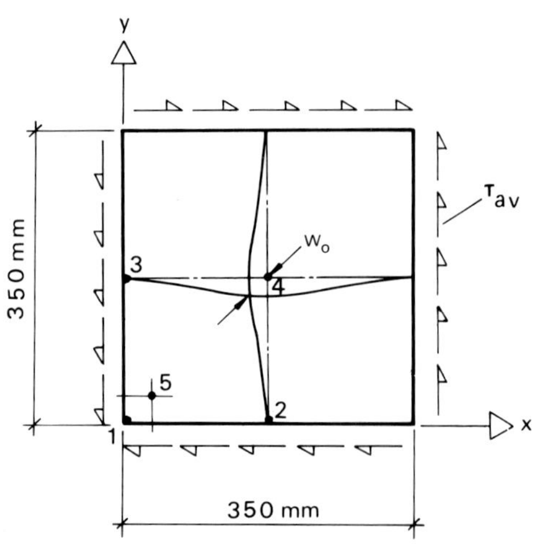


Fig. 5.3 Aircraft beam example

Material properties—aluminium alloy

$E = 74$ kN/mm$^2$ (10 730 ksi), $\nu = 0{\cdot}32$ (assume 0·3), $\sigma_y = 420$ N/mm$^2$ (60·9 ksi)

Critical buckling load

$$\sigma_b = \pi^2 E/[12(1-\nu^2)(b/t)^2] = 2{\cdot}18 \text{ N/mm}^2 \ (0{\cdot}320 \text{ ksi}) \qquad \text{(eq. 3.2)}$$

$$k = 9{\cdot}34 \qquad \text{(Fig. 4.40)}$$

$$\tau_{cr} = k\sigma_b = 20{\cdot}4 \text{ N/mm}^2 \ (2{\cdot}96 \text{ ksi}) \qquad \text{(eq. 3.1)}$$

Applied load

$$\tau_{av} = 21 \times 10^3/(350 \times 2) = 30{\cdot}0 \text{ N/mm}^2 \ (4{\cdot}35 \text{ ksi})$$
$$P/P_{cr} = \tau_{av}/\tau_{cr} = 1{\cdot}47$$

Initial geometric imperfection—the steel box girder imperfection formula is not applicable, assume $w_g = t$

Total imperfection—no residual welding stresses, therefore:

$$w_0/t = w_g/t = 1$$

**Deflection, effectiveness and stresses**

(i) Panel fully restrained—Fig. 4.40

deflection

$$W_4 = 1{\cdot}51, w_4 = W_4 \times t = 3{\cdot}0 \text{ mm } (0{\cdot}118 \text{ in}) \qquad (3.14)$$

stiffness effectiveness (to determine beam deflection)

$$\lambda_{\tau s} = s_f/s = 1/1{\cdot}075 = 0{\cdot}93 \qquad \text{(Table 16)}$$

equivalent surface stresses

$$\sigma_{eR} = (3)^{1/2}\tau_{av} = 52{\cdot}0 \text{ N/mm}^2 \ (7{\cdot}54 \text{ ksi}) \qquad \text{(Fig. 4.38)}$$
$$\bar{\sigma}_{es1} = 2{\cdot}26, \sigma_{es1} = \bar{\sigma}_{es1} \times \sigma_{eR} = 118 \text{ N/mm}^2 \ (17{\cdot}1 \text{ ksi}) \qquad \text{(eq. 3.15)}$$
$$\bar{\sigma}_{es4} = 1{\cdot}59, \sigma_{es4} = \bar{\sigma}_{es4} \times \sigma_{eR} = 83 \text{ N/mm}^2 \ (12{\cdot}0 \text{ ksi})$$
$$\bar{\sigma}_{es5} = 1{\cdot}40, \sigma_{es5} = \bar{\sigma}_{es5} \times \sigma_{eR} = 73 \text{ N/mm}^2 \ (10{\cdot}6 \text{ ksi})$$

equivalent membrane stresses

$$\sigma_{em1} = (3\tau_{m1}^2)^{1/2} = (3 \times 1{\cdot}072^2)^{1/2}\tau_{av} = 56 \text{ N/mm}^2 \ (8{\cdot}12 \text{ ksi}) \qquad \text{(Table 16)}$$
$$\sigma_{em4} = (\sigma_{mx4}^2 + \sigma_{my4}^2 - \sigma_{mx4}\sigma_{my4} + 3\tau_{m4}^2)^{1/2}$$
$$= (0{\cdot}41^2 + 3 \times 0{\cdot}99^2)^{1/2}\tau_{av} = 53 \text{ N/mm}^2 \ (7{\cdot}69 \text{ ksi})$$

edge forces

$$\sigma_{Rx} = \tau_{av} = 30 \text{ N/mm}^2 \ (4{\cdot}35 \text{ ksi})$$
$$\bar{\sigma}_{x\,av} = \bar{\sigma}_{y\,av} = 0{\cdot}24$$
$$\sigma_{x\,av} = \bar{\sigma}_{x\,av} \times \sigma_{Rx} = 7{\cdot}2 \text{ N/mm}^2 \ (1{\cdot}04 \text{ ksi}) \qquad \text{(eq. 3.17)}$$

average distributed load on edge framing member

$$q_x = 7{\cdot}2 \times 2 = 14{\cdot}4 \text{ N/mm } (82{\cdot}2 \text{ lb/in})$$

(ii) Panel unrestrained—Fig. 4.37

deflection

$$W_4 = 3{\cdot}02, w_4 = W_4 \times t = 6{\cdot}0 \text{ mm } (0{\cdot}236 \text{ in}) \qquad \text{(eq. 3.14)}$$

stiffness effectiveness (to determine beam deflection)

$$\lambda_{\tau s} = s_f/s = 1/2{\cdot}05 = 0{\cdot}49 \qquad \text{(Table 15)}$$

equivalent surface stresses

$$\sigma_{eR} = (3)^{1/2}\tau_{av} = 52{\cdot}0 \text{ N/mm}^2 \ (7{\cdot}54 \text{ ksi}) \qquad \text{(Fig. 4.36)}$$

$$\bar{\sigma}_{es1} = 5{\cdot}61, \sigma_{es1} = \bar{\sigma}_{es1} \times \sigma_{eR} = 292 \text{ N/mm}^2 \ (42{\cdot}3 \text{ ksi}) \qquad \text{(eq. 3.15)}$$

$$\bar{\sigma}_{es4} = \bar{\sigma}_{es5} = 2{\cdot}94, \sigma_{es4} = \bar{\sigma}_{es4} \times \sigma_{eR} = 153 \text{ N/mm}^2 \ (22{\cdot}2 \text{ ksi})$$

equivalent membrane stress

$$\bar{\sigma}_{em4} = 1{\cdot}26, \sigma_{em2} = \bar{\sigma}_{em2} \times \sigma_{eR} = 66 \text{ N/mm}^2 \ (9{\cdot}57 \text{ ksi}) \qquad \text{(eq. 3.16)}$$

Both surface and membrane equivalent stresses are well below yield in the above example, even for unrestrained sides. This is to be expected in such a thin plate. These stresses are however of interest in regard to fatigue calculations. The deflection, loss of in-plane stiffness and edge forces on framing members will also be of interest to aircraft designers.

### Example 5.4 Web panel in the double bottom of a welded steel bulk carrier ship

Appraise the deformations and stresses in the double bottom centre girder web panel of the bulk carrier ship under dry dock loading as illustrated in Fig. 5.4(i). The dead weight of the ship is $3{\cdot}5 \times 10^5$ kN ($78{\cdot}6 \times 10^6$ lbf) assumed equally distributed on three timber baulks over a length of 60 m (197 ft). Fig. 5.4(ii) shows the plate panel idealisation.

#### Plate parameters

Geometry

$$a = 825 \text{ mm } (32{\cdot}5 \text{ in}), b = 1250 \text{ mm } (49{\cdot}2 \text{ in}), t = 15 \text{ mm } (0{\cdot}591 \text{ in})$$

$$a/b = 825/1250 = 0{\cdot}66, b/t = 1250/15 = 83{\cdot}3$$

Material properties—mild steel

$$E = 210 \text{ kN/mm}^2 \ (30\,450 \text{ ksi}), \nu = 0{\cdot}3, \sigma_y = 248 \text{ N/mm}^2 \ (36{\cdot}0 \text{ ksi})$$

Critical buckling load

$$\sigma_b = \pi^2 E/[12(1-\nu^2)(b/t)^2] = 27{\cdot}35 \text{ N/mm}^2 \ (3{\cdot}97 \text{ ksi}) \qquad \text{(eq. 3.2)}$$

$$k = 12{\cdot}8 \qquad \text{(Fig. 4.87)}$$

$$\sigma_{y1cr} = k\sigma_b = 350 \text{ N/mm}^2 \ (50.8 \text{ ksi}) \qquad \text{(eq. 3.1)}$$

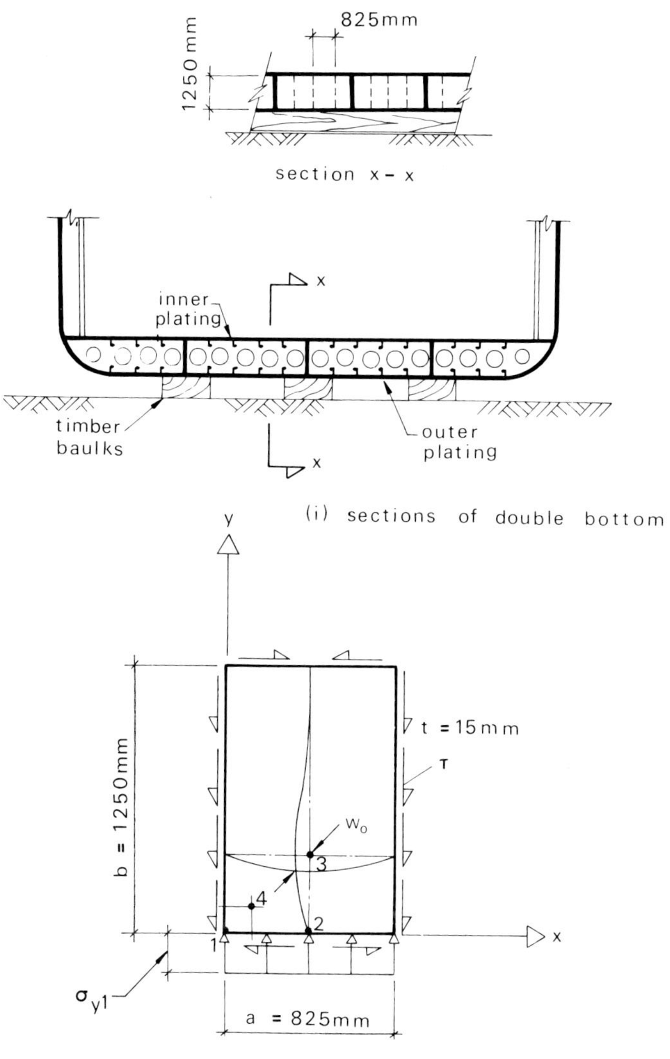


Fig. 5.4 Ship's double bottom example

Unfactored applied load

$$\sigma_{y1} = 3{\cdot}5 \times 10^8/(3 \times 60 \times 10^3 \times 15) = 130 \text{ N/mm}^2 \ (18{\cdot}9 \text{ ksi})$$

$$P/P_{cr} = \sigma_{y1}/\sigma_{y1\,cr} = 0{\cdot}37$$

Initial geometric imperfection (apply bridge formula) (Fig. 2.1)

$a < 3b, t < 25$ mm (0·984 in)

$G = a = 825$ mm (32·5 in)

$\Delta_x = G(1 + b/5000)/(30t) = 2{\cdot}29$ mm (0·090 in), use 3 mm (0·118 in)

$w_g = 1{\cdot}2\Delta_x = 3{\cdot}6$ mm (0·142 in)

Total imperfection

Assume residual stresses contribute effective imperfections of 3 mm in direction of both axes.

$$w_0 = (3{\cdot}6^2 + 3^2 + 3^2)^{1/2} = 5{\cdot}6 \text{ mm (0·220 in)}$$

$$w_0/t = 0{\cdot}37$$

## Serviceability

serviceability load factor = 1·00 (say)
serviceability factored load = 1·00 x $P/P_{cr} = 0{\cdot}37$
deflection–Fig. 4·87

$$W_3 = 0{\cdot}58, w_3 = W_3 \times t = 8{\cdot}7 \text{ mm (0·343 in)} \qquad \text{(eq. 3.14)}$$

equivalent surface stresses–Fig. 4.87

$$\sigma_{eR} = 1{\cdot}00(1 + 0{\cdot}75(a/b)^2)^{1/2}\sigma_{y1} = 150 \text{ N/mm}^2 \text{ (21·8 ksi)} \qquad \text{(Fig. 4.85)}$$

$$\bar{\sigma}_{es1} = 1{\cdot}90, \sigma_{es1} = \bar{\sigma}_{es1} \times \sigma_{eR} = 285 \text{ N/mm}^2 \text{ (41·3 ksi)} > \sigma_y \qquad \text{(eq. 3.15)}$$

$$\bar{\sigma}_{es4} = 1{\cdot}41, \sigma_{es4} = \bar{\sigma}_{es4} \times \sigma_{eR} = 212 \text{ N/mm}^2 \text{ (30·7 ksi)}$$

The maximum equivalent surface stress is greater than yield, but this is a case where limited surface yield would be acceptable.

$$\sigma_{es1} = 1{\cdot}15\sigma_y$$

$$\sigma_{es4} = 0{\cdot}85\sigma_y$$

Thus for the 10 ($x$-axis) by 14 ($y$-axis) finite-difference mesh grid used, it can be estimated that yield, with limited penetration, will spread over an area of less than 40 mm by 50 mm and will not therefore result in significant permanent set.

## Strength

collapse load factor = 1·50 (say)
collapse factored load = 1·50 x $P/P_{cr} = 0{\cdot}56$

equivalent membrane stress–Fig. 4.87

$$\sigma_{eR} = 1{\cdot}50\,(1 + 0{\cdot}75(a/b)^2)^{1/2}\,\sigma_{y1} = 225 \text{ N/mm}^2 \text{ (32·6 ksi)} \qquad \text{(Fig. 4.85)}$$

$\bar{\sigma}_{em2} = 0{\cdot}76 < \bar{\sigma}_{eR}$, hence use $\sigma_{eR} < \sigma_y$

**Example 5.5 Interaction–biaxial compression**

The following example illustrates the use of the rudimentary interaction procedure described in Section 4.9 of Chapter 4 for the case of biaxial compression. The plate panel shown in Figure 5.5(i) will be appraised for an applied load

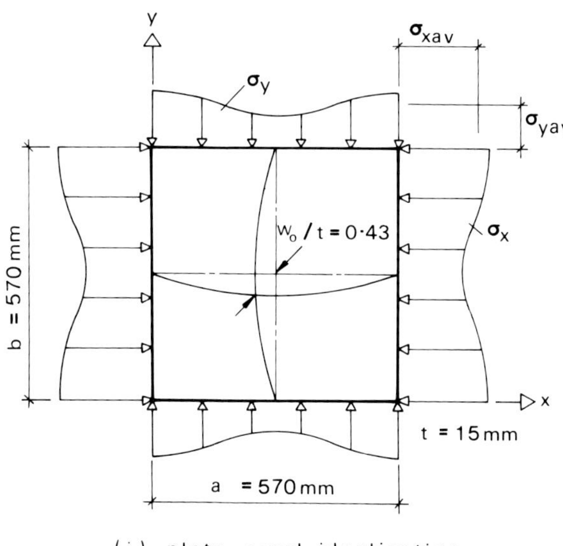

(i) plate panel idealisation

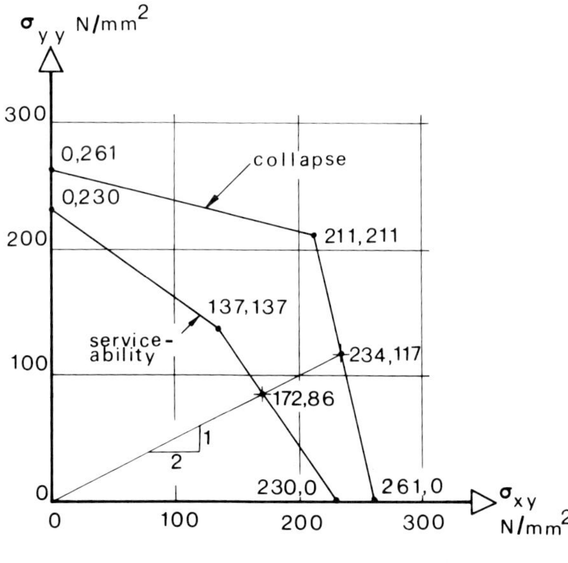

(ii) interaction diagram

Fig. 5.5 Interaction example–biaxial compression

combination of $\sigma_{xav} = 2\sigma_{yav} = 136$ N/mm$^2$ (19·7 ksi), for an initial imperfection of $w_0/t = 0{\cdot}43$. Note that geometry and material properties are identical to panel in Example 5.1.

Serviceability

serviceability load factor = 1·10

(i) Loads applied independently to cause surface yield

(a) $\sigma_{xav}$: $k = 4$, $\sigma_{xcr} = 526$ N/mm$^2$ (76·3 ksi), $\sigma_{eR} = \sigma_{xav}$

by trial: $\sigma_{xav} = 230$ N/mm$^2$ (33·4 ksi)

$P/P_{cr} = \sigma_{xav}/\sigma_{xcr} = 0{\cdot}44$

$\bar{\sigma}_{es1} = 1{\cdot}52$ (Fig. 4.17)

$\sigma_{es1} = \bar{\sigma}_{es1} \times \sigma_{eR} = 350$ N/mm$^2$ (50·8 ksi)($=\sigma_y$)

$\sigma_{xy0} = 230$ N/mm$^2$ (33·4 ksi)

(b) $\sigma_{yav}$: as above, giving $\sigma_{yy0} = 230$ N/mm$^2$ (33·4 ksi)

(ii) Combined load causing surface yield

$\sigma_{xav} = \sigma_{yav}$: $k = 2$, $\sigma_{xcr} = 263$ N/mm$^2$ (38·1 ksi), $\sigma_{eR} = \sigma_{xav}$

by trial: $\sigma_{xav} = 137$ N/mm$^2$ (19.9 ksi)

$P/P_{cr} = \sigma_{xav}/\sigma_{xcr} = 0{\cdot}52$

$\bar{\sigma}_{es1} = 2{\cdot}56$ (Fig. 4.24)

$\sigma_{es1} = \bar{\sigma}_{es1} \times \sigma_{eR} = 350$ N/mm$^2$ (50·8 ksi) ($=\sigma_y$)

$\sigma_{xy1} = \sigma_{yy1} = 137$ N/mm$^2$ (19·9 ksi)

(iii) Interaction solution for $\sigma_{xav} = 2\sigma_{yav} = 1{\cdot}10 \times 136 = 150$ N/mm$^2$ (21·8 ksi)

Refer to Fig. 5.5(ii). Cases (i)(a),(b) and (ii) are used to plot the serviceability interaction diagram. The slope of the load combination for which a solution is required is

$$\sigma_{x2}/\sigma_{y2} = \sigma_{xav}/\sigma_{yav} = 2$$

for which the diagram gives

$$\sigma_{xy2} = 2\sigma_{yy2} = 172 \text{ N/mm}^2 \text{ (24·9 ksi)}$$

the serviceability safety factor is therefore

$$\sigma_{xy2}/\sigma_{x2} = 172/150 = 1{\cdot}15$$

(iv) Check solution

The above solution can in fact be obtained directly from the graphs as follows:

$\sigma_{xav} = 2\sigma_{yav}$: $k = 8/3$, $\sigma_{xcr} = 350$ N/mm$^2$ (50.8 ksi), $\sigma_{eR} = (3/4)^{1/2}\sigma_{xav}$

by trial: $\sigma_{xav} = 175$ N/mm$^2$ (25·4 ksi)

$P/P_{cr} = \sigma_{xav}/\sigma_{xcr} = 0{\cdot}50$

$\bar{\sigma}_{es1} = 2{\cdot}31$ (Fig. 4.31)

$\sigma_{es1} = \bar{\sigma}_{es1} \times \sigma_{eR} = 350$ N/mm$^2$ (50·8 ksi) $(=\sigma_y)$

$\sigma_{xy2} = 175$ N/mm$^2$ (25·4 ksi)

This agrees very closely with the solution obtained from the interaction diagram.

Collapse

collapse load factor = 1·55

(i) Loads applied independently to cause membrane yield

(a) $\sigma_{xav}$:

by trial: $\sigma_{xav} = 261$ N/mm$^2$ (37·8 ksi)

$P/P_{cr} = \sigma_{xav}/\sigma_{xcr} = 0{\cdot}50$

$\bar{\sigma}_{em2} = 1{\cdot}34$ (Fig. 4.17)

$\sigma_{em2} = \bar{\sigma}_{em2} \times \sigma_{eR} = 350$ N/mm$^2$ (50·8 ksi) $(=\sigma_y)$

$\sigma_{xy0} = 261$ N/mm$^2$ (37·8 ksi)

(b) $\sigma_{yav}$: as above, giving $\sigma_{yy0} = 261$ N/mm$^2$ (37·8 ksi)

(ii) Combined load causing membrane yield

$\sigma_{xav} = \sigma_{yav}$:

by trial: $\sigma_{xav} = 211$ N/mm$^2$ (30·6 ksi)

$P/P_{cr} = 0{\cdot}80$

$\bar{\sigma}_{em1} = 1{\cdot}66$ (Fig. 4.24)

$\sigma_{em1} = \bar{\sigma}_{em1} \times \sigma_{eR} = 350$ N/mm$^2$ (50·8 ksi) $(=\sigma_y)$

$\sigma_{xy1} = \sigma_{yy1} = 211$ N/mm$^2$ (30·6 ksi)

(iii) Interaction solution for $\sigma_{xav} = 2\sigma_{yav} = 1{\cdot}55 \times 136 = 211$ N/mm$^2$ (30·6 ksi)

Solution will be conservative since maximum membrane stress solutions for interaction diagram are not at identical locations.

Refer to Fig. 5.5(ii). Cases (i) (a),(b) and (ii) are used to plot the collapse interaction diagram. The slope of the load combination for which a solution is required is

$$\sigma_{x2}/\sigma_{y2} = \sigma_{xav}/\sigma_{yav} = 2$$

for which the diagram gives

$$\sigma_{xy2} = 2\sigma_{yy2} = 234 \text{ N/mm}^2 \text{ (33·9 ksi)}$$

the collapse safety factor is therefore

$$\sigma_{xy2}/\sigma_{x2} = 234/211 = 1{\cdot}11$$

(iv) check solution

$\sigma_{x\,\mathrm{av}} = 2\sigma_{y\,\mathrm{av}}$:

by trial: $\sigma_{x\,\mathrm{av}} = 250$ N/mm$^2$ (36·3 ksi)

$P/P_{\mathrm{cr}} = \sigma_{x\,\mathrm{av}}/\sigma_{x\,\mathrm{cr}} = 0{\cdot}71$

$\bar{\sigma}_{\mathrm{em2}} = 1{\cdot}62$ (Fig. 4.31)

$\sigma_{\mathrm{em2}} = \bar{\sigma}_{\mathrm{em2}} \times \sigma_{\mathrm{eR}} = 350$ N/mm$^2$ (50·8 ksi) ($=\sigma_y$)

$\sigma_{xy2} = 250$ N/mm$^2$ (36·3 ksi)

This agrees quite closely with the solution obtained from the interaction diagram.

### Example 5.6 Interaction—combined uniaxial compression and shear

The following example illustrates the use of the rudimentary interaction procedure described in Section 4.9 of Chapter 4 for the case of combined uniaxial compression and shear. The plate panel shown in Fig. 5.6(i) will be appraised for an applied loading of $\sigma_{x\,\mathrm{av}} = 2\tau_{\mathrm{av}} = 120$ N/mm$^2$ (17·4 ksi), for an initial imperfection of $w_0/t = 0{\cdot}43$. Note that geometry and material properties are identical to the panel in Example 5.1.

#### Serviceability

serviceability load factor = 1·00

(i) Loads applied independently to cause surface yield

(a) $\sigma_{x\,\mathrm{av}}$: as for Example 5.5

$$\sigma_{xy0} = 230 \text{ N/mm}^2 \text{ (33·4 ksi)}$$

(b) $\tau_{\mathrm{av}}$: $k = 9{\cdot}34$, $\tau_{\mathrm{cr}} = 1227$ N/mm$^2$ (178 ksi), $\sigma_{\mathrm{eR}} = (3)^{1/2}\tau_{\mathrm{av}}$

by trial: $\tau_{\mathrm{av}} = 134$ N/mm$^2$ (19·4 ksi)

$P/P_{\mathrm{cr}} = \tau_{\mathrm{av}}/\tau_{\mathrm{cr}} = 0{\cdot}11$

$\bar{\sigma}_{\mathrm{es1}} = 1{\cdot}51$ (Fig. 4.40)

$\sigma_{\mathrm{es1}} = \bar{\sigma}_{\mathrm{es1}} \times \sigma_{\mathrm{eR}} = 350$ N/mm$^2$ (50·8 ksi) ($=\sigma_y$)

$\tau_{y0} = 134$ N/mm$^2$ (19·4 ksi)

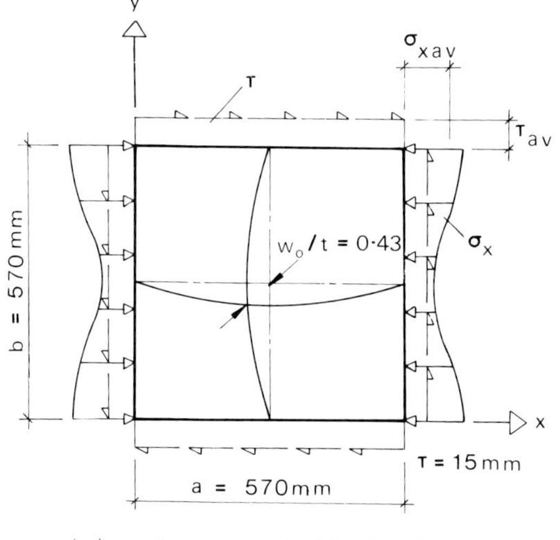


(i) plate panel idealisation

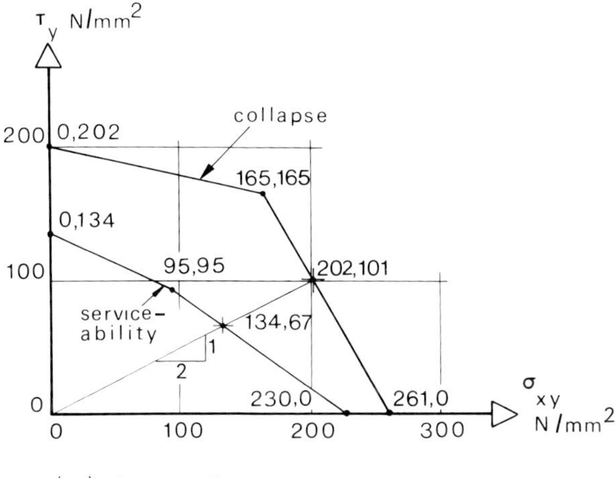


(ii) interaction diagram

Fig. 5.6 Interaction example–combined uniaxial compression and shear

## (ii) Combined load causing surface yield

$\sigma_{xav} = \tau_{av}$: $k = 3{\cdot}45$, $\sigma_{xcr} = 453$ N/mm$^2$ (65·7 ksi), $\sigma_{eR} = 2\sigma_{xav}$

by trial: $\sigma_{xav} = 95$ N/mm$^2$ (13·8 ksi)

$P/P_{cr} = \sigma_{xav}/\sigma_{xcr} = 0{\cdot}21$

$\bar{\sigma}_{es1} = 1{\cdot}84$ (Fig. 4.47)

$\sigma_{es1} = \bar{\sigma}_{es1} \times \sigma_{eR} = 350$ N/mm$^2$ (50·8 ksi) $(=\sigma_y)$

$\sigma_{xy1} = \tau_{y1} = 95$ N/mm$^2$ (13·8 ksi)

(iii) Interaction solution for $\sigma_{xav} = 2\tau_{av} = 1{\cdot}00 \times 120 = 120$ N/mm$^2$ (17·4 ksi)

Refer to Fig. 5.6(ii). Cases (i) (a),(b) and (ii) are used to plot the serviceability interaction diagram. The slope of this load combination for which a solution is required is

$$\sigma_{x2}/\tau_2 = \sigma_{xav}/\tau_{av} = 2$$

for which the diagram gives

$$\sigma_{xy2} = 2\tau_{y2} = 134 \text{ N/mm}^2 \text{ (19·4 ksi)}$$

the serviceability safety factor is therefore

$$\sigma_{xy2}/\sigma_{x2} = 134/120 = 1{\cdot}12$$

### Collapse

collapse load factor = 1·50

(i) Loads applied independently to cause membrane yield

(a) $\sigma_{xav}$: as for Example 5.5—collapse calculation

$$\sigma_{xyo} = 261 \text{ N/mm}^2 \text{ (37·8 ksi)}$$

(b) $\tau_{av}$: membrane yield not plotted, use $\sigma_{eR}$

$$\tau_{yo} = 350/(3)^{1/2} = 202 \text{ N/mm}^2 \text{ (29·3 ksi)}$$

(ii) Combined load causing membrane yield

$\sigma_{xav} = \tau_{av}$:

by trial: $\sigma_{xav} = 165$ N/mm$^2$ (23·9 ksi)

$P/P_{cr} = 0{\cdot}36$

$\bar{\sigma}_{em1} = 1{\cdot}06$ (Fig. 4.47)

$\sigma_{em1} = \bar{\sigma}_{em1} \times \sigma_{eR} = 350$ N/mm$^2$ (50·8 ksi) ($= \sigma_y$)

$\sigma_{xy1} = \tau_{y1} = 165$ N/mm$^2$ (23·9 ksi)

(iii) Interaction solution for $\sigma_{xav} = 2\tau_{av} = 1{\cdot}50 \times 120 = 180$ N/mm$^2$ (26·1 ksi)

Solution will be conservative since maximum membrane stress solutions for interaction diagram are not at identical locations.

Refer to Fig. 5.6(ii). Cases (i) (a),(b) and (ii) are used to plot the collapse interaction diagram. The slope of the load combination for which a solution is required is

$$\sigma_{x2}/\tau_2 = \sigma_{xav}/\tau_{av} = 2$$

for which the diagram gives

$$\sigma_{xy2} = 2\tau_{y2} = 202 \text{ N/mm}^2 \ (29{\cdot}3 \text{ ksi})$$

the collapse safety factor is therefore

$$\sigma_{xy2}/\sigma_{x2} = 202/180 = 1{\cdot}12$$

APPENDIX

# Tabulated Data

The surface and membrane stresses given in the graphs in Chapter 4 are equivalent stresses and incorporate all stress components acting at each location. There may be instances where a designer wishes to calculate the uniaxial stress in any given direction, or the principal stresses. This could arise in connection with fatigue calculations, or in cases where tensile yield is a consideration in connection with brittle materials or surface coating. For this purpose the tables give uniaxial stress components and bending moments at critical locations in non-dimensional form as defined in Section 3.1 of Chapter 3.

The edge stresses given in the graphs in Chapter 4 for unloaded constrained sides are average values. These will give an underestimate of the forces acting on side framing members and if a designer requires a more accurate estimate of these forces the maximum values given in the tables for such cases can be used.

There are restrained unloaded side cases where the maximum equivalent membrane stress is not plotted in the results given in Chapter 3. At lower loads ($< \sigma_{cr}/2$) the maximum membrane stress is in fact only marginally greater than the applied value but at the top of the load range it can exceed the applied value by up to 10%. If the maximum membrane stress is required it can be computed from the tabulated results by substituting the appropriate axial stress components into the expression for equivalent stress.

## Table 1

## Uniform uniaxial compressive stress in direction of $y$-axis Unloaded sides stress-free

Side ratio $a/b = 1{\cdot}0$

| W | P | W . | $\bar{\sigma}_{MX}$ | | | | $\bar{\sigma}_{MY}$ | | | | $\bar{M}_Y$ | $\bar{M}_X$ | $\bar{M}_{XY}$ |
|---|---|---|---|---|---|---|---|---|---|---|---|---|---|
| | | 4 | 1 | 2 | 3 | 4 | 1 | 2 | 3 | 4 | 4 | 4 | 1 |
| .10 | .25 | .123 | 0.000 | .006 | 0.000 | -.006 | 1.000 | 1.000 | 1.008 | .995 | .028 | .028 | .015 |
| | .50 | .217 | 0.000 | .039 | 0.000 | -.021 | 1.000 | 1.000 | 1.040 | .981 | .135 | .135 | .075 |
| | .75 | .385 | 0.000 | .088 | 0.000 | -.050 | 1.000 | 1.000 | 1.087 | .951 | .327 | .327 | .182 |
| | 1.00 | .885 | 0.000 | .402 | 0.000 | -.211 | 1.000 | 1.000 | 1.396 | .786 | .826 | .843 | .552 |
| | 1.25 | 1.552 | 0.000 | .913 | 0.000 | -.429 | 1.000 | 1.000 | 1.870 | .549 | 1.369 | 1.457 | 1.154 |
| | 1.50 | 2.055 | 0.000 | 1.453 | 0.000 | -.555 | 1.000 | 1.000 | 2.296 | .374 | 1.474 | 1.741 | 1.888 |
| .25 | .25 | .321 | 0.000 | .081 | 0.000 | -.047 | 1.000 | 1.000 | 1.080 | 1.013 | .084 | .084 | .044 |
| | .50 | .492 | 0.000 | .173 | 0.000 | -.098 | 1.000 | 1.000 | 1.172 | .965 | .271 | .273 | .160 |
| | .75 | .767 | 0.000 | .338 | 0.000 | -.186 | 1.000 | 1.000 | 1.336 | .812 | .563 | .570 | .352 |
| | 1.00 | 1.199 | 0.000 | .682 | 0.000 | -.343 | 1.000 | 1.000 | 1.667 | .639 | .936 | .969 | .716 |
| | 1.25 | 1.718 | 0.000 | 1.153 | 0.000 | -.516 | 1.000 | 1.000 | 2.088 | .466 | 1.265 | 1.378 | 1.266 |
| | 1.50 | 2.192 | 0.000 | 1.642 | 0.000 | -.607 | 1.000 | 1.000 | 2.454 | .310 | 1.354 | 1.644 | 1.995 |
| .50 | .25 | .642 | 0.000 | .314 | 0.000 | -.178 | 1.000 | 1.000 | 1.313 | .822 | .162 | .160 | .092 |
| | .50 | .867 | 0.000 | .485 | 0.000 | -.263 | 1.000 | 1.000 | 1.479 | .737 | .384 | .389 | .261 |
| | .75 | 1.177 | 0.000 | .738 | 0.000 | -.382 | 1.000 | 1.000 | 1.726 | .612 | .668 | .687 | .511 |
| | 1.00 | 1.554 | 0.000 | 1.076 | 0.000 | -.507 | 1.000 | 1.000 | 2.036 | .474 | .922 | .992 | .890 |
| | 1.25 | 1.974 | 0.000 | 1.480 | 0.000 | -.613 | 1.000 | 1.000 | 2.372 | .337 | 1.103 | 1.273 | 1.429 |
| | 1.50 | 2.404 | 0.000 | 1.919 | 0.000 | -.660 | 1.000 | 1.000 | 2.682 | .221 | 1.161 | 1.493 | 2.167 |
| 1.00 | .25 | 1.209 | 0.000 | .898 | 0.000 | -.470 | 1.000 | 1.000 | 1.893 | .526 | .171 | .205 | .160 |
| | .50 | 1.461 | 0.000 | 1.116 | 0.000 | -.552 | 1.000 | 1.000 | 2.098 | .439 | .369 | .418 | .391 |
| | .75 | 1.755 | 0.000 | 1.374 | 0.000 | -.635 | 1.000 | 1.000 | 2.336 | .347 | .595 | .638 | .696 |
| | 1.00 | 2.079 | 0.000 | 1.671 | 0.000 | -.700 | 1.000 | 1.000 | 2.583 | .259 | .729 | .836 | 1.118 |
| | 1.25 | 2.429 | 0.000 | 2.007 | 0.000 | -.740 | 1.000 | 1.000 | 2.825 | .176 | .779 | 1.016 | 1.694 |
| | 1.50 | 2.806 | 0.000 | 2.384 | 0.000 | -.751 | 1.000 | 1.000 | 3.049 | .100 | .788 | 1.192 | 2.468 |

## Table 2

## Uniform equal biaxial compressive stress ($\sigma_x = \sigma_y$)

Side ratio $a/b = 1{\cdot}0$

| W | P | W . | $\bar{\sigma}_{MX}$ | | | | $\bar{\sigma}_{MY}$ | | | | $\bar{M}_X$ | $\bar{M}_Y$ | $\bar{M}_{XY}$ |
|---|---|---|---|---|---|---|---|---|---|---|---|---|---|
| | | 4 | 1 | 2 | 3 | 4 | 1 | 2 | 3 | 4 | 4 | 4 | 1 |
| .10 | .25 | .123 | 1.000 | 1.019 | 1.000 | .986 | 1.000 | 1.000 | 1.019 | .986 | .028 | .028 | .015 |
| | .50 | .217 | 1.000 | 1.070 | 1.000 | .956 | 1.000 | 1.000 | 1.070 | .956 | .135 | .135 | .075 |
| | .75 | .385 | 1.000 | 1.175 | 1.000 | .900 | 1.000 | 1.000 | 1.175 | .900 | .327 | .327 | .182 |
| | 1.00 | .885 | 1.000 | 1.796 | 1.000 | .557 | 1.000 | 1.000 | 1.796 | .557 | .832 | .832 | .350 |
| | 1.25 | 1.548 | 1.000 | 2.765 | 1.000 | .171 | 1.000 | 1.000 | 2.765 | .171 | 1.420 | 1.420 | 1.131 |
| | 1.50 | 2.039 | 1.000 | 3.700 | 1.000 | .198 | 1.000 | 1.000 | 3.700 | .198 | 1.656 | 1.656 | 1.802 |
| .25 | .25 | .321 | 1.000 | 1.162 | 1.000 | .904 | 1.000 | 1.000 | 1.162 | .904 | .085 | .085 | .044 |
| | .50 | .492 | 1.000 | 1.345 | 1.000 | .804 | 1.000 | 1.000 | 1.345 | .804 | .273 | .273 | .160 |
| | .75 | .767 | 1.000 | 1.674 | 1.000 | .626 | 1.000 | 1.000 | 1.674 | .626 | .568 | .568 | .351 |
| | 1.00 | 1.199 | 1.000 | 2.344 | 1.000 | .303 | 1.000 | 1.000 | 2.344 | .303 | .958 | .958 | .711 |
| | 1.25 | 1.713 | 1.000 | 3.221 | 1.000 | -.070 | 1.000 | 1.000 | 3.221 | -.070 | 1.342 | 1.342 | 1.239 |
| | 1.50 | 2.176 | 1.000 | 4.037 | 1.000 | -.320 | 1.000 | 1.000 | 4.037 | -.320 | 1.558 | 1.558 | 1.910 |
| .50 | .25 | .642 | 1.000 | 1.625 | 1.000 | .644 | 1.000 | 1.000 | 1.625 | .644 | .161 | .161 | .092 |
| | .50 | .867 | 1.000 | 1.964 | 1.000 | .473 | 1.000 | 1.000 | 1.964 | .473 | .388 | .388 | .261 |
| | .75 | 1.177 | 1.000 | 2.466 | 1.000 | .229 | 1.000 | 1.000 | 2.466 | .229 | .680 | .680 | .509 |
| | 1.00 | 1.553 | 1.000 | 3.103 | 1.000 | -.037 | 1.000 | 1.000 | 3.103 | -.037 | .965 | .965 | .879 |
| | 1.25 | 1.970 | 1.000 | 3.820 | 1.000 | -.290 | 1.000 | 1.000 | 3.820 | -.290 | 1.212 | 1.212 | 1.392 |
| | 1.50 | 2.388 | 1.000 | 4.532 | 1.000 | -.481 | 1.000 | 1.000 | 4.532 | -.481 | 1.386 | 1.386 | 2.071 |
| 1.00 | .25 | 1.209 | 1.000 | 2.790 | 1.000 | .055 | 1.000 | 1.000 | 2.790 | .055 | .205 | .205 | .160 |
| | .50 | 1.461 | 1.000 | 3.214 | 1.000 | -.116 | 1.000 | 1.000 | 3.214 | -.116 | .412 | .412 | .390 |
| | .75 | 1.755 | 1.000 | 3.706 | 1.000 | -.293 | 1.000 | 1.000 | 3.706 | -.293 | .620 | .620 | .691 |
| | 1.00 | 2.077 | 1.000 | 4.237 | 1.000 | -.452 | 1.000 | 1.000 | 4.237 | -.452 | .795 | .795 | 1.092 |
| | 1.25 | 2.422 | 1.000 | 4.794 | 1.000 | -.586 | 1.000 | 1.000 | 4.794 | -.586 | .935 | .935 | 1.531 |
| | 1.50 | 2.788 | 1.000 | 5.364 | 1.000 | -.695 | 1.000 | 1.000 | 5.364 | -.695 | 1.040 | 1.040 | 2.360 |

## Table 3

**Uniform uniaxial compressive displacement in direction of *y*-axis**
**Unloaded sides stress-free**

Side ratio $a/b = 1{\cdot}0$

| W | P | V | W | $\bar{\sigma}_{MX}$ | | | | $\bar{\sigma}_{MY}$ | | | | $\bar{M}_{Y}$ | $\bar{M}_{X}$ | $\bar{M}_{XY}$ |
|---|---|---|---|---|---|---|---|---|---|---|---|---|---|---|
| | | | 4 | 1 | 2 | 3 | 4 | 1 | 2 | 3 | 4 | 4 | 4 | 1 |
| .10 | .25 | 1.010 | .131 | 0.000 | .006 | 0.000 | -.006 | 1.002 | .977 | 1.020 | .987 | .036 | .036 | .019 |
| | .50 | 1.023 | .209 | 0.000 | .014 | 0.000 | -.014 | 1.004 | .992 | 1.042 | .973 | .126 | .126 | .069 |
| | .75 | 1.071 | .400 | 0.000 | .044 | 0.000 | -.043 | 1.011 | .944 | 1.131 | .918 | .346 | .346 | .192 |
| | 1.00 | 1.225 | .810 | 0.000 | .135 | 0.000 | -.125 | 1.036 | .824 | 1.415 | .754 | .785 | .781 | .480 |
| | 1.25 | 1.457 | 1.268 | 0.000 | .256 | 0.000 | -.223 | 1.078 | .650 | 1.822 | .533 | 1.207 | 1.198 | .870 |
| | 1.50 | 1.684 | 1.652 | 0.000 | .353 | 0.000 | -.288 | 1.125 | .487 | 2.197 | .347 | 1.490 | 1.471 | 1.294 |
| .25 | .25 | 1.060 | .325 | 0.000 | .037 | 0.000 | -.036 | 1.009 | .952 | 1.111 | .930 | .088 | .087 | .047 |
| | .50 | 1.114 | .479 | 0.000 | .070 | 0.000 | -.067 | 1.018 | .909 | 1.210 | .871 | .260 | .259 | .150 |
| | .75 | 1.215 | .726 | 0.000 | .129 | 0.000 | -.121 | 1.035 | .832 | 1.393 | .765 | .528 | .525 | .321 |
| | 1.00 | 1.391 | 1.078 | 0.000 | .225 | 0.000 | -.203 | 1.064 | .699 | 1.712 | .589 | .874 | .864 | .599 |
| | 1.25 | 1.592 | 1.447 | 0.000 | .322 | 0.000 | -.274 | 1.104 | .552 | 2.054 | .410 | 1.190 | 1.169 | .947 |
| | 1.50 | 1.777 | 1.771 | 0.000 | .391 | 0.000 | -.314 | 1.149 | .424 | 2.343 | .274 | 1.412 | 1.376 | 1.339 |
| .50 | .25 | 1.216 | .638 | 0.000 | .131 | 0.000 | -.125 | 1.032 | .833 | 1.396 | .761 | .155 | .154 | .093 |
| | .50 | 1.313 | .838 | 0.000 | .187 | 0.000 | -.175 | 1.051 | .756 | 1.571 | .659 | .367 | .362 | .234 |
| | .75 | 1.448 | 1.096 | 0.000 | .259 | 0.000 | -.235 | 1.075 | .653 | 1.810 | .528 | .628 | .608 | .439 |
| | 1.00 | 1.608 | 1.389 | 0.000 | .335 | 0.000 | -.292 | 1.103 | .538 | 2.082 | .389 | .878 | .853 | .705 |
| | 1.25 | 1.772 | 1.685 | 0.000 | .401 | 0.000 | -.333 | 1.135 | .425 | 2.352 | .263 | 1.104 | 1.060 | 1.025 |
| | 1.50 | 1.928 | 1.965 | 0.000 | .448 | 0.000 | -.353 | 1.174 | .323 | 2.592 | .163 | 1.280 | 1.218 | 1.401 |
| 1.00 | .25 | 1.578 | 1.187 | 0.000 | .332 | 0.000 | -.302 | 1.094 | .557 | 2.049 | .397 | .188 | .179 | .143 |
| | .50 | 1.691 | 1.399 | 0.000 | .384 | 0.000 | -.339 | 1.114 | .477 | 2.244 | .302 | .386 | .363 | .326 |
| | .75 | 1.811 | 1.627 | 0.000 | .431 | 0.000 | -.368 | 1.139 | .395 | 2.445 | .215 | .572 | .532 | .550 |
| | 1.00 | 1.934 | 1.864 | 0.000 | .471 | 0.000 | -.388 | 1.168 | .314 | 2.646 | .137 | .743 | .683 | .819 |
| | 1.25 | 2.057 | 2.102 | 0.000 | .503 | 0.000 | -.396 | 1.200 | .238 | 2.832 | .069 | .892 | .809 | 1.139 |
| | 1.50 | 2.177 | 2.332 | 0.000 | .523 | 0.000 | -.395 | 1.231 | .173 | 2.986 | .011 | 1.023 | .911 | 1.517 |

## Table 4

**Uniform uniaxial compressive displacement in direction of *y*-axis**
**Unloaded sides stress-free**

Side ratio $a/b = 2{\cdot}0$

| W | P | V | W | $\bar{\sigma}_{MX}$ | | | | $\bar{\sigma}_{MY}$ | | | | $\bar{M}_{X}$ | $\bar{M}_{Y}$ | $\bar{M}_{XY}$ |
|---|---|---|---|---|---|---|---|---|---|---|---|---|---|---|
| | | | 4 | 1 | 2 | 3 | 4 | 1 | 2 | 3 | 4 | 4 | 4 | 1 |
| .10 | .25 | 1.030 | .134 | 0.000 | .010 | 0.000 | -.006 | 1.037 | .981 | 1.037 | .978 | .016 | .032 | .011 |
| | .50 | 1.059 | .207 | 0.000 | .017 | 0.000 | -.013 | 1.056 | .952 | 1.072 | .947 | .052 | .103 | .035 |
| | .75 | 1.132 | .348 | 0.000 | .031 | 0.000 | -.031 | 1.098 | .874 | 1.163 | .858 | .116 | .237 | .083 |
| | 1.00 | 1.346 | .627 | 0.000 | .074 | 0.000 | -.071 | 1.260 | .696 | 1.431 | .691 | .228 | .498 | .193 |
| | 1.25 | 1.661 | .955 | 0.000 | .121 | 0.000 | -.116 | 1.487 | .483 | 1.838 | .489 | .335 | .796 | .352 |
| | 1.50 | 2.024 | 1.232 | 0.000 | .145 | 0.000 | -.143 | 1.713 | .313 | 2.333 | .345 | .378 | 1.035 | .546 |
| .25 | .25 | 1.151 | .326 | 0.000 | .040 | 0.000 | -.035 | 1.137 | .863 | 1.185 | .857 | .036 | .072 | .025 |
| | .50 | 1.248 | .451 | 0.000 | .060 | 0.000 | -.055 | 1.213 | .778 | 1.305 | .772 | .091 | .191 | .071 |
| | .75 | 1.405 | .631 | 0.000 | .090 | 0.000 | -.083 | 1.335 | .652 | 1.503 | .647 | .162 | .359 | .142 |
| | 1.00 | 1.646 | .863 | 0.000 | .125 | 0.000 | -.116 | 1.513 | .494 | 1.817 | .495 | .235 | .570 | .254 |
| | 1.25 | 1.944 | 1.119 | 0.000 | .157 | 0.000 | -.144 | 1.734 | .345 | 2.217 | .356 | .296 | .796 | .407 |
| | 1.50 | 2.257 | 1.357 | 0.000 | .176 | 0.000 | -.157 | 1.982 | .254 | 2.655 | .280 | .331 | 1.001 | .594 |
| .50 | .25 | 1.463 | .616 | 0.000 | .100 | 0.000 | -.098 | 1.344 | .583 | 1.581 | .584 | .049 | .109 | .043 |
| | .50 | 1.611 | .760 | 0.000 | .117 | 0.000 | -.117 | 1.430 | .478 | 1.777 | .491 | .099 | .241 | .108 |
| | .75 | 1.794 | .934 | 0.000 | .136 | 0.000 | -.139 | 1.538 | .357 | 2.023 | .388 | .155 | .400 | .191 |
| | 1.00 | 2.030 | 1.132 | 0.000 | .163 | 0.000 | -.154 | 1.759 | .271 | 2.333 | .297 | .196 | .574 | .312 |
| | 1.25 | 2.301 | 1.343 | 0.000 | .190 | 0.000 | -.169 | 2.037 | .212 | 2.700 | .232 | .238 | .758 | .469 |
| | 1.50 | 2.585 | 1.553 | 0.000 | .205 | 0.000 | -.188 | 2.298 | .167 | 3.115 | .216 | .302 | .947 | .657 |
| 1.00 | .25 | 2.095 | 1.131 | 0.000 | .172 | 0.000 | -.172 | 1.769 | .170 | 2.410 | .214 | .038 | .118 | .068 |
| | .50 | 2.249 | 1.273 | 0.000 | .171 | 0.000 | -.192 | 1.836 | .113 | 2.633 | .232 | .094 | .250 | .153 |
| | .75 | 2.427 | 1.426 | 0.000 | .179 | 0.000 | -.223 | 1.982 | .081 | 2.880 | .291 | .167 | .395 | .258 |
| | 1.00 | 2.654 | 1.601 | 0.000 | .205 | 0.000 | -.252 | 2.291 | .081 | 3.190 | .277 | .253 | .559 | .396 |
| | 1.25 | 2.917 | 1.793 | 0.000 | .242 | 0.000 | -.287 | 2.728 | .103 | 3.562 | .238 | .351 | .737 | .566 |
| | 1.50 | 3.199 | 1.998 | 0.000 | .285 | 0.000 | -.334 | 3.256 | .139 | 3.993 | .226 | .464 | .927 | .766 |

# Table 5

## Uniform uniaxial compressive displacement in direction of $y$-axis Unloaded sides stress-free

### Side ratio $a/b = 3{\cdot}0$

| W | P | V | W | $\bar{\sigma}_{MX}$ | | | | $\bar{\sigma}_{MY}$ | | | | $\bar{M}_X$ | $\bar{M}_Y$ | $\bar{M}_{XY}$ |
|---|---|---|---|---|---|---|---|---|---|---|---|---|---|---|
| | | | 4 | 1 | 2 | 3 | 4 | 1 | 2 | 3 | 4 | 4 | 4 | 1 |
| .10 | .25 | 1.033 | .133 | 0.000 | .004 | 0.000 | -.004 | 1.029 | .962 | 1.037 | .967 | .012 | .030 | .007 |
| | .50 | 1.067 | .202 | 0.000 | .007 | 0.000 | -.007 | 1.060 | .936 | 1.075 | .936 | .037 | .095 | .023 |
| | .75 | 1.157 | .335 | 0.000 | .015 | 0.000 | -.014 | 1.141 | .862 | 1.176 | .861 | .079 | .215 | .057 |
| | 1.00 | 1.409 | .584 | 0.000 | .027 | 0.000 | -.026 | 1.355 | .700 | 1.467 | .700 | .149 | .440 | .141 |
| | 1.25 | 1.839 | .890 | 0.000 | .033 | 0.000 | -.032 | 1.685 | .533 | 1.996 | .537 | .221 | .711 | .291 |
| | 1.50 | 2.451 | 1.197 | 0.000 | .023 | 0.000 | -.022 | 2.112 | .431 | 2.789 | .447 | .278 | .980 | .516 |
| .25 | .25 | 1.182 | .322 | 0.000 | .019 | 0.000 | -.018 | 1.163 | .830 | 1.202 | .828 | .025 | .066 | .017 |
| | .50 | 1.294 | .437 | 0.000 | .026 | 0.000 | -.025 | 1.259 | .749 | 1.329 | .748 | .060 | .171 | .050 |
| | .75 | 1.481 | .597 | 0.000 | .033 | 0.000 | -.030 | 1.416 | .640 | 1.546 | .639 | .103 | .315 | .106 |
| | 1.00 | 1.795 | .811 | 0.000 | .035 | 0.000 | -.032 | 1.663 | .522 | 1.932 | .525 | .150 | .503 | .207 |
| | 1.25 | 2.222 | 1.064 | 0.000 | .032 | 0.000 | -.055 | 1.969 | .427 | 2.481 | .445 | .227 | .732 | .361 |
| | 1.50 | 2.733 | 1.339 | 0.000 | .023 | 0.000 | -.133 | 2.289 | .390 | 3.175 | .445 | .364 | .996 | .574 |
| .50 | .25 | 1.555 | .606 | 0.000 | .042 | 0.000 | -.041 | 1.484 | .557 | 1.622 | .555 | .031 | .096 | .032 |
| | .50 | 1.741 | .738 | 0.000 | .043 | 0.000 | -.042 | 1.633 | .484 | 1.847 | .484 | .064 | .214 | .085 |
| | .75 | 1.989 | .898 | 0.000 | .040 | 0.000 | -.052 | 1.825 | .416 | 2.161 | .448 | .111 | .359 | .164 |
| | 1.00 | 2.323 | 1.103 | 0.000 | .033 | 0.000 | -.120 | 2.055 | .366 | 2.593 | .555 | .224 | .558 | .282 |
| | 1.25 | 2.723 | 1.339 | 0.000 | .024 | 0.000 | -.203 | 2.306 | .333 | 3.133 | .640 | .359 | .788 | .444 |
| | 1.50 | 3.154 | 1.580 | 0.000 | .019 | 0.000 | -.219 | 2.555 | .312 | 3.756 | .409 | .426 | 1.001 | .649 |
| 1.00 | .25 | 2.334 | 1.125 | 0.000 | .041 | 0.000 | -.121 | 2.110 | .251 | 2.556 | .556 | .051 | .117 | .058 |
| | .50 | 2.559 | 1.271 | 0.000 | .033 | 0.000 | -.152 | 2.260 | .238 | 2.855 | .566 | .119 | .256 | .137 |
| | .75 | 2.828 | 1.440 | 0.000 | .025 | 0.000 | -.190 | 2.431 | .231 | 3.225 | .546 | .202 | .418 | .242 |
| | 1.00 | 3.145 | 1.632 | 0.000 | .017 | 0.000 | -.233 | 2.625 | .224 | 3.669 | .482 | .295 | .599 | .380 |
| | 1.25 | 3.493 | 1.840 | 0.000 | .012 | 0.000 | -.286 | 2.818 | .217 | 4.175 | .412 | .405 | .796 | .553 |
| | 1.50 | 3.848 | 2.057 | 0.000 | .012 | 0.000 | -.353 | 2.974 | .210 | 4.719 | .397 | .533 | 1.002 | .763 |

# Table 6

## Uniform uniaxial compressive displacement in direction of $y$-axis Unloaded sides straight

### Side ratio $a/b = 1{\cdot}0$

| W | P | V | U | W | $\bar{\sigma}_{MX}$ | | | | $\bar{\sigma}_{MY}$ | | | | $\bar{M}_Y$ | $\bar{M}_X$ | $\bar{M}_{XY}$ |
|---|---|---|---|---|---|---|---|---|---|---|---|---|---|---|---|
| | | | | 4 | 1 | 2 | 3 | 4 | 1 | 2 | 3 | 4 | 4 | 4 | 1 |
| .10 | .25 | 1.010 | .966 | .131 | .010 | .010 | -.009 | -.010 | 1.008 | .990 | 1.012 | .990 | .036 | .036 | .020 |
| | .50 | 1.021 | .926 | .205 | .022 | .022 | -.021 | -.021 | 1.020 | .979 | 1.024 | .979 | .122 | .123 | .067 |
| | .75 | 1.061 | .829 | .365 | .051 | .051 | -.066 | -.066 | 1.060 | .942 | 1.059 | .942 | .303 | .305 | .170 |
| | 1.00 | 1.165 | .464 | .691 | .161 | .160 | -.165 | -.160 | 1.166 | .838 | 1.162 | .842 | .657 | .664 | .391 |
| | 1.25 | 1.301 | -.026 | 1.071 | .312 | .303 | -.291 | -.273 | 1.306 | .699 | 1.294 | .716 | 1.039 | 1.060 | .669 |
| | 1.50 | 1.422 | -.409 | 1.338 | .436 | .406 | -.407 | -.363 | 1.435 | .578 | 1.397 | .619 | 1.243 | 1.294 | .913 |
| .25 | .25 | 1.059 | .797 | .327 | .060 | .061 | -.058 | -.058 | 1.057 | .940 | 1.062 | .940 | .089 | .089 | .049 |
| | .50 | 1.101 | .614 | .468 | .115 | .116 | -.092 | -.091 | 1.102 | .895 | 1.106 | .897 | .248 | .249 | .141 |
| | .75 | 1.175 | .354 | .676 | .194 | .193 | -.162 | -.158 | 1.179 | .821 | 1.179 | .827 | .475 | .479 | .282 |
| | 1.00 | 1.280 | .055 | .938 | .286 | .280 | -.273 | -.260 | 1.284 | .721 | 1.276 | .734 | .740 | .753 | .472 |
| | 1.25 | 1.396 | -.264 | 1.215 | .388 | .369 | -.395 | -.365 | 1.401 | .610 | 1.380 | .638 | .992 | 1.026 | .694 |
| | 1.50 | 1.487 | -.591 | 1.451 | .495 | .455 | -.472 | -.418 | 1.506 | .518 | 1.457 | .569 | 1.157 | 1.231 | .919 |
| .50 | .25 | 1.193 | .312 | .629 | .206 | .206 | -.184 | -.181 | 1.196 | .807 | 1.197 | .807 | .144 | .145 | .085 |
| | .50 | 1.264 | .048 | .804 | .285 | .283 | -.249 | -.241 | 1.270 | .733 | 1.268 | .741 | .332 | .334 | .206 |
| | .75 | 1.345 | -.215 | 1.006 | .366 | .357 | -.330 | -.311 | 1.355 | .652 | 1.344 | .671 | .533 | .540 | .353 |
| | 1.00 | 1.425 | -.426 | 1.212 | .436 | .415 | -.414 | -.381 | 1.439 | .576 | 1.411 | .609 | .719 | .739 | .519 |
| | 1.25 | 1.497 | -.590 | 1.412 | .495 | .458 | -.493 | -.440 | 1.517 | .511 | 1.467 | .559 | .874 | .920 | .699 |
| | 1.50 | 1.560 | -.760 | 1.599 | .557 | .498 | -.553 | -.478 | 1.586 | .451 | 1.515 | .518 | .994 | 1.078 | .892 |
| 1.00 | .25 | 1.459 | -.444 | 1.152 | .440 | .424 | -.466 | -.439 | 1.456 | .552 | 1.442 | .579 | .154 | .155 | .111 |
| | .50 | 1.518 | -.722 | 1.318 | .528 | .501 | -.512 | -.467 | 1.533 | .487 | 1.506 | .527 | .309 | .315 | .240 |
| | .75 | 1.574 | -.954 | 1.487 | .604 | .563 | -.557 | -.494 | 1.603 | .428 | 1.556 | .484 | .451 | .467 | .385 |
| | 1.00 | 1.624 | -1.080 | 1.653 | .651 | .591 | -.604 | -.525 | 1.658 | .382 | 1.587 | .457 | .574 | .607 | .542 |
| | 1.25 | 1.664 | -1.099 | 1.810 | .667 | .586 | -.651 | -.556 | 1.695 | .350 | 1.599 | .447 | .670 | .731 | .707 |
| | 1.50 | 1.698 | -1.206 | 1.952 | .710 | .606 | -.676 | -.558 | 1.740 | .314 | 1.621 | .431 | .739 | .838 | .879 |

## Table 7

### Uniform uniaxial compressive displacement in direction of $y$-axis
### Unloaded sides straight

Side ratio $a/b$ = 2·0

| W | P | U | V | W 4 | $\bar{\sigma}_{MX}$ 1 | $\bar{\sigma}_{MX}$ 2 | $\bar{\sigma}_{MX}$ 3 | $\bar{\sigma}_{MX}$ 4 | $\bar{\sigma}_{MY}$ 1 | $\bar{\sigma}_{MY}$ 2 | $\bar{\sigma}_{MY}$ 3 | $\bar{\sigma}_{MY}$ 4 | $\bar{M}_X$ 4 | $\bar{M}_Y$ 4 | $\bar{M}_{XY}$ 1 |
|---|---|---|---|---|---|---|---|---|---|---|---|---|---|---|---|
| .10 | .25 | .984 | 1.025 | .129 | .005 | .006 | -.007 | -.007 | 1.023 | .977 | 1.023 | .976 | .015 | .028 | .009 |
| | .50 | .963 | 1.054 | .203 | .012 | .012 | -.016 | -.015 | 1.053 | .948 | 1.053 | .947 | .050 | .099 | .033 |
| | .75 | .904 | 1.133 | .345 | .029 | .027 | -.041 | -.037 | 1.133 | .875 | 1.133 | .875 | .115 | .234 | .082 |
| | 1.00 | .742 | 1.340 | .615 | .086 | .065 | -.109 | -.085 | 1.338 | .700 | 1.336 | .701 | .224 | .487 | .186 |
| | 1.25 | .477 | 1.634 | .940 | .188 | .117 | -.211 | -.135 | 1.633 | .482 | 1.623 | .493 | .337 | .786 | .332 |
| | 1.50 | .135 | 1.923 | 1.184 | .339 | .164 | -.325 | -.145 | 1.935 | .337 | 1.903 | .378 | .380 | .998 | .488 |
| .25 | .25 | .890 | 1.146 | .323 | .033 | .033 | -.043 | -.041 | 1.140 | .858 | 1.140 | .857 | .035 | .070 | .024 |
| | .50 | .815 | 1.239 | .447 | .057 | .051 | -.072 | -.064 | 1.234 | .774 | 1.234 | .773 | .089 | .188 | .068 |
| | .75 | .695 | 1.391 | .623 | .100 | .079 | -.122 | -.097 | 1.388 | .647 | 1.387 | .649 | .160 | .353 | .137 |
| | 1.00 | .493 | 1.611 | .847 | .180 | .116 | -.199 | -.131 | 1.611 | .492 | 1.602 | .506 | .234 | .557 | .240 |
| | 1.25 | .218 | 1.869 | 1.091 | .295 | .159 | -.295 | -.154 | 1.876 | .339 | 1.852 | .373 | .298 | .773 | .371 |
| | 1.50 | -.102 | 2.110 | 1.309 | .443 | .199 | -.398 | -.152 | 2.134 | .235 | 2.092 | .294 | .336 | .965 | .517 |
| .50 | .25 | .657 | 1.459 | .615 | .113 | .090 | -.141 | -.117 | 1.463 | .581 | 1.451 | .590 | .049 | .109 | .042 |
| | .50 | .527 | 1.599 | .756 | .166 | .114 | -.187 | -.135 | 1.608 | .480 | 1.589 | .501 | .099 | .239 | .103 |
| | .75 | .363 | 1.775 | .926 | .236 | .140 | -.248 | -.153 | 1.781 | .362 | 1.764 | .401 | .155 | .395 | .180 |
| | 1.00 | .128 | 1.966 | 1.115 | .334 | .175 | -.325 | -.164 | 1.979 | .264 | 1.954 | .309 | .200 | .562 | .286 |
| | 1.25 | -.115 | 2.164 | 1.309 | .443 | .204 | -.419 | -.174 | 2.186 | .186 | 2.145 | .238 | .237 | .730 | .411 |
| | 1.50 | -.269 | 2.350 | 1.483 | .536 | .201 | -.531 | -.194 | 2.377 | .128 | 2.316 | .205 | .268 | .884 | .538 |
| 1.00 | .25 | .121 | 2.065 | 1.129 | .337 | .177 | -.355 | -.194 | 2.070 | .168 | 2.051 | .224 | .037 | .116 | .063 |
| | .50 | -.010 | 2.204 | 1.265 | .407 | .184 | -.420 | -.204 | 2.213 | .113 | 2.180 | .200 | .073 | .238 | .139 |
| | .75 | -.206 | 2.339 | 1.410 | .499 | .206 | -.484 | -.207 | 2.368 | .070 | 2.315 | .186 | .111 | .368 | .228 |
| | 1.00 | -.331 | 2.473 | 1.560 | .570 | .205 | -.572 | -.263 | 2.505 | .040 | 2.439 | .217 | .188 | .513 | .331 |
| | 1.25 | -.434 | 2.605 | 1.712 | .638 | .197 | -.670 | -.339 | 2.635 | .022 | 2.556 | .263 | .286 | .665 | .444 |
| | 1.50 | -.591 | 2.733 | 1.865 | .730 | .200 | -.755 | -.383 | 2.776 | .013 | 2.672 | .278 | .379 | .813 | .565 |

## Table 8

### Uniform uniaxial compressive displacement in direction of $y$-axis
### Unloaded sides straight

Side ratio $a/b$ = 3·0

| W | P | U | V | W 4 | $\bar{\sigma}_{MX}$ 1 | $\bar{\sigma}_{MX}$ 2 | $\bar{\sigma}_{MX}$ 3 | $\bar{\sigma}_{MX}$ 4 | $\bar{\sigma}_{MY}$ 1 | $\bar{\sigma}_{MY}$ 2 | $\bar{\sigma}_{MY}$ 3 | $\bar{\sigma}_{MY}$ 4 | $\bar{M}_X$ 4 | $\bar{M}_Y$ 4 | $\bar{M}_{XY}$ 1 |
|---|---|---|---|---|---|---|---|---|---|---|---|---|---|---|---|
| .10 | .25 | .987 | 1.030 | .129 | .004 | .004 | -.003 | -.003 | 1.030 | .971 | 1.030 | .970 | .011 | .027 | .006 |
| | .50 | .974 | 1.065 | .200 | .008 | .008 | -.008 | -.007 | 1.066 | .937 | 1.066 | .935 | .036 | .093 | .022 |
| | .75 | .944 | 1.157 | .333 | .019 | .013 | -.022 | -.015 | 1.158 | .858 | 1.158 | .859 | .078 | .214 | .057 |
| | 1.00 | .857 | 1.411 | .579 | .061 | .020 | -.075 | -.031 | 1.410 | .700 | 1.408 | .703 | .146 | .436 | .141 |
| | 1.25 | .683 | 1.820 | .886 | .163 | .023 | -.189 | -.055 | 1.821 | .528 | 1.809 | .539 | .229 | .704 | .279 |
| | 1.50 | .391 | 2.344 | 1.162 | .363 | .014 | -.389 | -.083 | 2.358 | .447 | 2.323 | .479 | .310 | .928 | .471 |
| .25 | .25 | .928 | 1.148 | .321 | .023 | .020 | -.021 | -.018 | 1.178 | .831 | 1.178 | .829 | .025 | .093 | .016 |
| | .50 | .885 | 1.267 | .436 | .040 | .027 | -.041 | -.025 | 1.292 | .750 | 1.292 | .748 | .060 | .084 | .049 |
| | .75 | .817 | 1.485 | .598 | .074 | .031 | -.081 | -.034 | 1.480 | .639 | 1.479 | .635 | .103 | .074 | .104 |
| | 1.00 | .689 | 1.783 | .805 | .149 | .031 | -.162 | -.037 | 1.779 | .524 | 1.770 | .529 | .149 | .404 | .200 |
| | 1.25 | .489 | 2.152 | 1.044 | .278 | .029 | -.291 | -.058 | 2.170 | .427 | 2.146 | .451 | .218 | .863 | .335 |
| | 1.50 | .211 | 2.587 | 1.294 | .470 | .030 | -.471 | -.136 | 2.612 | .376 | 2.562 | .446 | .346 | .982 | .504 |
| .50 | .25 | .795 | 1.553 | .606 | .078 | .040 | -.086 | -.045 | 1.557 | .557 | 1.545 | .556 | .031 | .127 | .031 |
| | .50 | .717 | 1.734 | .737 | .122 | .041 | -.131 | -.047 | 1.739 | .483 | 1.726 | .484 | .064 | .112 | .083 |
| | .75 | .598 | 1.974 | .895 | .193 | .041 | -.199 | -.054 | 1.979 | .414 | 1.970 | .435 | .108 | .071 | .158 |
| | 1.00 | .424 | 2.277 | 1.089 | .307 | .040 | -.305 | -.098 | 2.289 | .361 | 2.256 | .457 | .191 | .407 | .264 |
| | 1.25 | .212 | 2.621 | 1.306 | .458 | .037 | -.450 | -.169 | 2.647 | .324 | 2.594 | .501 | .299 | .919 | .398 |
| | 1.50 | -.011 | 2.969 | 1.527 | .633 | .031 | -.624 | -.233 | 3.011 | .297 | 2.926 | .465 | .397 | 1.142 | .553 |
| 1.00 | .25 | .461 | 2.316 | 1.124 | .268 | .047 | -.265 | -.128 | 2.324 | .249 | 2.310 | .545 | .049 | .115 | .055 |
| | .50 | .367 | 2.525 | 1.266 | .346 | .036 | -.353 | -.172 | 2.536 | .234 | 2.512 | .557 | .115 | .251 | .129 |
| | .75 | .247 | 2.765 | 1.427 | .448 | .023 | -.465 | -.228 | 2.786 | .226 | 2.737 | .564 | .194 | .407 | .223 |
| | 1.00 | .075 | 3.026 | 1.604 | .583 | .020 | -.596 | -.278 | 3.057 | .216 | 2.986 | .518 | .277 | .575 | .336 |
| | 1.25 | -.133 | 3.296 | 1.792 | .742 | .024 | -.740 | -.325 | 3.338 | .204 | 3.247 | .440 | .364 | .753 | .468 |
| | 1.50 | -.342 | 3.562 | 1.983 | .911 | .026 | -.894 | -.383 | 3.620 | .191 | 3.500 | .379 | .460 | .933 | .615 |

## Table 9

### Uniform equal biaxial compressive displacement ($\sigma_{x\text{av}} = \sigma_{y\text{av}}$)

Side ratio $a/b = 1{\cdot}0$

| W | P | U | V | W 4 | $\bar{\sigma}_{MX}$ 1 | $\bar{\sigma}_{MX}$ 2 | $\bar{\sigma}_{MX}$ 3 | $\bar{\sigma}_{MX}$ 4 | $\bar{\sigma}_{MY}$ 1 | $\bar{\sigma}_{MY}$ 2 | $\bar{\sigma}_{MY}$ 3 | $\bar{\sigma}_{MY}$ 4 | $\bar{M}_X$ 4 | $\bar{M}_Y$ 4 | $\bar{M}_{XY}$ 1 |
|---|---|---|---|---|---|---|---|---|---|---|---|---|---|---|---|
| .10 | .25 | 1.026 | 1.026 | .129 | 1.022 | 1.021 | .978 | .978 | 1.022 | .978 | 1.021 | .978 | .034 | .034 | .018 |
| | .50 | 1.064 | 1.064 | .208 | 1.047 | 1.049 | .953 | .953 | 1.047 | .953 | 1.049 | .953 | .126 | .126 | .069 |
| | .75 | 1.165 | 1.165 | .370 | 1.113 | 1.118 | .887 | .887 | 1.113 | .887 | 1.118 | .887 | .312 | .312 | .173 |
| | 1.00 | 1.462 | 1.462 | .697 | 1.322 | 1.323 | .681 | .689 | 1.322 | .681 | 1.323 | .689 | .572 | .672 | .393 |
| | 1.25 | 1.838 | 1.838 | 1.035 | 1.595 | 1.575 | .422 | .456 | 1.595 | .422 | 1.575 | .456 | 1.012 | 1.012 | .645 |
| | 1.50 | 2.169 | 2.169 | 1.323 | 1.845 | 1.785 | .196 | .269 | 1.845 | .196 | 1.785 | .269 | 1.262 | 1.262 | .889 |
| .25 | .25 | 1.169 | 1.169 | .329 | 1.118 | 1.121 | .881 | .879 | 1.118 | .881 | 1.121 | .879 | .094 | .094 | .048 |
| | .50 | 1.288 | 1.288 | .465 | 1.201 | 1.205 | .798 | .797 | 1.201 | .798 | 1.205 | .797 | .251 | .251 | .137 |
| | .75 | 1.510 | 1.510 | .676 | 1.358 | 1.356 | .645 | .653 | 1.358 | .645 | 1.356 | .653 | .479 | .479 | .282 |
| | 1.00 | 1.806 | 1.806 | .940 | 1.570 | 1.555 | .443 | .469 | 1.570 | .443 | 1.555 | .469 | .751 | .751 | .472 |
| | 1.25 | 2.094 | 2.094 | 1.198 | 1.783 | 1.744 | .246 | .303 | 1.783 | .246 | 1.744 | .303 | .993 | .993 | .679 |
| | 1.50 | 2.359 | 2.359 | 1.439 | 1.984 | 1.903 | .067 | .166 | 1.984 | .067 | 1.903 | .166 | 1.197 | 1.197 | .897 |
| .50 | .25 | 1.561 | 1.561 | .633 | 1.390 | 1.392 | .610 | .616 | 1.390 | .610 | 1.392 | .616 | .150 | .150 | .086 |
| | .50 | 1.751 | 1.751 | .800 | 1.524 | 1.520 | .479 | .494 | 1.524 | .479 | 1.520 | .494 | .329 | .329 | .202 |
| | .75 | 1.973 | 1.973 | .998 | 1.686 | 1.670 | .326 | .359 | 1.686 | .326 | 1.670 | .359 | .531 | .531 | .347 |
| | 1.00 | 2.215 | 2.215 | 1.214 | 1.867 | 1.827 | .162 | .224 | 1.867 | .162 | 1.827 | .224 | .735 | .735 | .518 |
| | 1.25 | 2.442 | 2.442 | 1.426 | 2.045 | 1.968 | .009 | .109 | 2.045 | .009 | 1.968 | .109 | .919 | .919 | .705 |
| | 1.50 | 2.610 | 2.610 | 1.613 | 2.190 | 2.060 | -.101 | .037 | 2.190 | -.101 | 2.050 | .037 | 1.065 | 1.065 | .892 |
| 1.00 | .25 | 2.323 | 2.323 | 1.157 | 1.936 | 1.910 | .085 | .142 | 1.936 | .085 | 1.910 | .142 | .159 | .159 | .113 |
| | .50 | 2.485 | 2.485 | 1.321 | 2.061 | 2.012 | -.025 | .058 | 2.061 | -.025 | 2.012 | .058 | .314 | .314 | .242 |
| | .75 | 2.639 | 2.639 | 1.489 | 2.185 | 2.101 | -.128 | -.013 | 2.185 | -.128 | 2.101 | -.013 | .462 | .462 | .384 |
| | 1.00 | 2.786 | 2.786 | 1.656 | 2.305 | 2.184 | -.225 | -.073 | 2.305 | -.225 | 2.184 | -.073 | .598 | .598 | .540 |
| | 1.25 | 2.919 | 2.919 | 1.818 | 2.417 | 2.254 | -.311 | -.119 | 2.417 | -.311 | 2.254 | -.119 | .718 | .718 | .707 |
| | 1.50 | 3.023 | 3.023 | 1.969 | 2.515 | 2.301 | -.375 | -.145 | 2.515 | -.375 | 2.301 | -.145 | .820 | .820 | .883 |

## Table 10

### Uniform equal biaxial compressive displacement ($\sigma_{x\text{av}} = \sigma_{y\text{av}}$)

Side ratio $a/b = 2{\cdot}0$

| W | P | U | V | W 4 | $\bar{\sigma}_{MX}$ 1 | $\bar{\sigma}_{MX}$ 2 | $\bar{\sigma}_{MX}$ 3 | $\bar{\sigma}_{MX}$ 4 | $\bar{\sigma}_{MY}$ 1 | $\bar{\sigma}_{MY}$ 2 | $\bar{\sigma}_{MY}$ 3 | $\bar{\sigma}_{MY}$ 4 | $\bar{M}_X$ 4 | $\bar{M}_Y$ 4 | $\bar{M}_{XY}$ 1 |
|---|---|---|---|---|---|---|---|---|---|---|---|---|---|---|---|
| .10 | .25 | 1.013 | 1.046 | .133 | 1.009 | 1.009 | .992 | .992 | 1.032 | .967 | 1.032 | .967 | .016 | .032 | .011 |
| | .50 | 1.027 | 1.097 | .206 | 1.017 | 1.017 | .982 | .982 | 1.066 | .933 | 1.066 | .933 | .051 | .102 | .034 |
| | .75 | 1.066 | 1.235 | .348 | 1.041 | 1.039 | .955 | .959 | 1.162 | .843 | 1.163 | .843 | .115 | .237 | .084 |
| | 1.00 | 1.179 | 1.610 | .622 | 1.126 | 1.094 | .870 | .910 | 1.432 | .628 | 1.428 | .635 | .210 | .491 | .196 |
| | 1.25 | 1.357 | 2.182 | .942 | 1.286 | 1.147 | .710 | .858 | 1.846 | .381 | 1.824 | .416 | .302 | .770 | .367 |
| | 1.50 | 1.588 | 2.899 | 1.227 | 1.532 | 1.169 | .465 | .823 | 2.371 | .189 | 2.385 | .286 | .316 | 1.017 | .590 |
| .25 | .25 | 1.074 | 1.262 | .325 | 1.048 | 1.048 | .951 | .955 | 1.184 | .821 | 1.184 | .819 | .035 | .072 | .025 |
| | .50 | 1.121 | 1.430 | .450 | 1.081 | 1.074 | .915 | .928 | 1.304 | .717 | 1.300 | .717 | .088 | .189 | .071 |
| | .75 | 1.199 | 1.706 | .627 | 1.142 | 1.111 | .852 | .890 | 1.581 | .562 | 1.490 | .569 | .155 | .353 | .143 |
| | 1.00 | 1.333 | 2.132 | .856 | 1.256 | 1.148 | .740 | .853 | 1.808 | .374 | 1.787 | .402 | .216 | .558 | .261 |
| | 1.25 | 1.500 | 2.646 | 1.100 | 1.425 | 1.175 | .574 | .825 | 2.182 | .209 | 2.139 | .275 | .257 | .770 | .419 |
| | 1.50 | 1.676 | 3.188 | 1.323 | 1.645 | 1.184 | .352 | .812 | 2.578 | .121 | 2.496 | .242 | .267 | .958 | .610 |
| .50 | .25 | 1.225 | 1.815 | .615 | 1.159 | 1.129 | .841 | .873 | 1.575 | .476 | 1.571 | .486 | .048 | .108 | .043 |
| | .50 | 1.305 | 2.078 | .759 | 1.227 | 1.152 | .775 | .850 | 1.766 | .351 | 1.755 | .379 | .096 | .240 | .108 |
| | .75 | 1.407 | 2.409 | .931 | 1.321 | 1.175 | .685 | .825 | 2.008 | .218 | 1.983 | .266 | .144 | .395 | .195 |
| | 1.00 | 1.524 | 2.792 | 1.118 | 1.448 | 1.196 | .547 | .807 | 2.283 | .118 | 2.240 | .186 | .172 | .556 | .316 |
| | 1.25 | 1.657 | 3.205 | 1.313 | 1.612 | 1.209 | .376 | .761 | 2.584 | .049 | 2.512 | .162 | .237 | .734 | .466 |
| | 1.50 | 1.807 | 3.625 | 1.508 | 1.813 | 1.202 | .190 | .644 | 2.980 | .005 | 2.781 | .220 | .402 | .941 | .635 |
| 1.00 | .25 | 1.455 | 2.909 | 1.128 | 1.449 | 1.236 | .551 | .767 | 2.355 | -.028 | 2.317 | .040 | .035 | .115 | .064 |
| | .50 | 1.548 | 3.175 | 1.268 | 1.542 | 1.235 | .459 | .736 | 2.553 | -.084 | 2.493 | .075 | .086 | .244 | .146 |
| | .75 | 1.728 | 3.457 | 1.420 | 1.638 | 1.225 | .360 | .684 | 2.772 | -.127 | 2.588 | .145 | .160 | .389 | .249 |
| | 1.00 | 1.831 | 3.755 | 1.583 | 1.774 | 1.224 | .220 | .602 | 2.991 | -.141 | 2.885 | .184 | .269 | .553 | .373 |
| | 1.25 | 1.899 | 4.056 | 1.751 | 1.937 | 1.227 | .059 | .511 | 3.206 | -.141 | 3.073 | .193 | .392 | .723 | .514 |
| | 1.50 | 2.019 | 4.339 | 1.913 | 2.101 | 1.222 | -.095 | .439 | 3.414 | -.147 | 3.242 | .184 | .497 | .882 | .667 |

## Table 11

## Uniform equal biaxial compressive displacement ($\sigma_{xav} = \sigma_{yav}$)

Side ratio $a/b$ = 3·0

| W | P | U | V | W | $\bar{\sigma}_{MX}$ | | | | $\bar{\sigma}_{MY}$ | | | | $\bar{M}_X$ | $\bar{M}_Y$ | $\bar{M}_{XY}$ |
|---|---|---|---|---|---|---|---|---|---|---|---|---|---|---|---|
| | | | | 4 | 1 | 2 | 3 | 4 | 1 | 2 | 3 | 4 | 4 | 4 | 1 |
| .10 | .25 | 1.004 | 1.052 | .133 | 1.003 | 1.003 | .995 | .997 | 1.036 | .964 | 1.036 | .964 | .012 | .030 | .007 |
| | .50 | 1.011 | 1.108 | .203 | 1.007 | 1.007 | .990 | .992 | 1.074 | .928 | 1.074 | .928 | .036 | .095 | .023 |
| | .75 | 1.032 | 1.251 | .333 | 1.021 | 1.015 | .979 | .987 | 1.175 | .849 | 1.175 | .851 | .075 | .212 | .059 |
| | 1.00 | 1.103 | 1.681 | .584 | 1.088 | 1.023 | .913 | .980 | 1.480 | .690 | 1.476 | .696 | .137 | .435 | .158 |
| | 1.25 | 1.239 | 2.479 | .914 | 1.282 | 1.022 | .718 | .932 | 2.052 | .548 | 2.026 | .543 | .244 | .736 | .344 |
| | 1.50 | 1.451 | 3.696 | 1.284 | 1.659 | 1.005 | .337 | .815 | 2.928 | .503 | 2.851 | .459 | .410 | 1.088 | .634 |
| .25 | .25 | 1.027 | 1.291 | .322 | 1.026 | 1.022 | .975 | .979 | 1.203 | .813 | 1.203 | .811 | .025 | .066 | .017 |
| | .50 | 1.048 | 1.472 | .436 | 1.049 | 1.026 | .951 | .973 | 1.330 | .728 | 1.330 | .727 | .058 | .169 | .052 |
| | .75 | 1.099 | 1.782 | .598 | 1.097 | 1.029 | .902 | .964 | 1.549 | .620 | 1.548 | .632 | .106 | .315 | .114 |
| | 1.00 | 1.191 | 2.334 | .818 | 1.216 | 1.030 | .782 | .928 | 1.943 | .525 | 1.930 | .582 | .183 | .518 | .233 |
| | 1.25 | 1.317 | 3.087 | 1.088 | 1.434 | 1.024 | .563 | .851 | 2.484 | .458 | 2.441 | .529 | .295 | .772 | .412 |
| | 1.50 | 1.465 | 3.980 | 1.393 | 1.780 | 1.008 | .217 | .716 | 3.128 | .437 | 3.026 | .410 | .445 | 1.065 | .645 |
| .50 | .25 | 1.104 | 1.876 | .604 | 1.086 | 1.042 | .904 | .956 | 1.619 | .688 | 1.607 | .496 | .026 | .093 | .031 |
| | .50 | 1.144 | 2.183 | .738 | 1.163 | 1.058 | .859 | .954 | 1.836 | .358 | 1.820 | .482 | .072 | .216 | .088 |
| | .75 | 1.211 | 2.607 | .905 | 1.292 | 1.080 | .784 | .940 | 2.135 | .406 | 2.117 | .529 | .138 | .373 | .175 |
| | 1.00 | 1.304 | 3.188 | 1.122 | 1.435 | 1.039 | .584 | .844 | 2.549 | .383 | 2.511 | .504 | .239 | .578 | .308 |
| | 1.25 | 1.426 | 3.872 | 1.371 | 1.628 | .978 | .267 | .698 | 3.042 | .299 | 2.966 | .423 | .365 | .817 | .484 |
| | 1.50 | 1.589 | 4.560 | 1.623 | 1.944 | .987 | -.125 | .580 | 3.544 | .301 | 3.416 | .357 | .502 | 1.063 | .693 |
| 1.00 | .25 | 1.268 | 3.106 | 1.125 | 1.299 | 1.044 | .695 | .846 | 2.490 | .179 | 2.461 | .492 | .050 | .117 | .057 |
| | .50 | 1.320 | 3.479 | 1.275 | 1.413 | 1.033 | .578 | .790 | 2.758 | .183 | 2.713 | .524 | .123 | .261 | .139 |
| | .75 | 1.391 | 3.927 | 1.450 | 1.571 | 1.022 | .419 | .712 | 3.078 | .198 | 3.019 | .539 | .222 | .433 | .248 |
| | 1.00 | 1.501 | 4.448 | 1.651 | 1.783 | 1.014 | .208 | .613 | 3.460 | .207 | 3.365 | .497 | .340 | .631 | .390 |
| | 1.25 | 1.636 | 5.007 | 1.867 | 2.039 | 1.009 | -.039 | .499 | 3.876 | .208 | 3.723 | .416 | .471 | .844 | .563 |
| | 1.50 | 1.775 | 5.546 | 2.085 | 2.313 | 1.006 | -.299 | .379 | 4.282 | .203 | 4.056 | .333 | .605 | 1.054 | .759 |

## Table 12

## Uniform unequal biaxial compressive displacement ($\sigma_{xav} = 2\sigma_{yav}$)

Side ratio $a/b$ = 1·0

| W | P | U | V | W | $\bar{\sigma}_{MX}$ | | | | $\bar{\sigma}_{MY}$ | | | | $\bar{M}_X$ | $\bar{M}_Y$ | $\bar{M}_{XY}$ |
|---|---|---|---|---|---|---|---|---|---|---|---|---|---|---|---|
| | | | | 4 | 1 | 2 | 3 | 4 | 1 | 2 | 3 | 4 | 4 | 4 | 1 |
| .10 | .25 | 1.013 | 1.071 | .128 | 1.015 | 1.014 | .989 | .989 | 1.027 | .974 | 1.027 | .974 | .833 | .033 | .018 |
| | .50 | 1.045 | 1.166 | .210 | 1.035 | 1.036 | .968 | .968 | 1.068 | .932 | 1.068 | .932 | .127 | .127 | .070 |
| | .75 | 1.121 | 1.435 | .374 | 1.087 | 1.091 | .911 | .912 | 1.180 | .818 | 1.181 | .819 | .316 | .317 | .176 |
| | 1.00 | 1.282 | 2.218 | .693 | 1.244 | 1.243 | .757 | .765 | 1.486 | .518 | 1.483 | .529 | .666 | .668 | .392 |
| | 1.25 | 1.489 | 3.190 | 1.029 | 1.444 | 1.429 | .566 | .591 | 1.873 | .152 | 1.845 | .200 | 1.006 | 1.012 | .639 |
| | 1.50 | 1.724 | 4.091 | 1.319 | 1.633 | 1.591 | .392 | .448 | 2.266 | -.200 | 2.181 | -.083 | 1.252 | 1.272 | .889 |
| .25 | .25 | 1.112 | 1.418 | .327 | 1.084 | 1.086 | .915 | .914 | 1.168 | .828 | 1.171 | .824 | .094 | .094 | .045 |
| | .50 | 1.192 | 1.746 | .465 | 1.149 | 1.151 | .850 | .852 | 1.301 | .698 | 1.303 | .697 | .248 | .249 | .137 |
| | .75 | 1.304 | 2.324 | .675 | 1.267 | 1.265 | .734 | .740 | 1.526 | .477 | 1.526 | .487 | .477 | .477 | .278 |
| | 1.00 | 1.484 | 3.090 | .935 | 1.423 | 1.411 | .585 | .604 | 1.842 | .178 | 1.823 | .215 | .743 | .746 | .468 |
| | 1.25 | 1.707 | 3.907 | 1.208 | 1.592 | 1.560 | .429 | .472 | 2.198 | -.148 | 2.137 | -.058 | .996 | 1.007 | .688 |
| | 1.50 | 1.869 | 4.609 | 1.449 | 1.743 | 1.680 | .292 | .367 | 2.504 | -.413 | 2.384 | -.258 | 1.193 | 1.213 | .908 |
| .50 | .25 | 1.339 | 2.469 | .631 | 1.292 | 1.296 | .708 | .713 | 1.584 | .421 | 1.586 | .428 | .148 | .148 | .085 |
| | .50 | 1.434 | 2.955 | .794 | 1.392 | 1.392 | .610 | .624 | 1.776 | .236 | 1.766 | .258 | .321 | .322 | .198 |
| | .75 | 1.511 | 3.462 | .974 | 1.511 | 1.500 | .496 | .519 | 1.966 | .054 | 1.949 | .093 | .510 | .510 | .325 |
| | 1.00 | 1.705 | 4.141 | 1.201 | 1.649 | 1.618 | .371 | .416 | 2.269 | -.222 | 2.212 | -.133 | .720 | .726 | .507 |
| | 1.25 | 1.932 | 4.840 | 1.439 | 1.786 | 1.726 | .252 | .327 | 2.600 | -.513 | 2.481 | -.354 | .920 | .936 | .717 |
| | 1.50 | 1.966 | 5.196 | 1.604 | 1.882 | 1.789 | .168 | .271 | 2.745 | -.622 | 2.575 | -.416 | 1.043 | 1.069 | .885 |
| 1.00 | .25 | 1.812 | 4.468 | 1.156 | 1.707 | 1.679 | .313 | .355 | 2.401 | -.369 | 2.357 | -.284 | .158 | .159 | .113 |
| | .50 | 1.918 | 4.901 | 1.320 | 1.799 | 1.755 | .230 | .292 | 2.591 | -.537 | 2.514 | -.411 | .313 | .315 | .242 |
| | .75 | 2.059 | 5.357 | 1.497 | 1.886 | 1.822 | .154 | .242 | 2.812 | -.719 | 2.687 | -.542 | .466 | .472 | .392 |
| | 1.00 | 2.103 | 5.687 | 1.656 | 1.978 | 1.888 | .078 | .191 | 2.954 | -.828 | 2.778 | -.602 | .591 | .602 | .539 |
| | 1.25 | 2.102 | 5.939 | 1.803 | 2.068 | 1.946 | .007 | .147 | 3.053 | -.894 | 2.823 | -.619 | .694 | .713 | .691 |
| | 1.50 | 2.174 | 6.220 | 1.953 | 2.140 | 1.979 | -.044 | .125 | 3.193 | -.990 | 2.901 | -.655 | .791 | .820 | .867 |

## Table 13

### Uniform unequal biaxial compressive displacement ($\sigma_{xav} = 2\sigma_{yav}$)

Side ratio $a/b = 2{\cdot}0$

| W | P | U | V | W 4 | $\bar{\sigma}_{MX}$ 1 | $\bar{\sigma}_{MX}$ 2 | $\bar{\sigma}_{MX}$ 3 | $\bar{\sigma}_{MX}$ 4 | $\bar{\sigma}_{MY}$ 1 | $\bar{\sigma}_{MY}$ 2 | $\bar{\sigma}_{MY}$ 3 | $\bar{\sigma}_{MY}$ 4 | $\bar{M}_X$ 4 | $\bar{M}_Y$ 4 | $\bar{M}_{XY}$ 1 |
|---|---|---|---|---|---|---|---|---|---|---|---|---|---|---|---|
| .10 | .25 | 1.004 | 1.100 | .133 | 1.005 | 1.005 | .995 | .995 | 1.040 | .960 | 1.040 | .960 | .016 | .032 | .011 |
| | .50 | 1.009 | 1.209 | .206 | 1.012 | 1.012 | .989 | .991 | 1.083 | .917 | 1.083 | .917 | .051 | .102 | .035 |
| | .75 | 1.028 | 1.506 | .349 | 1.028 | 1.026 | .971 | .978 | 1.201 | .807 | 1.201 | .809 | .111 | .236 | .085 |
| | 1.00 | 1.092 | 2.339 | .628 | 1.084 | 1.054 | .916 | .948 | 1.542 | .555 | 1.533 | .570 | .205 | .491 | .209 |
| | 1.25 | 1.198 | 3.660 | .952 | 1.206 | 1.079 | .794 | .908 | 2.092 | .294 | 2.051 | .355 | .294 | .783 | .411 |
| | 1.50 | 1.343 | 5.397 | 1.236 | 1.414 | 1.086 | .584 | .872 | 2.826 | .153 | 2.715 | .307 | .341 | 1.032 | .692 |
| .25 | .25 | 1.020 | 1.564 | .325 | 1.029 | 1.029 | .970 | .973 | 1.220 | .786 | 1.220 | .784 | .035 | .072 | .025 |
| | .50 | 1.040 | 1.919 | .448 | 1.050 | 1.044 | .948 | .958 | 1.363 | .665 | 1.358 | .665 | .086 | .188 | .071 |
| | .75 | 1.086 | 2.505 | .625 | 1.091 | 1.064 | .907 | .940 | 1.604 | .486 | 1.591 | .498 | .144 | .349 | .147 |
| | 1.00 | 1.173 | 3.459 | .855 | 1.173 | 1.083 | .826 | .919 | 2.007 | .276 | 1.977 | .324 | .191 | .551 | .281 |
| | 1.25 | 1.279 | 4.676 | 1.108 | 1.304 | 1.095 | .696 | .878 | 2.519 | .114 | 2.449 | .214 | .280 | .783 | .470 |
| | 1.50 | 1.376 | 6.035 | 1.348 | 1.488 | 1.096 | .512 | .795 | 3.077 | .080 | 2.931 | .237 | .463 | 1.033 | .710 |
| .50 | .25 | 1.078 | 2.729 | .614 | 1.098 | 1.080 | .902 | .921 | 1.679 | .378 | 1.680 | .390 | .048 | .108 | .042 |
| | .50 | 1.120 | 3.287 | .757 | 1.142 | 1.092 | .859 | .908 | 1.909 | .238 | 1.902 | .273 | .092 | .237 | .109 |
| | .75 | 1.183 | 3.996 | .927 | 1.202 | 1.103 | .798 | .897 | 2.212 | .089 | 2.183 | .153 | .131 | .388 | .201 |
| | 1.00 | 1.239 | 4.864 | 1.114 | 1.301 | 1.112 | .696 | .863 | 2.572 | -.001 | 2.509 | .154 | .207 | .562 | .337 |
| | 1.25 | 1.306 | 5.840 | 1.318 | 1.435 | 1.116 | .563 | .804 | 2.980 | -.042 | 2.867 | .188 | .331 | .763 | .512 |
| | 1.50 | 1.411 | 6.848 | 1.541 | 1.591 | 1.111 | .412 | .725 | 3.421 | -.059 | 3.235 | .108 | .512 | .994 | .719 |
| 1.00 | .25 | 1.249 | 5.017 | 1.129 | 1.269 | 1.139 | .733 | .866 | 2.628 | -.233 | 2.592 | -.157 | .031 | .115 | .065 |
| | .50 | 1.283 | 5.603 | 1.268 | 1.335 | 1.137 | .666 | .836 | 2.875 | -.281 | 2.809 | -.067 | .095 | .248 | .151 |
| | .75 | 1.328 | 6.234 | 1.420 | 1.417 | 1.132 | .581 | .779 | 3.147 | -.299 | 3.037 | .109 | .193 | .400 | .257 |
| | 1.00 | 1.371 | 6.904 | 1.586 | 1.523 | 1.129 | .475 | .720 | 3.424 | -.286 | 3.276 | .135 | .311 | .569 | .393 |
| | 1.25 | 1.418 | 7.601 | 1.764 | 1.648 | 1.128 | .354 | .659 | 3.714 | -.257 | 3.518 | .061 | .446 | .748 | .556 |
| | 1.50 | 1.488 | 8.304 | 1.946 | 1.779 | 1.122 | .229 | .589 | 4.035 | -.234 | 3.749 | .009 | .591 | .930 | .741 |

## Table 14

### Uniform unequal biaxial compressive displacement ($\sigma_{xav} = 2\sigma_{yav}$)

Side ratio $a/b = 3{\cdot}0$

| W | P | U | V | W 4 | $\bar{\sigma}_{MX}$ 1 | $\bar{\sigma}_{MX}$ 2 | $\bar{\sigma}_{MX}$ 3 | $\bar{\sigma}_{MX}$ 4 | $\bar{\sigma}_{MY}$ 1 | $\bar{\sigma}_{MY}$ 2 | $\bar{\sigma}_{MY}$ 3 | $\bar{\sigma}_{MY}$ 4 | $\bar{M}_X$ 4 | $\bar{M}_Y$ 4 | $\bar{M}_{XY}$ 1 |
|---|---|---|---|---|---|---|---|---|---|---|---|---|---|---|---|
| .10 | .25 | 1.001 | 1.102 | .133 | 1.002 | 1.002 | .997 | .997 | 1.039 | .962 | 1.039 | .959 | .012 | .030 | .007 |
| | .50 | 1.004 | 1.208 | .202 | 1.004 | 1.004 | .995 | .995 | 1.083 | .924 | 1.083 | .921 | .036 | .094 | .024 |
| | .75 | 1.016 | 1.496 | .332 | 1.022 | 1.007 | .987 | .993 | 1.202 | .844 | 1.201 | .853 | .074 | .211 | .064 |
| | 1.00 | 1.054 | 2.445 | .590 | 1.066 | 1.008 | .932 | .983 | 1.588 | .711 | 1.577 | .729 | .150 | .444 | .189 |
| | 1.25 | 1.129 | 4.393 | .966 | 1.241 | 1.003 | .757 | .918 | 2.381 | .625 | 2.326 | .537 | .317 | .800 | .440 |
| | 1.50 | 1.246 | 7.461 | 1.454 | 1.565 | .991 | .432 | .782 | 3.632 | .630 | 3.492 | .271 | .597 | 1.278 | .832 |
| .25 | .25 | 1.009 | 1.564 | .322 | 1.014 | 1.012 | .986 | .989 | 1.222 | .795 | 1.222 | .793 | .025 | .066 | .017 |
| | .50 | 1.021 | 1.919 | .435 | 1.028 | 1.013 | .972 | .985 | 1.365 | .709 | 1.365 | .708 | .056 | .168 | .054 |
| | .75 | 1.047 | 2.561 | .599 | 1.062 | 1.013 | .939 | .977 | 1.625 | .612 | 1.622 | .623 | .107 | .318 | .125 |
| | 1.00 | 1.088 | 3.820 | .845 | 1.157 | 1.008 | .842 | .944 | 2.136 | .562 | 2.108 | .588 | .217 | .550 | .277 |
| | 1.25 | 1.171 | 5.624 | 1.163 | 1.341 | 1.001 | .658 | .867 | 2.885 | .542 | 2.797 | .491 | .387 | .857 | .514 |
| | 1.50 | 1.318 | 7.886 | 1.542 | 1.634 | .993 | .370 | .728 | 3.850 | .533 | 3.655 | .232 | .615 | 1.229 | .837 |
| .50 | .25 | 1.051 | 2.689 | .604 | 1.050 | 1.028 | .949 | .974 | 1.679 | .469 | 1.676 | .461 | .027 | .093 | .032 |
| | .50 | 1.067 | 3.349 | .746 | 1.092 | 1.027 | .905 | .958 | 1.944 | .417 | 1.934 | .449 | .082 | .225 | .098 |
| | .75 | 1.113 | 4.231 | .922 | 1.154 | 1.017 | .841 | .934 | 2.308 | .390 | 2.289 | .548 | .163 | .392 | .195 |
| | 1.00 | 1.171 | 5.588 | 1.169 | 1.292 | 1.005 | .703 | .865 | 2.870 | .407 | 2.806 | .491 | .306 | .637 | .366 |
| | 1.25 | 1.249 | 7.222 | 1.458 | 1.492 | .996 | .508 | .769 | 3.555 | .430 | 3.416 | .342 | .483 | .922 | .597 |
| | 1.50 | 1.358 | 8.816 | 1.739 | 1.727 | .996 | .284 | .675 | 4.241 | .404 | 4.009 | .211 | .651 | 1.187 | .865 |
| 1.00 | .25 | 1.116 | 5.087 | 1.126 | 1.167 | 1.026 | .827 | .913 | 2.648 | .112 | 2.630 | .487 | .052 | .118 | .058 |
| | .50 | 1.142 | 5.894 | 1.283 | 1.244 | 1.020 | .748 | .875 | 2.979 | .137 | 2.938 | .505 | .135 | .272 | .150 |
| | .75 | 1.198 | 6.890 | 1.473 | 1.356 | 1.011 | .639 | .820 | 3.400 | .182 | 3.314 | .494 | .248 | .461 | .276 |
| | 1.00 | 1.267 | 8.121 | 1.700 | 1.518 | 1.003 | .480 | .737 | 3.923 | .211 | 3.773 | .422 | .401 | .688 | .452 |
| | 1.25 | 1.342 | 9.458 | 1.947 | 1.715 | .997 | .288 | .638 | 4.495 | .226 | 4.262 | .310 | .577 | .935 | .671 |
| | 1.50 | 1.413 | 10.697 | 2.192 | 1.922 | .993 | .091 | .538 | 5.032 | .227 | 4.695 | .191 | .752 | 1.173 | .920 |

## Table 15

### Uniform shear stress

### Side ratio $a/b = 1 \cdot 0$

| W | P | S | W | $\bar{\tau}_{MXY}$ | | $\bar{\sigma}_{MX}$ | | $\bar{\sigma}_{MY}$ | | $\bar{M}_X$ | $\bar{M}_Y$ | $\bar{M}_{XY}$ | |
|---|---|---|---|---|---|---|---|---|---|---|---|---|---|
| | | | 4 | 1 | 4 | 3 | 4 | 2 | 4 | 4 | 4 | 1 | 4 |
| .10 | .25 | 1.002 | .137 | 1.000 | 1.000 | 0.000 | .005 | 0.000 | .005 | .069 | .038 | .060 | .021 |
| | .50 | 1.007 | .205 | 1.000 | 1.002 | 0.000 | .009 | 0.000 | .009 | .197 | .196 | .173 | .060 |
| | .75 | 1.035 | .463 | 1.000 | 1.011 | 0.000 | .042 | 0.000 | .042 | .674 | .692 | .595 | .211 |
| | 1.00 | 1.120 | .957 | 1.000 | 1.034 | 0.000 | .121 | 0.000 | .121 | 1.638 | 1.639 | 1.630 | .550 |
| | 1.25 | 1.308 | 1.644 | 1.000 | 1.078 | 0.000 | .257 | 0.000 | .257 | 3.111 | 3.083 | 3.652 | 1.140 |
| | 1.50 | 1.662 | 2.395 | 1.000 | 1.148 | 0.000 | .452 | 0.000 | .452 | 4.999 | 4.995 | 7.110 | 2.040 |
| .25 | .25 | 1.020 | .338 | 1.000 | 1.005 | 0.000 | .028 | 0.000 | .028 | .166 | .166 | .148 | .051 |
| | .50 | 1.028 | .491 | 1.000 | 1.012 | 0.000 | .048 | 0.000 | .048 | .453 | .453 | .415 | .142 |
| | .75 | 1.063 | .820 | 1.000 | 1.030 | 0.000 | .111 | 0.000 | .111 | 1.078 | 1.078 | 1.066 | .362 |
| | 1.00 | 1.151 | 1.332 | 1.000 | 1.060 | 0.000 | .220 | 0.000 | .220 | 2.109 | 2.109 | 2.332 | .759 |
| | 1.25 | 1.323 | 1.955 | 1.000 | 1.104 | 0.000 | .361 | 0.000 | .361 | 3.501 | 3.501 | 4.444 | 1.352 |
| | 1.50 | 1.618 | 2.526 | 1.000 | 1.161 | 0.000 | .493 | 0.000 | .493 | 5.092 | 5.092 | 7.633 | 2.133 |
| .50 | .25 | 1.085 | .654 | 1.000 | 1.027 | 0.000 | .095 | 0.000 | .095 | .292 | .292 | .283 | .095 |
| | .50 | 1.137 | .887 | 1.000 | 1.038 | 0.000 | .142 | 0.000 | .142 | .742 | .742 | .758 | .252 |
| | .75 | 1.238 | 1.239 | 1.000 | 1.064 | 0.000 | .221 | 0.000 | .221 | 1.448 | 1.448 | 1.631 | .527 |
| | 1.00 | 1.398 | 1.703 | 1.000 | 1.104 | 0.000 | .329 | 0.000 | .329 | 2.439 | 2.439 | 3.084 | .951 |
| | 1.25 | 1.624 | 2.227 | 1.000 | 1.151 | 0.000 | .451 | 0.000 | .451 | 3.699 | 3.699 | 5.305 | 1.533 |
| | 1.50 | 1.919 | 2.718 | 1.000 | 1.187 | 0.000 | .557 | 0.000 | .557 | 5.158 | 5.158 | 8.479 | 2.262 |
| 1.00 | .25 | 1.268 | 1.218 | 1.000 | 1.072 | 0.000 | .244 | 0.000 | .244 | .434 | .434 | .517 | .168 |
| | .50 | 1.363 | 1.498 | 1.000 | 1.095 | 0.000 | .306 | 0.000 | .306 | 1.009 | 1.009 | 1.292 | .406 |
| | .75 | 1.502 | 1.837 | 1.000 | 1.124 | 0.000 | .381 | 0.000 | .381 | 1.769 | 1.769 | 2.469 | .742 |
| | 1.00 | 1.679 | 2.225 | 1.000 | 1.158 | 0.000 | .463 | 0.000 | .463 | 2.734 | 2.734 | 4.190 | 1.189 |
| | 1.25 | 1.878 | 2.641 | 1.000 | 1.189 | 0.000 | .545 | 0.000 | .545 | 3.906 | 3.906 | 6.619 | 1.752 |
| | 1.50 | 2.066 | 3.055 | 1.000 | 1.209 | 0.000 | .613 | 0.000 | .613 | 5.264 | 5.264 | 9.935 | 2.429 |

# Table 16

## Uniform shear displacement

Side ratio $a/b = 1 \cdot 0$

| W | P | S | W | $\bar{\tau}$ MXY | | $\bar{\sigma}$ MX | | $\bar{\sigma}$ MY | | $\bar{M}$ X | $\bar{M}$ Y | $\bar{M}$ XY | |
|---|---|---|---|---|---|---|---|---|---|---|---|---|---|
| | | | 4 | 1 | 4 | 3 | 4 | 2 | 4 | 4 | 4 | 1 | 4 |
| .10 | .25 | 1.001 | .120 | 1.000 | 1.002 | .005 | .005 | .005 | .005 | .023 | .023 | .017 | .003 |
| | .50 | 1.002 | .192 | 1.001 | 1.004 | .016 | .018 | .016 | .018 | .167 | .167 | .147 | .050 |
| | .75 | 1.007 | .342 | 1.006 | 1.006 | .047 | .054 | .047 | .054 | .472 | .473 | .423 | .150 |
| | 1.00 | 1.014 | .540 | 1.014 | 1.010 | .092 | .107 | .092 | .107 | .913 | .913 | .832 | .305 |
| | 1.25 | 1.023 | .748 | 1.024 | 1.013 | .145 | .171 | .145 | .171 | 1.448 | 1.448 | 1.348 | .508 |
| | 1.50 | 1.034 | .946 | 1.035 | 1.014 | .198 | .239 | .198 | .239 | 2.038 | 2.038 | 1.946 | .744 |
| | 1.75 | 1.045 | 1.122 | 1.046 | 1.013 | .247 | .301 | .247 | .301 | 2.637 | 2.637 | 2.594 | .996 |
| | 2.00 | 1.054 | 1.268 | 1.055 | 1.007 | .285 | .351 | .285 | .351 | 3.195 | 3.196 | 3.256 | 1.238 |
| .25 | .25 | 1.007 | .324 | 1.006 | 1.009 | .054 | .061 | .054 | .061 | .134 | .134 | .118 | .040 |
| | .50 | 1.011 | .426 | 1.010 | 1.011 | .078 | .089 | .078 | .089 | .356 | .356 | .326 | .115 |
| | .75 | 1.018 | .568 | 1.017 | 1.013 | .114 | .132 | .114 | .132 | .682 | .682 | .637 | .233 |
| | 1.00 | 1.025 | .730 | 1.026 | 1.013 | .156 | .184 | .156 | .184 | 1.092 | 1.092 | 1.041 | .389 |
| | 1.25 | 1.034 | .894 | 1.035 | 1.012 | .199 | .237 | .199 | .237 | 1.561 | 1.561 | 1.522 | .575 |
| | 1.50 | 1.043 | 1.050 | 1.044 | 1.010 | .240 | .289 | .240 | .289 | 2.065 | 2.065 | 2.067 | .783 |
| | 1.75 | 1.051 | 1.193 | 1.052 | 1.007 | .275 | .336 | .275 | .336 | 2.578 | 2.578 | 2.659 | 1.002 |
| | 2.00 | 1.059 | 1.323 | 1.060 | 1.003 | .305 | .375 | .305 | .375 | 3.075 | 3.075 | 3.283 | 1.221 |
| .50 | .25 | 1.022 | .596 | 1.021 | 1.018 | .138 | .157 | .138 | .157 | .210 | .210 | .208 | .075 |
| | .50 | 1.027 | .704 | 1.026 | 1.019 | .166 | .190 | .166 | .190 | .478 | .478 | .482 | .176 |
| | .75 | 1.033 | .827 | 1.033 | 1.021 | .197 | .229 | .197 | .229 | .809 | .809 | .823 | .304 |
| | 1.00 | 1.040 | .956 | 1.041 | 1.020 | .229 | .270 | .229 | .270 | 1.190 | 1.190 | 1.231 | .457 |
| | 1.25 | 1.048 | 1.084 | 1.049 | 1.015 | .259 | .310 | .259 | .310 | 1.604 | 1.604 | 1.701 | .629 |
| | 1.50 | 1.055 | 1.208 | 1.056 | 1.006 | .287 | .348 | .287 | .348 | 2.039 | 2.039 | 2.225 | .816 |
| | 1.75 | 1.062 | 1.325 | 1.063 | .997 | .313 | .382 | .313 | .382 | 2.483 | 2.483 | 2.789 | 1.011 |
| | 2.00 | 1.069 | 1.434 | 1.069 | .992 | .336 | .412 | .336 | .412 | 2.923 | 2.924 | 3.377 | 1.206 |
| 1.00 | .25 | 1.049 | 1.078 | 1.048 | 1.013 | .236 | .271 | .236 | .271 | .256 | .256 | .310 | .114 |
| | .50 | 1.054 | 1.161 | 1.053 | 1.007 | .259 | .303 | .259 | .303 | .544 | .544 | .664 | .240 |
| | .75 | 1.059 | 1.249 | 1.059 | 1.003 | .281 | .334 | .281 | .334 | .866 | .866 | 1.064 | .380 |
| | 1.00 | 1.065 | 1.340 | 1.066 | .998 | .302 | .365 | .302 | .365 | 1.212 | 1.212 | 1.504 | .531 |
| | 1.25 | 1.071 | 1.432 | 1.071 | .992 | .322 | .392 | .322 | .392 | 1.573 | 1.573 | 1.981 | .691 |
| | 1.50 | 1.076 | 1.523 | 1.077 | .986 | .341 | .417 | .341 | .417 | 1.944 | 1.944 | 2.495 | .858 |
| | 1.75 | 1.081 | 1.611 | 1.082 | .980 | .359 | .439 | .359 | .439 | 2.320 | 2.320 | 3.046 | 1.028 |
| | 2.00 | 1.086 | 1.693 | 1.087 | .977 | .376 | .461 | .376 | .461 | 2.699 | 2.699 | 3.637 | 1.201 |

## Table 17

## Uniform shear displacement

Side ratio $a/b = 1\cdot 5$

| W | P | S | W | $\bar{T}$ MXY | | $\bar{\sigma}$ MX | | $\bar{\sigma}$ MY | | $\bar{M}$ X | $\bar{M}$ Y | $\bar{M}$ XY | |
|---|---|---|---|---|---|---|---|---|---|---|---|---|---|
| | | | .4 | 1 | 4 | 3 | 4 | 2 | 4 | 4 | 4 | 1 | 4 |
| .10 | .25 | 1.000 | .122 | 1.000 | 1.000 | .049 | .007 | .002 | .004 | .024 | .017 | .013 | .003 |
| | .50 | 1.002 | .195 | 1.002 | 1.002 | .027 | .023 | .009 | .015 | .144 | .114 | .087 | .038 |
| | .75 | 1.006 | .348 | 1.005 | 1.006 | .055 | .066 | .027 | .044 | .394 | .315 | .247 | .110 |
| | 1.00 | 1.011 | .543 | 1.011 | 1.009 | .112 | .126 | .057 | .089 | .748 | .605 | .471 | .221 |
| | 1.25 | 1.019 | .745 | 1.019 | 1.008 | .177 | .195 | .099 | .145 | 1.172 | .959 | .739 | .368 |
| | 1.50 | 1.028 | .935 | 1.030 | 1.003 | .239 | .265 | .149 | .208 | 1.634 | 1.352 | 1.036 | .541 |
| | 1.75 | 1.037 | 1.100 | 1.040 | .995 | .292 | .327 | .201 | .270 | 2.097 | 1.751 | 1.354 | .724 |
| | 2.00 | 1.045 | 1.238 | 1.047 | .988 | .334 | .374 | .248 | .320 | 2.523 | 2.121 | 1.685 | .898 |
| .25 | .25 | 1.006 | .325 | 1.006 | 1.006 | .066 | .074 | .030 | .050 | .112 | .090 | .070 | .030 |
| | .50 | 1.009 | .428 | 1.009 | 1.008 | .095 | .106 | .046 | .073 | .293 | .236 | .186 | .085 |
| | .75 | 1.014 | .567 | 1.014 | 1.010 | .139 | .154 | .073 | .110 | .555 | .451 | .358 | .170 |
| | 1.00 | 1.021 | .726 | 1.020 | 1.010 | .189 | .210 | .108 | .156 | .881 | .721 | .574 | .281 |
| | 1.25 | 1.028 | .885 | 1.028 | 1.007 | .239 | .266 | .148 | .206 | 1.249 | 1.029 | .821 | .415 |
| | 1.50 | 1.035 | 1.037 | 1.037 | 1.001 | .286 | .317 | .190 | .256 | 1.639 | 1.361 | 1.092 | .565 |
| | 1.75 | 1.042 | 1.174 | 1.045 | .994 | .326 | .361 | .232 | .303 | 2.032 | 1.700 | 1.379 | .723 |
| | 2.00 | 1.049 | 1.295 | 1.052 | .987 | .361 | .396 | .272 | .344 | 2.411 | 2.026 | 1.677 | .879 |
| .50 | .25 | 1.018 | .595 | 1.017 | 1.010 | .166 | .182 | .090 | .132 | .170 | .139 | .117 | .055 |
| | .50 | 1.022 | .701 | 1.022 | 1.010 | .199 | .218 | .114 | .162 | .383 | .314 | .267 | .128 |
| | .75 | 1.027 | .821 | 1.026 | 1.008 | .235 | .258 | .143 | .198 | .643 | .529 | .453 | .222 |
| | 1.00 | 1.033 | .946 | 1.032 | 1.005 | .272 | .299 | .176 | .238 | .938 | .777 | .670 | .332 |
| | 1.25 | 1.039 | 1.068 | 1.039 | 1.000 | .308 | .337 | .211 | .277 | 1.258 | 1.046 | .910 | .454 |
| | 1.50 | 1.045 | 1.186 | 1.047 | .993 | .340 | .372 | .245 | .315 | 1.593 | 1.328 | 1.171 | .584 |
| | 1.75 | 1.051 | 1.296 | 1.054 | .986 | .371 | .402 | .278 | .350 | 1.933 | 1.616 | 1.447 | .721 |
| | 2.00 | 1.057 | 1.400 | 1.059 | .981 | .399 | .428 | .306 | .381 | 2.269 | 1.902 | 1.737 | .860 |
| 1.00 | .25 | 1.040 | 1.074 | 1.039 | .999 | .284 | .287 | .208 | .248 | .195 | .154 | .171 | .083 |
| | .50 | 1.044 | 1.153 | 1.044 | .995 | .312 | .319 | .234 | .280 | .414 | .347 | .363 | .174 |
| | .75 | 1.048 | 1.237 | 1.047 | .989 | .339 | .349 | .262 | .311 | .659 | .553 | .577 | .273 |
| | 1.00 | 1.053 | 1.324 | 1.052 | .984 | .365 | .376 | .288 | .340 | .923 | .773 | .811 | .379 |
| | 1.25 | 1.057 | 1.410 | 1.058 | .980 | .390 | .401 | .313 | .368 | 1.198 | 1.002 | 1.061 | .490 |
| | 1.50 | 1.062 | 1.494 | 1.064 | .976 | .413 | .422 | .336 | .393 | 1.481 | 1.236 | 1.328 | .606 |
| | 1.75 | 1.066 | 1.576 | 1.069 | .972 | .435 | .441 | .359 | .417 | 1.768 | 1.475 | 1.610 | .726 |
| | 2.00 | 1.071 | 1.655 | 1.073 | .966 | .455 | .459 | .384 | .439 | 2.057 | 1.717 | 1.908 | .847 |

# Table 18

## Uniform shear displacement

### Side ratio $a/b = 2{\cdot}0$

| W | P | S | W | $\bar{\tau}$ | | $\bar{\sigma}$ | | $\bar{\sigma}$ | | $\bar{M}$ | $\bar{M}$ | $\bar{M}$ | |
|---|---|---|---|---|---|---|---|---|---|---|---|---|---|
| | | | | MXY | | MX | | MY | | X | Y | XY | |
| | | | 4 | 1 | 4 | 3 | 4 | 2 | 4 | 4 | 4 | 1 | 4 |
| .10 | .25 | 1.005 | .121 | .999 | 1.000 | .008 | .009 | .002 | .004 | .024 | .017 | -.003 | .004 |
| | .50 | 1.005 | .189 | 1.000 | 1.000 | .022 | .024 | .008 | .014 | .130 | .095 | .016 | .032 |
| | .75 | 1.005 | .323 | 1.003 | 1.002 | .054 | .060 | .020 | .035 | .328 | .242 | .090 | .085 |
| | 1.00 | 1.007 | .495 | 1.007 | 1.003 | .105 | .115 | .045 | .075 | .625 | .472 | .153 | .175 |
| | 1.25 | 1.012 | .682 | 1.012 | 1.002 | .168 | .181 | .084 | .132 | .999 | .776 | .179 | .302 |
| | 1.50 | 1.020 | .860 | 1.019 | .997 | .236 | .251 | .135 | .202 | 1.411 | 1.123 | .171 | .454 |
| | 1.75 | 1.029 | 1.003 | 1.026 | .989 | .297 | .310 | .187 | .271 | 1.796 | 1.460 | .162 | .608 |
| | 2.00 | 1.035 | 1.086 | 1.032 | .978 | .335 | .341 | .227 | .322 | 2.072 | 1.715 | .210 | .729 |
| .25 | .25 | 1.008 | .318 | 1.003 | 1.002 | .066 | .074 | .025 | .045 | .099 | .072 | .031 | .025 |
| | .50 | 1.010 | .413 | 1.005 | 1.003 | .095 | .104 | .039 | .066 | .255 | .190 | .068 | .070 |
| | .75 | 1.011 | .542 | 1.009 | 1.003 | .138 | .150 | .062 | .100 | .477 | .351 | .122 | .138 |
| | 1.00 | 1.014 | .685 | 1.013 | 1.002 | .187 | .201 | .093 | .144 | .753 | .579 | .163 | .230 |
| | 1.25 | 1.019 | .827 | 1.019 | .999 | .239 | .252 | .131 | .195 | 1.068 | .834 | .175 | .343 |
| | 1.50 | 1.025 | .960 | 1.024 | .993 | .287 | .299 | .172 | .250 | 1.404 | 1.114 | .154 | .472 |
| | 1.75 | 1.032 | 1.078 | 1.029 | .986 | .330 | .338 | .214 | .302 | 1.741 | 1.401 | .103 | .606 |
| | 2.00 | 1.037 | 1.176 | 1.034 | .978 | .365 | .366 | .250 | .346 | 2.053 | 1.674 | .033 | .735 |
| .50 | .25 | 1.016 | .586 | 1.010 | 1.005 | .168 | .179 | .079 | .123 | .146 | .110 | .045 | .046 |
| | .50 | 1.018 | .682 | 1.013 | 1.004 | .201 | .212 | .100 | .152 | .328 | .249 | .087 | .106 |
| | .75 | 1.020 | .790 | 1.018 | 1.001 | .238 | .249 | .129 | .188 | .547 | .423 | .131 | .182 |
| | 1.00 | 1.022 | .901 | 1.022 | .996 | .275 | .286 | .160 | .227 | .796 | .622 | .163 | .271 |
| | 1.25 | 1.027 | 1.009 | 1.027 | .989 | .311 | .319 | .193 | .268 | 1.066 | .840 | .172 | .372 |
| | 1.50 | 1.033 | 1.111 | 1.031 | .981 | .344 | .349 | .225 | .308 | 1.349 | 1.071 | .154 | .481 |
| | 1.75 | 1.039 | 1.205 | 1.036 | .974 | .375 | .374 | .257 | .347 | 1.637 | 1.310 | .105 | .595 |
| | 2.00 | 1.043 | 1.290 | 1.040 | .969 | .406 | .395 | .286 | .382 | 1.922 | 1.549 | .024 | .711 |
| 1.00 | .25 | 1.031 | 1.063 | 1.025 | .989 | .284 | .267 | .189 | .243 | .164 | .129 | .056 | .057 |
| | .50 | 1.034 | 1.132 | 1.028 | .985 | .312 | .295 | .215 | .274 | .348 | .274 | .104 | .125 |
| | .75 | 1.035 | 1.205 | 1.033 | .979 | .341 | .322 | .240 | .305 | .554 | .436 | .146 | .214 |
| | 1.00 | 1.037 | 1.280 | 1.037 | .974 | .368 | .346 | .265 | .335 | .774 | .609 | .175 | .308 |
| | 1.25 | 1.041 | 1.353 | 1.041 | .970 | .393 | .368 | .289 | .364 | 1.005 | .793 | .187 | .401 |
| | 1.50 | 1.046 | 1.424 | 1.044 | .966 | .416 | .387 | .311 | .391 | 1.243 | .983 | .178 | .494 |
| | 1.75 | 1.051 | 1.493 | 1.047 | .961 | .438 | .404 | .332 | .416 | 1.485 | 1.178 | .142 | .587 |
| | 2.00 | 1.053 | 1.559 | 1.050 | .952 | .458 | .419 | .353 | .441 | 1.729 | 1.377 | .074 | .689 |

## Table 19

## Uniform shear displacement

### Side ratio $a/b = 3{\cdot}0$

| W | P | S | W | $\bar{\tau}_{MXY}$ | | $\bar{\sigma}_{MX}$ | | $\bar{\sigma}_{MY}$ | | $\bar{M}_X$ | $\bar{M}_Y$ | $\bar{M}_{XY}$ | |
|---|---|---|---|---|---|---|---|---|---|---|---|---|---|
| | | | 4 | 1 | 4 | 3 | 4 | 2 | 4 | 4 | 4 | 1 | 4 |
| .10 | .25 | 1.004 | .122 | 1.000 | 1.000 | .008 | .010 | .004 | .007 | .024 | .018 | .011 | .004 |
| | .50 | 1.006 | .189 | 1.001 | 1.002 | .022 | .026 | .009 | .020 | .125 | .091 | .046 | .031 |
| | .75 | 1.008 | .317 | 1.005 | 1.004 | .055 | .064 | .022 | .049 | .319 | .231 | .132 | .083 |
| | 1.00 | 1.011 | .480 | 1.010 | 1.007 | .102 | .118 | .044 | .094 | .587 | .427 | .220 | .159 |
| | 1.25 | 1.016 | .649 | 1.016 | 1.007 | .157 | .176 | .076 | .148 | .901 | .662 | .264 | .257 |
| | 1.50 | 1.023 | .807 | 1.023 | 1.005 | .215 | .232 | .115 | .208 | 1.240 | .925 | .258 | .371 |
| | 1.75 | 1.031 | .945 | 1.029 | 1.000 | .270 | .281 | .159 | .268 | 1.584 | 1.203 | .216 | .496 |
| | 2.00 | 1.038 | 1.057 | 1.035 | .993 | .318 | .318 | .202 | .323 | 1.913 | 1.480 | .175 | .624 |
| .25 | .25 | 1.009 | .320 | 1.005 | 1.005 | .069 | .081 | .030 | .063 | .099 | .082 | .040 | .025 |
| | .50 | 1.013 | .411 | 1.008 | 1.006 | .097 | .112 | .042 | .088 | .245 | .192 | .092 | .066 |
| | .75 | 1.016 | .529 | 1.013 | 1.009 | .134 | .153 | .062 | .124 | .444 | .331 | .167 | .125 |
| | 1.00 | 1.019 | .660 | 1.019 | 1.010 | .177 | .199 | .088 | .166 | .685 | .499 | .242 | .198 |
| | 1.25 | 1.024 | .792 | 1.024 | 1.008 | .222 | .244 | .117 | .212 | .952 | .692 | .294 | .285 |
| | 1.50 | 1.030 | .917 | 1.029 | 1.003 | .267 | .284 | .150 | .257 | 1.236 | .907 | .309 | .381 |
| | 1.75 | 1.036 | 1.031 | 1.034 | .997 | .310 | .318 | .185 | .302 | 1.527 | 1.137 | .278 | .486 |
| | 2.00 | 1.043 | 1.133 | 1.040 | .992 | .349 | .346 | .222 | .345 | 1.818 | 1.372 | .193 | .594 |
| .50 | .25 | 1.021 | .585 | 1.016 | 1.011 | .167 | .191 | .079 | .157 | .141 | .100 | .062 | .042 |
| | .50 | 1.026 | .677 | 1.020 | 1.011 | .197 | .221 | .098 | .185 | .307 | .218 | .127 | .094 |
| | .75 | 1.028 | .778 | 1.025 | 1.011 | .229 | .253 | .121 | .218 | .503 | .358 | .199 | .156 |
| | 1.00 | 1.031 | .882 | 1.030 | 1.009 | .262 | .284 | .146 | .252 | .721 | .514 | .270 | .226 |
| | 1.25 | 1.034 | .985 | 1.035 | 1.006 | .295 | .313 | .173 | .285 | .954 | .683 | .327 | .303 |
| | 1.50 | 1.039 | 1.083 | 1.039 | 1.001 | .328 | .339 | .201 | .318 | 1.197 | .852 | .357 | .385 |
| | 1.75 | 1.045 | 1.176 | 1.042 | .996 | .359 | .362 | .230 | .349 | 1.446 | 1.047 | .346 | .471 |
| | 2.00 | 1.049 | 1.262 | 1.046 | .991 | .388 | .382 | .260 | .380 | 1.695 | 1.238 | .280 | .560 |
| 1.00 | .25 | 1.041 | 1.066 | 1.036 | 1.005 | .276 | .292 | .098 | .274 | .157 | .107 | .085 | .057 |
| | .50 | 1.045 | 1.135 | 1.040 | 1.003 | .304 | .317 | .066 | .301 | .327 | .225 | .167 | .118 |
| | .75 | 1.047 | 1.208 | 1.044 | 1.000 | .329 | .339 | .039 | .326 | .512 | .352 | .249 | .181 |
| | 1.00 | 1.048 | 1.282 | 1.047 | .997 | .352 | .358 | .053 | .351 | .708 | .488 | .324 | .247 |
| | 1.25 | 1.050 | 1.355 | 1.051 | .995 | .375 | .376 | .120 | .374 | .911 | .629 | .387 | .316 |
| | 1.50 | 1.054 | 1.427 | 1.054 | .993 | .397 | .392 | .226 | .396 | 1.119 | .774 | .431 | .387 |
| | 1.75 | 1.059 | 1.497 | 1.056 | .991 | .418 | .407 | .335 | .417 | 1.329 | .923 | .449 | .460 |
| | 2.00 | 1.062 | 1.565 | 1.058 | .988 | .438 | .420 | .391 | .437 | 1.541 | 1.076 | .436 | .534 |

## Table 20

### Combined uniform uniaxial compressive and shear displacement ($\sigma_{x\,av} = \tau_{av}$)

Side ratio $a/b = 1{\cdot}0$

| W | P | U | V | S | W | $\bar{\sigma}_{MX}$ 1 | $\bar{\sigma}_{MX}$ 2 | $\bar{\sigma}_{MY}$ 1 | $\bar{\sigma}_{MY}$ 2 | $\bar{\tau}_{MXY}$ 1 | $\bar{\tau}_{MXY}$ 2 | $\bar{M}_X$ 2 | $\bar{M}_Y$ 2 | $\bar{M}_{XY}$ 1 | $\bar{M}_{XY}$ 2 |
|---|---|---|---|---|---|---|---|---|---|---|---|---|---|---|---|
| .10 | .25 | 1.009 | .989 | 1.002 | .123 | 1.013 | .990 | .008 | -.014 | 1.002 | 1.002 | .030 | .029 | .024 | .005 |
| | .50 | 1.025 | .920 | 1.005 | .209 | 1.034 | .970 | .034 | -.031 | 1.005 | 1.006 | .141 | .139 | .121 | .023 |
| | .75 | 1.056 | .823 | 1.008 | .357 | 1.078 | .933 | .076 | -.059 | 1.007 | 1.015 | .331 | .327 | .288 | .056 |
| | 1.00 | 1.165 | .386 | 1.027 | .696 | 1.235 | .799 | .256 | -.190 | 1.027 | 1.052 | .770 | .760 | .730 | .145 |
| | 1.25 | 1.294 | .073 | 1.056 | 1.007 | 1.424 | .661 | .404 | -.361 | 1.056 | 1.103 | 1.195 | 1.170 | 1.257 | .258 |
| | 1.50 | 1.441 | -.109 | 1.094 | 1.289 | 1.643 | .519 | .519 | -.583 | 1.093 | 1.167 | 1.604 | 1.552 | 1.865 | .394 |
| .25 | .25 | 1.061 | .801 | 1.009 | .322 | 1.088 | .927 | .087 | -.075 | 1.009 | 1.014 | .093 | .091 | .076 | .014 |
| | .50 | 1.105 | .660 | 1.017 | .459 | 1.154 | .875 | .148 | -.132 | 1.015 | 1.029 | .270 | .266 | .247 | .049 |
| | .75 | 1.177 | .416 | 1.028 | .663 | 1.252 | .788 | .250 | -.220 | 1.028 | 1.053 | .534 | .526 | .514 | .104 |
| | 1.00 | 1.290 | .017 | 1.052 | .928 | 1.413 | .663 | .418 | -.337 | 1.053 | 1.096 | .886 | .871 | .939 | .193 |
| | 1.25 | 1.398 | -.340 | 1.084 | 1.191 | 1.572 | .547 | .573 | -.459 | 1.083 | 1.149 | 1.259 | 1.237 | 1.475 | .308 |
| | 1.50 | 1.496 | -.634 | 1.124 | 1.445 | 1.722 | .445 | .704 | -.586 | 1.118 | 1.211 | 1.647 | 1.617 | 2.115 | .448 |
| .50 | .25 | 1.205 | .355 | 1.030 | .626 | 1.284 | .762 | .279 | -.249 | 1.029 | 1.058 | .161 | .158 | .151 | .030 |
| | .50 | 1.269 | .143 | 1.044 | .786 | 1.375 | .690 | .370 | -.324 | 1.044 | 1.083 | .370 | .364 | .390 | .080 |
| | .75 | 1.346 | -.137 | 1.064 | .981 | 1.490 | .600 | .487 | -.412 | 1.063 | 1.117 | .630 | .618 | .716 | .149 |
| | 1.00 | 1.433 | -.453 | 1.090 | 1.197 | 1.615 | .508 | .620 | -.501 | 1.090 | 1.162 | .930 | .914 | 1.151 | .240 |
| | 1.25 | 1.526 | -.783 | 1.124 | 1.422 | 1.748 | .417 | .760 | -.590 | 1.124 | 1.212 | 1.276 | 1.248 | 1.716 | .360 |
| | 1.50 | 1.617 | -1.103 | 1.167 | 1.643 | 1.886 | .330 | .897 | -.677 | 1.163 | 1.257 | 1.681 | 1.616 | 2.426 | .515 |
| 1.00 | .25 | 1.465 | -.552 | 1.088 | 1.136 | 1.658 | .480 | .663 | -.529 | 1.093 | 1.170 | .187 | .183 | .249 | .052 |
| | .50 | 1.524 | -.765 | 1.111 | 1.294 | 1.745 | .422 | .750 | -.584 | 1.113 | 1.205 | .399 | .390 | .571 | .120 |
| | .75 | 1.584 | -.987 | 1.138 | 1.473 | 1.841 | .362 | .842 | -.639 | 1.136 | 1.243 | .636 | .621 | .969 | .202 |
| | 1.00 | 1.645 | -1.126 | 1.164 | 1.640 | 1.923 | .311 | .912 | -.707 | 1.163 | 1.277 | .911 | .883 | 1.463 | .308 |
| | 1.25 | 1.705 | -1.297 | 1.193 | 1.810 | 2.002 | .260 | .989 | -.774 | 1.196 | 1.311 | 1.235 | 1.187 | 2.077 | .436 |
| | 1.50 | 1.766 | -1.552 | 1.227 | 1.989 | 2.085 | .204 | 1.086 | -.833 | 1.235 | 1.347 | 1.611 | 1.536 | 2.818 | .588 |

## Table 21

### Combined uniform uniaxial compressive and shear displacement ($\sigma_{x\,av} = \tau_{av}$)

Side ratio $a/b = 2{\cdot}0$

| W | P | U | V | S | W | $\bar{\sigma}_{MX}$ 1 | $\bar{\sigma}_{MX}$ 2 | $\bar{\sigma}_{MY}$ 1 | $\bar{\sigma}_{MY}$ 2 | $\bar{\tau}_{MXY}$ 1 | $\bar{\tau}_{MXY}$ 2 | $\bar{M}_X$ 2 | $\bar{M}_Y$ 2 | $\bar{M}_{XY}$ 1 | $\bar{M}_{XY}$ 2 |
|---|---|---|---|---|---|---|---|---|---|---|---|---|---|---|---|
| .10 | .25 | 1.014 | .984 | 1.000 | .121 | 1.011 | .989 | .021 | -.016 | 1.001 | 1.002 | .026 | .027 | .019 | .004 |
| | .50 | 1.028 | .894 | 1.001 | .208 | 1.040 | .964 | .049 | -.026 | .999 | 1.004 | .133 | .138 | .108 | .026 |
| | .75 | 1.053 | .748 | 1.011 | .356 | 1.084 | .922 | .102 | -.057 | 1.010 | 1.015 | .315 | .326 | .258 | .061 |
| | 1.00 | 1.172 | .276 | 1.038 | .683 | 1.261 | .767 | .294 | -.183 | 1.038 | 1.047 | .715 | .747 | .658 | .156 |
| | 1.25 | 1.264 | -.213 | 1.086 | 1.007 | 1.433 | .561 | .531 | -.274 | 1.078 | 1.098 | 1.114 | 1.172 | 1.177 | .267 |
| | 1.50 | 1.316 | -.688 | 1.156 | 1.311 | 1.585 | .308 | .806 | -.317 | 1.130 | 1.157 | 1.489 | 1.378 | 1.808 | .389 |
| .25 | .25 | 1.058 | .720 | 1.011 | .321 | 1.120 | .909 | .210 | -.066 | 1.017 | 1.016 | .087 | .090 | .064 | .016 |
| | .50 | 1.110 | .560 | 1.022 | .457 | 1.164 | .849 | .178 | -.124 | 1.021 | 1.029 | .254 | .264 | .222 | .053 |
| | .75 | 1.188 | .238 | 1.032 | .663 | 1.284 | .744 | .312 | -.208 | 1.035 | 1.042 | .507 | .530 | .473 | .113 |
| | 1.00 | 1.296 | -.277 | 1.088 | .922 | 1.449 | .604 | .519 | -.303 | 1.092 | 1.106 | .824 | .863 | .861 | .198 |
| | 1.25 | 1.429 | -.827 | 1.137 | 1.200 | 1.649 | .452 | .740 | -.413 | 1.139 | 1.159 | 1.185 | 1.250 | 1.421 | .317 |
| | 1.50 | 1.583 | -1.284 | 1.132 | 1.474 | 1.882 | .303 | .950 | -.548 | 1.128 | 1.142 | 1.579 | 1.683 | 2.172 | .477 |
| .50 | .25 | 1.210 | .146 | 1.043 | .623 | 1.311 | .719 | .355 | -.222 | 1.066 | 1.055 | .150 | .156 | .136 | .033 |
| | .50 | 1.280 | -.210 | 1.066 | .786 | 1.425 | .628 | .488 | -.282 | 1.059 | 1.081 | .349 | .365 | .360 | .084 |
| | .75 | 1.365 | -.639 | 1.095 | .988 | 1.570 | .513 | .662 | -.355 | 1.095 | 1.114 | .597 | .628 | .674 | .155 |
| | 1.00 | 1.459 | -.908 | 1.125 | 1.193 | 1.686 | .413 | .778 | -.453 | 1.128 | 1.149 | .864 | .911 | 1.088 | .245 |
| | 1.25 | 1.542 | -1.329 | 1.175 | 1.404 | 1.830 | .315 | .947 | -.508 | 1.174 | 1.197 | 1.142 | 1.207 | 1.596 | .341 |
| | 1.50 | 1.614 | -1.890 | 1.244 | 1.621 | 2.000 | .220 | 1.163 | -.525 | 1.233 | 1.256 | 1.433 | 1.517 | 2.196 | .445 |
| 1.00 | .25 | 1.470 | -1.052 | 1.132 | 1.147 | 1.785 | .376 | .847 | .388 | 1.132 | 1.175 | .175 | .186 | .232 | .054 |
| | .50 | 1.553 | -1.365 | 1.160 | 1.305 | 1.821 | .307 | .959 | .360 | 1.160 | 1.207 | .374 | .395 | .543 | .121 |
| | .75 | 1.615 | -1.684 | 1.193 | 1.472 | 1.941 | .231 | 1.088 | -.590 | 1.197 | 1.226 | .591 | .627 | .932 | .200 |
| | 1.00 | 1.655 | -1.975 | 1.232 | 1.637 | 2.023 | .157 | 1.200 | -.632 | 1.238 | 1.261 | .821 | .870 | 1.401 | .290 |
| | 1.25 | 1.744 | -2.367 | 1.295 | 1.826 | 2.147 | .078 | 1.356 | -.551 | 1.297 | 1.316 | 1.106 | 1.169 | 2.061 | .408 |
| | 1.50 | 1.903 | -2.900 | 1.386 | 2.047 | 2.346 | -.009 | 1.579 | -.705 | 1.378 | 1.389 | 1.463 | 1.538 | 2.948 | .560 |

## Table 22

### Combined uniform biaxial compressive and shear displacement ($\sigma_{x\,av} = \sigma_{y\,av} = \tau_{av}/2$)

Side ratio $a/b = 1{\cdot}0$

| W | P | U | V | S | H | $\bar{\sigma}_{MX}$ 1 | $\bar{\sigma}_{MX}$ 2 | $\bar{\sigma}_{MY}$ 1 | $\bar{\sigma}_{MY}$ 2 | $\bar{\tau}_{MXY}$ 1 | $\bar{\tau}_{MXY}$ 2 | $\bar{M}_X$ 2 | $\bar{M}_Y$ 2 | $\bar{M}_{XY}$ 1 | $\bar{M}_{XY}$ 2 |
|---|---|---|---|---|---|---|---|---|---|---|---|---|---|---|---|
| .10 | .25 | 1.027 | 1.027 | 1.000 | .125 | 1.024 | .978 | 1.024 | .978 | 1.001 | 1.000 | .032 | .032 | .026 | .005 |
| | .50 | 1.071 | 1.071 | 1.002 | .210 | 1.075 | .939 | 1.075 | .939 | 1.003 | 1.005 | .141 | .141 | .122 | .024 |
| | .75 | 1.160 | 1.160 | 1.008 | .359 | 1.179 | .862 | 1.179 | .862 | 1.009 | 1.016 | .333 | .333 | .294 | .058 |
| | 1.00 | 1.469 | 1.469 | 1.027 | .672 | 1.482 | .608 | 1.482 | .608 | 1.027 | 1.051 | .738 | .738 | .722 | .147 |
| | 1.25 | 1.897 | 1.897 | 1.058 | 1.055 | 1.891 | .267 | 1.891 | .267 | 1.058 | 1.107 | 1.250 | 1.250 | 1.345 | .279 |
| | 1.50 | 2.233 | 2.233 | 1.097 | 1.327 | 2.218 | .022 | 2.218 | .022 | 1.099 | 1.172 | 1.657 | 1.657 | 2.010 | .424 |
| .25 | .25 | 1.180 | 1.180 | 1.008 | .324 | 1.156 | .851 | 1.156 | .851 | 1.009 | 1.019 | .095 | .095 | .080 | .015 |
| | .50 | 1.312 | 1.312 | 1.017 | .462 | 1.293 | .740 | 1.293 | .740 | 1.016 | 1.032 | .273 | .273 | .254 | .051 |
| | .75 | 1.522 | 1.522 | 1.029 | .667 | 1.523 | .562 | 1.523 | .563 | 1.029 | 1.054 | .537 | .537 | .525 | .106 |
| | 1.00 | 1.819 | 1.819 | 1.052 | .924 | 1.828 | .325 | 1.828 | .314 | 1.052 | 1.097 | .876 | .876 | .946 | .195 |
| | 1.25 | 2.149 | 2.149 | 1.084 | 1.198 | 2.146 | .080 | 2.146 | .116 | 1.084 | 1.145 | 1.260 | 1.260 | 1.502 | .315 |
| | 1.50 | 2.449 | 2.449 | 1.123 | 1.449 | 2.400 | -.113 | 2.400 | .117 | 1.124 | 1.180 | 1.654 | 1.654 | 2.163 | .458 |
| .50 | .25 | 1.570 | 1.570 | 1.030 | .624 | 1.585 | .518 | 1.585 | .518 | 1.029 | 1.054 | .158 | .158 | .154 | .031 |
| | .50 | 1.758 | 1.758 | 1.044 | .785 | 1.773 | .370 | 1.773 | .370 | 1.044 | 1.077 | .365 | .365 | .393 | .080 |
| | .75 | 1.994 | 1.994 | 1.064 | .984 | 2.015 | .188 | 2.015 | .188 | 1.064 | 1.119 | .624 | .624 | .720 | .149 |
| | 1.00 | 2.254 | 2.254 | 1.091 | 1.199 | 2.246 | .005 | 2.246 | .005 | 1.091 | 1.164 | .921 | .921 | 1.158 | .242 |
| | 1.25 | 2.513 | 2.513 | 1.123 | 1.419 | 2.481 | -.172 | 2.481 | -.172 | 1.124 | 1.214 | 1.255 | 1.255 | 1.720 | .362 |
| | 1.50 | 2.745 | 2.745 | 1.161 | 1.634 | 2.746 | -.339 | 2.746 | -.339 | 1.162 | 1.272 | 1.629 | 1.629 | 2.417 | .514 |
| 1.00 | .25 | 2.319 | 2.319 | 1.093 | 1.147 | 2.334 | -.050 | 2.334 | -.050 | 1.089 | 1.170 | .185 | .185 | .248 | .052 |
| | .50 | 2.482 | 2.482 | 1.113 | 1.304 | 2.500 | -.163 | 2.500 | -.163 | 1.110 | 1.205 | .393 | .393 | .571 | .120 |
| | .75 | 2.653 | 2.653 | 1.137 | 1.469 | 2.677 | -.278 | 2.677 | -.278 | 1.137 | 1.242 | .626 | .626 | .967 | .203 |
| | 1.00 | 2.825 | 2.825 | 1.167 | 1.639 | 2.832 | -.392 | 2.832 | -.392 | 1.166 | 1.282 | .891 | .891 | 1.469 | .307 |
| | 1.25 | 2.998 | 2.998 | 1.201 | 1.812 | 2.991 | -.505 | 2.991 | -.505 | 1.199 | 1.322 | 1.197 | 1.197 | 2.087 | .435 |
| | 1.50 | 3.174 | 3.174 | 1.235 | 1.985 | 3.209 | -.616 | 3.209 | -.616 | 1.241 | 1.356 | 1.556 | 1.556 | 2.822 | .588 |

## Table 23

### Combined uniform biaxial compressive and shear displacement ($\sigma_{x\,av} = \sigma_{y\,av} = \tau_{av}/2$)

Side ratio $a/b = 2{\cdot}0$

| W | P | U | V | S | H | $\bar{\sigma}_{MX}$ 1 | $\bar{\sigma}_{MX}$ 2 | $\bar{\sigma}_{MY}$ 1 | $\bar{\sigma}_{MY}$ 2 | $\bar{\tau}_{MXY}$ 1 | $\bar{\tau}_{MXY}$ 2 | $\bar{M}_X$ 2 | $\bar{M}_Y$ 2 | $\bar{M}_{XY}$ 1 | $\bar{M}_{XY}$ 2 |
|---|---|---|---|---|---|---|---|---|---|---|---|---|---|---|---|
| .10 | .25 | 1.001 | 1.038 | 1.001 | .117 | 1.020 | .989 | 1.022 | .979 | 1.003 | 1.002 | .010 | .018 | .007 | .001 |
| | .50 | 1.015 | 1.121 | 1.002 | .206 | 1.045 | .963 | 1.089 | .919 | 1.005 | 1.005 | .056 | .107 | .056 | .009 |
| | .75 | 1.018 | 1.261 | 1.005 | .361 | 1.072 | .920 | 1.212 | .802 | 1.005 | 1.010 | .138 | .263 | .141 | .023 |
| | 1.00 | 1.079 | 1.688 | 1.015 | .636 | 1.228 | .800 | 1.549 | .541 | 1.016 | 1.035 | .259 | .538 | .342 | .050 |
| | 1.25 | 1.211 | 2.341 | 1.039 | .961 | 1.474 | .667 | 2.081 | .245 | 1.038 | 1.073 | .374 | .364 | .694 | .086 |
| | 1.50 | 1.417 | 3.185 | 1.078 | 1.306 | 1.788 | .543 | 2.794 | -.039 | 1.070 | 1.118 | .465 | 1.216 | 1.205 | .131 |
| .25 | .25 | 1.035 | 1.278 | 1.007 | .325 | 1.101 | .905 | 1.234 | .779 | 1.006 | 1.010 | .043 | .076 | .032 | .006 |
| | .50 | 1.077 | 1.473 | 1.012 | .455 | 1.172 | .849 | 1.399 | .652 | 1.011 | 1.020 | .105 | .206 | .119 | .019 |
| | .75 | 1.127 | 1.806 | 1.019 | .647 | 1.268 | .760 | 1.636 | .453 | 1.020 | 1.037 | .192 | .399 | .263 | .039 |
| | 1.00 | 1.184 | 2.245 | 1.036 | .870 | 1.433 | .668 | 1.999 | .249 | 1.036 | 1.066 | .275 | .623 | .481 | .064 |
| | 1.25 | 1.279 | 2.769 | 1.061 | 1.112 | 1.648 | .584 | 2.439 | .085 | 1.060 | 1.106 | .350 | .869 | .803 | .093 |
| | 1.50 | 1.457 | 3.362 | 1.092 | 1.369 | 1.868 | .505 | 2.867 | -.011 | 1.095 | 1.158 | .425 | 1.141 | 1.259 | .127 |
| .50 | .25 | 1.134 | 1.895 | 1.020 | .619 | 1.302 | .732 | 1.729 | .377 | 1.021 | 1.037 | .058 | .119 | .075 | .012 |
| | .50 | 1.190 | 2.180 | 1.030 | .768 | 1.405 | .670 | 1.959 | .236 | 1.031 | 1.057 | .121 | .269 | .203 | .030 |
| | .75 | 1.257 | 2.545 | 1.045 | .945 | 1.529 | .603 | 2.252 | .085 | 1.042 | 1.083 | .190 | .450 | .383 | .053 |
| | 1.00 | 1.335 | 2.951 | 1.063 | 1.139 | 1.692 | .540 | 2.564 | -.035 | 1.064 | 1.118 | .254 | .547 | .636 | .078 |
| | 1.25 | 1.408 | 3.361 | 1.090 | 1.332 | 1.880 | .481 | 2.891 | -.117 | 1.095 | 1.160 | .313 | .850 | .954 | .104 |
| | 1.50 | 1.468 | 3.768 | 1.130 | 1.519 | 2.088 | .425 | 3.230 | -.160 | 1.136 | 1.207 | .368 | 1.056 | 1.333 | .129 |
| 1.00 | .25 | 1.306 | 3.034 | 1.054 | 1.131 | 1.712 | .497 | 2.633 | -.135 | 1.069 | 1.122 | .049 | .130 | .131 | .020 |
| | .50 | 1.365 | 3.304 | 1.093 | 1.269 | 1.822 | .463 | 2.859 | -.195 | 1.083 | 1.149 | .099 | .272 | .307 | .042 |
| | .75 | 1.417 | 3.587 | 1.180 | 1.414 | 1.947 | .427 | 3.083 | -.240 | 1.098 | 1.182 | .147 | .426 | .525 | .065 |
| | 1.00 | 1.526 | 3.920 | 1.147 | 1.575 | 2.090 | .387 | 3.372 | -.302 | 1.123 | 1.217 | .206 | .607 | .830 | .095 |
| | 1.25 | 1.643 | 4.276 | 1.130 | 1.744 | 2.241 | .336 | 3.682 | -.362 | 1.154 | 1.258 | .273 | .820 | 1.207 | .128 |
| | 1.50 | 1.734 | 4.637 | 1.227 | 1.913 | 2.390 | .273 | 3.983 | -.405 | 1.189 | 1.307 | .346 | 1.068 | 1.645 | .163 |

# Table 24

## Varying uniaxial compressive displacement
## Unloaded sides restrained

Side ratio $a/b = 0{\cdot}667$

| W | P | U | W | $\bar{\sigma}_{MX}$ | | $\bar{\sigma}_{MY}$ | | $\bar{M}_X$ | $\bar{M}_Y$ | $\bar{M}_{XY}$ |
|---|---|---|---|---|---|---|---|---|---|---|
| | | | | 4 | 4 | 2 | 3 | 4 | 4 | 1 |
| .10 | .25 | 1.000 | .128 | .558 | -.037 | -.013 | -.009 | .065 | .046 | .033 |
| | .50 | 1.000 | .193 | .538 | -.018 | -.027 | -.017 | .215 | .150 | .112 |
| | .75 | 1.000 | .298 | .497 | -.045 | -.052 | -.033 | .455 | .316 | .239 |
| | 1.00 | 1.000 | .430 | .447 | -.081 | -.089 | -.055 | .759 | .526 | .409 |
| | 1.25 | 1.000 | .573 | .395 | -.117 | -.133 | -.079 | 1.089 | .755 | .606 |
| | 1.50 | 1.000 | .707 | .348 | -.147 | -.174 | -.101 | 1.390 | .961 | .800 |
| .25 | .25 | 1.000 | .316 | .471 | -.066 | -.074 | -.046 | .151 | .106 | .080 |
| | .50 | 1.000 | .399 | .439 | -.088 | -.099 | -.060 | .342 | .238 | .185 |
| | .75 | 1.000 | .501 | .396 | -.115 | -.130 | -.078 | .576 | .399 | .320 |
| | 1.00 | 1.000 | .607 | .360 | -.141 | -.160 | -.095 | .817 | .567 | .465 |
| | 1.25 | 1.000 | .711 | .330 | -.165 | -.189 | -.110 | 1.051 | .731 | .616 |
| | 1.50 | 1.000 | .809 | .304 | -.185 | -.214 | -.123 | 1.271 | .882 | .768 |
| .50 | .25 | 1.000 | .579 | .338 | -.155 | -.176 | -.105 | .180 | .125 | .104 |
| | .50 | 1.000 | .661 | .313 | -.173 | -.198 | -.116 | .365 | .251 | .217 |
| | .75 | 1.000 | .746 | .286 | -.189 | -.219 | -.126 | .556 | .378 | .338 |
| | 1.00 | 1.000 | .830 | .267 | -.204 | -.237 | -.135 | .743 | .505 | .464 |
| | 1.25 | 1.000 | .911 | .254 | -.218 | -.253 | -.143 | .923 | .632 | .593 |
| | 1.50 | 1.000 | .989 | .244 | -.230 | -.268 | -.150 | 1.094 | .752 | .723 |
| 1.00 | .25 | 1.000 | 1.066 | .220 | -.259 | -.301 | -.167 | .146 | .100 | .103 |
| | .50 | 1.000 | 1.129 | .216 | -.266 | -.311 | -.171 | .287 | .198 | .208 |
| | .75 | 1.000 | 1.191 | .210 | -.271 | -.318 | -.174 | .423 | .294 | .314 |
| | 1.00 | 1.000 | 1.252 | .206 | -.276 | -.326 | -.176 | .554 | .386 | .421 |
| | 1.25 | 1.000 | 1.311 | .201 | -.281 | -.333 | -.179 | .680 | .473 | .529 |
| | 1.50 | 1.000 | 1.369 | .198 | -.285 | -.340 | -.181 | .803 | .557 | .638 |

# Table 25

## Varying uniaxial compressive displacement
## Unloaded sides restrained

Side ratio $a/b = 1{\cdot}0$

| W | P | U | W | $\bar{\sigma}_{MX}$ | | $\bar{\sigma}_{MY}$ | | $\bar{M}_X$ | $\bar{M}_Y$ | $\bar{M}_{XY}$ |
|---|---|---|---|---|---|---|---|---|---|---|
| | | | | 4 | 4 | 2 | 3 | 4 | 4 | 1 |
| .10 | .25 | 1.000 | .128 | .533 | -.011 | -.011 | -.010 | .032 | .033 | .020 |
| | .50 | 1.000 | .194 | .521 | -.024 | -.024 | -.020 | .110 | .113 | .069 |
| | .75 | 1.000 | .303 | .502 | -.046 | -.047 | -.039 | .237 | .244 | .149 |
| | 1.00 | 1.000 | .441 | .472 | -.078 | -.081 | -.067 | .393 | .409 | .258 |
| | 1.25 | 1.000 | .592 | .437 | -.114 | -.120 | -.099 | .558 | .587 | .384 |
| | 1.50 | 1.000 | .732 | .406 | -.147 | -.156 | -.127 | .701 | .749 | .513 |
| .25 | .25 | 1.000 | .317 | .484 | -.063 | -.066 | -.055 | .078 | .080 | .049 |
| | .50 | 1.000 | .403 | .465 | -.085 | -.088 | -.073 | .176 | .182 | .116 |
| | .75 | 1.000 | .510 | .437 | -.111 | -.116 | -.095 | .296 | .310 | .201 |
| | 1.00 | 1.000 | .621 | .414 | -.136 | -.144 | -.117 | .418 | .441 | .294 |
| | 1.25 | 1.000 | .730 | .396 | -.158 | -.169 | -.136 | .531 | .569 | .394 |
| | 1.50 | 1.000 | .833 | .381 | -.177 | -.190 | -.152 | .630 | .686 | .498 |
| .50 | .25 | 1.000 | .583 | .408 | -.149 | -.156 | -.128 | .094 | .099 | .065 |
| | .50 | 1.000 | .668 | .393 | -.166 | -.175 | -.142 | .188 | .200 | .137 |
| | .75 | 1.000 | .758 | .378 | -.182 | -.193 | -.156 | .284 | .305 | .215 |
| | 1.00 | 1.000 | .845 | .366 | -.196 | -.209 | -.167 | .373 | .407 | .297 |
| | 1.25 | 1.000 | .929 | .356 | -.207 | -.223 | -.177 | .454 | .503 | .383 |
| | 1.50 | 1.000 | 1.009 | .348 | -.217 | -.235 | -.185 | .524 | .594 | .474 |
| 1.00 | .25 | 1.000 | 1.070 | .337 | -.244 | -.264 | -.205 | .073 | .082 | .066 |
| | .50 | 1.000 | 1.137 | .333 | -.249 | -.271 | -.209 | .140 | .150 | .133 |
| | .75 | 1.000 | 1.202 | .328 | -.252 | -.276 | -.212 | .201 | .236 | .202 |
| | 1.00 | 1.000 | 1.265 | .324 | -.255 | -.281 | -.215 | .256 | .307 | .273 |
| | 1.25 | 1.000 | 1.327 | .319 | -.258 | -.286 | -.219 | .305 | .374 | .349 |
| | 1.50 | 1.000 | 1.387 | .317 | -.260 | -.291 | -.222 | .346 | .438 | .431 |

## Table 26

### Varying uniaxial compressive displacement
### Unloaded sides restrained

Side ratio $a/b = 1{\cdot}5$

| W | P | U | W | $\bar{\sigma}_{MX}$ | | $\bar{\sigma}_{MY}$ | | $\bar{M}_X$ | $\bar{M}_Y$ | $\bar{M}_{XY}$ |
|---|---|---|---|---|---|---|---|---|---|---|
| | | | | 4 | 4 | 2 | 3 | 4 | 4 | 1 |
| .10 | .25 | 1.000 | .128 | .549 | -.012 | -.011 | -.011 | .052 | .041 | .028 |
| | .50 | 1.000 | .196 | .530 | -.025 | -.025 | -.021 | .179 | .139 | .100 |
| | .75 | 1.000 | .307 | .498 | -.049 | -.049 | -.039 | .388 | .302 | .218 |
| | 1.00 | 1.000 | .447 | .453 | -.083 | -.084 | -.065 | .645 | .503 | .377 |
| | 1.25 | 1.000 | .599 | .401 | -.122 | -.125 | -.094 | .921 | .718 | .560 |
| | 1.50 | 1.000 | .740 | .355 | -.157 | -.163 | -.120 | 1.172 | .913 | .743 |
| .25 | .25 | 1.000 | .318 | .475 | -.066 | -.066 | -.052 | .127 | .098 | .072 |
| | .50 | 1.000 | .404 | .445 | -.088 | -.089 | -.068 | .287 | .223 | .168 |
| | .75 | 1.000 | .513 | .407 | -.116 | -.118 | -.090 | .485 | .378 | .291 |
| | 1.00 | 1.000 | .625 | .372 | -.144 | -.147 | -.110 | .689 | .537 | .428 |
| | 1.25 | 1.000 | .735 | .341 | -.168 | -.174 | -.128 | .887 | .691 | .573 |
| | 1.50 | 1.000 | .839 | .316 | -.189 | -.197 | -.144 | 1.068 | .834 | .720 |
| .50 | .25 | 1.000 | .583 | .358 | -.155 | -.158 | -.119 | .153 | .119 | .095 |
| | .50 | 1.000 | .669 | .334 | -.173 | -.178 | -.133 | .309 | .240 | .200 |
| | .75 | 1.000 | .759 | .311 | -.190 | -.197 | -.145 | .470 | .365 | .314 |
| | 1.00 | 1.000 | .848 | .293 | -.206 | -.215 | -.157 | .626 | .488 | .434 |
| | 1.25 | 1.000 | .934 | .279 | -.220 | -.230 | -.166 | .776 | .605 | .558 |
| | 1.50 | 1.000 | 1.017 | .267 | -.231 | -.244 | -.174 | .915 | .717 | .686 |
| 1.00 | .25 | 1.000 | 1.070 | .250 | -.257 | -.269 | -.191 | .123 | .097 | .098 |
| | .50 | 1.000 | 1.138 | .243 | -.264 | -.277 | -.195 | .242 | .191 | .198 |
| | .75 | 1.000 | 1.204 | .241 | -.270 | -.285 | -.199 | .355 | .285 | .300 |
| | 1.00 | 1.000 | 1.269 | .235 | -.275 | -.291 | -.203 | .462 | .371 | .406 |
| | 1.25 | 1.000 | 1.332 | .228 | -.279 | -.298 | -.206 | .565 | .452 | .516 |
| | 1.50 | 1.000 | 1.394 | .221 | -.283 | -.303 | -.209 | .662 | .530 | .629 |

## Table 27

### Varying uniaxial compressive displacement
### Unloaded sides restrained

Side ratio $a/b = 2{\cdot}0$

| W | P | U | W | $\bar{\sigma}_{MX}$ | | $\bar{\sigma}_{MY}$ | | $\bar{M}_X$ | $\bar{M}_Y$ | $\bar{M}_{XY}$ |
|---|---|---|---|---|---|---|---|---|---|---|
| | | | | 4 | 4 | 2 | 3 | 4 | 4 | 1 |
| .10 | .25 | 1.000 | .127 | .536 | -.011 | -.011 | -.009 | .032 | .032 | .019 |
| | .50 | 1.000 | .195 | .525 | -.023 | -.024 | -.020 | .111 | .114 | .069 |
| | .75 | 1.000 | .305 | .507 | -.046 | -.047 | -.040 | .239 | .247 | .150 |
| | 1.00 | 1.000 | .445 | .475 | -.079 | -.082 | -.069 | .398 | .414 | .260 |
| | 1.25 | 1.000 | .598 | .437 | -.116 | -.122 | -.101 | .565 | .593 | .388 |
| | 1.50 | 1.000 | .738 | .405 | -.149 | -.157 | -.130 | .708 | .756 | .517 |
| .25 | .25 | 1.000 | .317 | .491 | -.062 | -.064 | -.054 | .078 | .080 | .049 |
| | .50 | 1.000 | .403 | .468 | -.084 | -.087 | -.073 | .176 | .183 | .116 |
| | .75 | 1.000 | .512 | .441 | -.111 | -.115 | -.096 | .300 | .313 | .202 |
| | 1.00 | 1.000 | .625 | .419 | -.137 | -.143 | -.119 | .423 | .447 | .296 |
| | 1.25 | 1.000 | .735 | .401 | -.160 | -.169 | -.139 | .538 | .577 | .397 |
| | 1.50 | 1.000 | .839 | .385 | -.179 | -.191 | -.155 | .637 | .696 | .500 |
| .50 | .25 | 1.000 | .583 | .406 | -.148 | -.154 | -.128 | .095 | .099 | .065 |
| | .50 | 1.000 | .670 | .398 | -.166 | -.174 | -.143 | .190 | .202 | .137 |
| | .75 | 1.000 | .760 | .383 | -.182 | -.192 | -.157 | .287 | .309 | .216 |
| | 1.00 | 1.000 | .850 | .371 | -.196 | -.209 | -.170 | .379 | .414 | .300 |
| | 1.25 | 1.000 | .937 | .362 | -.209 | -.224 | -.181 | .462 | .513 | .387 |
| | 1.50 | 1.000 | 1.018 | .354 | -.218 | -.236 | -.190 | .532 | .605 | .477 |
| 1.00 | .25 | 1.000 | 1.070 | .343 | -.244 | -.260 | -.207 | .074 | .084 | .078 |
| | .50 | 1.000 | 1.139 | .338 | -.249 | -.268 | -.212 | .143 | .165 | .131 |
| | .75 | 1.000 | 1.207 | .334 | -.254 | -.274 | -.216 | .207 | .243 | .203 |
| | 1.00 | 1.000 | 1.271 | .330 | -.257 | -.280 | -.220 | .263 | .317 | .277 |
| | 1.25 | 1.000 | 1.333 | .326 | -.259 | -.285 | -.223 | .313 | .386 | .352 |
| | 1.50 | 1.000 | 1.392 | .323 | -.261 | -.289 | -.226 | .353 | .451 | .431 |

## Table 28

### Varying uniaxial compressive displacement Unloaded sides stress-free

Side ratio $a/b = 0{\cdot}667$

| W | P | U | W | $\bar{\sigma}_{MX}$ | | $\bar{\sigma}_{MY}$ | | $\bar{M}_X$ | $\bar{M}_Y$ | $\bar{M}_{XY}$ |
|---|---|---|---|---|---|---|---|---|---|---|
| | | | | 4 | 2 | 4 | 2,3 | 4 | 4 | 1 |
| .10 | .25 | 1.006 | .125 | .562 | 1.010 | -.003 | 0.000 | .058 | .040 | .029 |
| | .50 | 1.013 | .201 | .543 | 1.024 | -.008 | 0.000 | .235 | .164 | .123 |
| | .75 | 1.024 | .328 | .510 | 1.046 | -.017 | 0.000 | .528 | .370 | .277 |
| | 1.00 | 1.052 | .525 | .442 | 1.100 | -.036 | 0.000 | .978 | .684 | .546 |
| | 1.25 | 1.091 | .752 | .356 | 1.175 | -.059 | 0.000 | 1.492 | 1.041 | .890 |
| | 1.50 | 1.124 | .942 | .290 | 1.247 | -.075 | 0.000 | 1.904 | 1.324 | 1.230 |
| .25 | .25 | 1.031 | .321 | .487 | 1.061 | -.022 | 0.000 | .165 | .115 | .086 |
| | .50 | 1.045 | .428 | .449 | 1.089 | -.033 | 0.000 | .411 | .285 | .227 |
| | .75 | 1.067 | .572 | .398 | 1.132 | -.047 | 0.000 | .739 | .514 | .423 |
| | 1.00 | 1.095 | .734 | .336 | 1.186 | -.063 | 0.000 | 1.104 | .765 | .671 |
| | 1.25 | 1.126 | .902 | .276 | 1.245 | -.077 | 0.000 | 1.474 | 1.020 | .957 |
| | 1.50 | 1.151 | 1.059 | .237 | 1.301 | -.090 | 0.000 | 1.813 | 1.263 | 1.263 |
| .50 | .25 | 1.083 | .607 | .347 | 1.166 | -.060 | 0.000 | .245 | .171 | .139 |
| | .50 | 1.103 | .722 | .310 | 1.205 | -.071 | 0.000 | .507 | .354 | .309 |
| | .75 | 1.125 | .848 | .270 | 1.251 | -.081 | 0.000 | .789 | .550 | .514 |
| | 1.00 | 1.147 | .981 | .232 | 1.299 | -.092 | 0.000 | 1.082 | .755 | .749 |
| | 1.25 | 1.170 | 1.115 | .200 | 1.343 | -.101 | 0.000 | 1.372 | .961 | 1.011 |
| | 1.50 | 1.194 | 1.247 | .175 | 1.383 | -.108 | 0.000 | 1.651 | 1.167 | 1.297 |
| 1.00 | .25 | 1.181 | 1.104 | .182 | 1.357 | -.109 | 0.000 | .231 | .162 | .173 |
| | .50 | 1.198 | 1.210 | .169 | 1.391 | -.115 | 0.000 | .465 | .337 | .367 |
| | .75 | 1.215 | 1.318 | .159 | 1.425 | -.123 | 0.000 | .703 | .531 | .584 |
| | 1.00 | 1.231 | 1.425 | .150 | 1.457 | -.130 | 0.000 | .939 | .731 | .819 |
| | 1.25 | 1.246 | 1.531 | .139 | 1.486 | -.134 | 0.000 | 1.166 | .925 | 1.072 |
| | 1.50 | 1.260 | 1.636 | .126 | 1.511 | -.137 | 0.000 | 1.382 | 1.113 | 1.345 |

## Table 29

### Varying uniaxial compressive displacement Unloaded sides stress-free

Side ratio $a/b = 1{\cdot}0$

| W | P | U | W | $\bar{\sigma}_{MX}$ | | $\bar{\sigma}_{MY}$ | | $\bar{M}_X$ | $\bar{M}_Y$ | $\bar{M}_{XY}$ |
|---|---|---|---|---|---|---|---|---|---|---|
| | | | | 4 | 2 | 4 | 2,3 | 4 | 4 | 1 |
| .10 | .25 | 1.003 | .120 | .537 | 1.007 | -.002 | 0.000 | .023 | .024 | .014 |
| | .50 | 1.011 | .210 | .525 | 1.023 | -.007 | 0.000 | .128 | .132 | .081 |
| | .75 | 1.023 | .362 | .505 | 1.043 | -.016 | 0.000 | .306 | .315 | .186 |
| | 1.00 | 1.070 | .661 | .428 | 1.139 | -.042 | 0.000 | .635 | .658 | .453 |
| | 1.25 | 1.135 | 1.016 | .329 | 1.270 | -.074 | 0.000 | 1.001 | 1.052 | .754 |
| | 1.50 | 1.187 | 1.277 | .266 | 1.362 | -.093 | 0.000 | 1.220 | 1.320 | .881 |
| .25 | .25 | 1.028 | .324 | .498 | 1.057 | -.017 | 0.000 | .087 | .089 | .053 |
| | .50 | 1.048 | .460 | .464 | 1.095 | -.029 | 0.000 | .241 | .248 | .160 |
| | .75 | 1.079 | .661 | .411 | 1.154 | -.048 | 0.000 | .465 | .482 | .323 |
| | 1.00 | 1.124 | .907 | .343 | 1.247 | -.071 | 0.000 | .723 | .757 | .556 |
| | 1.25 | 1.176 | 1.167 | .276 | 1.350 | -.092 | 0.000 | .971 | 1.036 | .847 |
| | 1.50 | 1.226 | 1.405 | .226 | 1.430 | -.106 | 0.000 | 1.169 | 1.283 | 1.181 |
| .50 | .25 | 1.092 | .626 | .398 | 1.184 | -.056 | 0.000 | .144 | .150 | .099 |
| | .50 | 1.123 | .787 | .353 | 1.242 | -.071 | 0.000 | .318 | .332 | .239 |
| | .75 | 1.159 | .981 | .298 | 1.310 | -.089 | 0.000 | .521 | .546 | .422 |
| | 1.00 | 1.196 | 1.184 | .251 | 1.385 | -.104 | 0.000 | .717 | .767 | .649 |
| | 1.25 | 1.236 | 1.389 | .210 | 1.457 | -.114 | 0.000 | .897 | .982 | .922 |
| | 1.50 | 1.281 | 1.597 | .172 | 1.519 | -.121 | 0.000 | 1.053 | 1.183 | 1.247 |
| 1.00 | .25 | 1.217 | 1.148 | .226 | 1.426 | -.118 | 0.000 | .156 | .166 | .142 |
| | .50 | 1.244 | 1.306 | .198 | 1.479 | -.126 | 0.000 | .311 | .337 | .312 |
| | .75 | 1.273 | 1.471 | .173 | 1.533 | -.132 | 0.000 | .463 | .513 | .512 |
| | 1.00 | 1.300 | 1.635 | .152 | 1.576 | -.136 | 0.000 | .603 | .688 | .748 |
| | 1.25 | 1.326 | 1.797 | .133 | 1.614 | -.136 | 0.000 | .728 | .860 | 1.021 |
| | 1.50 | 1.351 | 1.956 | .115 | 1.647 | -.134 | 0.000 | .837 | 1.030 | 1.333 |

# Table 30

## Varying uniaxial compressive displacement
## Unloaded sides stress-free

Side ratio $a/b = 1{\cdot}5$

| W | P | U | W | $\bar{\sigma}_{MX}$ | | $\bar{\sigma}_{MY}$ | | $\bar{M}_X$ | $\bar{M}_Y$ | $\bar{M}_{XY}$ |
|---|---|---|---|---|---|---|---|---|---|---|
| | | | | 4 | 2 | 4 | 2,3 | 4 | 4 | 1 |
| .10 | .25 | 1.005 | .123 | .554 | 1.006 | -.003 | 0.000 | .043 | .034 | .022 |
| | .50 | 1.012 | .207 | .534 | 1.019 | -.008 | 0.000 | .199 | .155 | .112 |
| | .75 | 1.024 | .351 | .500 | 1.046 | -.018 | 0.000 | .467 | .364 | .265 |
| | 1.00 | 1.050 | .570 | .426 | 1.098 | -.038 | 0.000 | .870 | .671 | .530 |
| | 1.25 | 1.087 | .822 | .334 | 1.166 | -.062 | 0.000 | 1.321 | 1.015 | .875 |
| | 1.50 | 1.125 | 1.033 | .270 | 1.235 | -.081 | 0.000 | 1.670 | 1.288 | 1.228 |
| .25 | .25 | 1.028 | .324 | .488 | 1.049 | -.021 | 0.000 | .138 | .107 | .077 |
| | .50 | 1.043 | .440 | .447 | 1.079 | -.031 | 0.000 | .352 | .272 | .210 |
| | .75 | 1.062 | .601 | .389 | 1.123 | -.047 | 0.000 | .647 | .499 | .402 |
| | 1.00 | 1.091 | .783 | .327 | 1.175 | -.065 | 0.000 | .969 | .749 | .651 |
| | 1.25 | 1.123 | .971 | .270 | 1.229 | -.082 | 0.000 | 1.291 | 1.001 | .945 |
| | 1.50 | 1.150 | 1.144 | .225 | 1.279 | -.094 | 0.000 | 1.583 | 1.228 | 1.265 |
| .50 | .25 | 1.081 | .611 | .354 | 1.149 | -.058 | 0.000 | .204 | .157 | .128 |
| | .50 | 1.099 | .739 | .311 | 1.183 | -.070 | 0.000 | .434 | .334 | .293 |
| | .75 | 1.115 | .883 | .265 | 1.220 | -.082 | 0.000 | .691 | .531 | .497 |
| | 1.00 | 1.140 | 1.033 | .227 | 1.265 | -.094 | 0.000 | .951 | .735 | .737 |
| | 1.25 | 1.169 | 1.184 | .197 | 1.313 | -.104 | 0.000 | 1.205 | .940 | 1.014 |
| | 1.50 | 1.195 | 1.334 | .171 | 1.356 | -.112 | 0.000 | 1.442 | 1.145 | 1.330 |
| 1.00 | .25 | 1.172 | 1.115 | .186 | 1.328 | -.110 | 0.000 | .231 | .160 | .169 |
| | .50 | 1.191 | 1.233 | .170 | 1.352 | -.115 | 0.000 | .394 | .327 | .364 |
| | .75 | 1.209 | 1.355 | .156 | 1.386 | -.120 | 0.000 | .618 | .505 | .586 |
| | 1.00 | 1.229 | 1.478 | .136 | 1.417 | -.120 | 0.000 | .828 | .701 | .835 |
| | 1.25 | 1.245 | 1.601 | .117 | 1.445 | -.120 | 0.000 | 1.019 | .891 | 1.110 |
| | 1.50 | 1.255 | 1.719 | .106 | 1.468 | -.123 | 0.000 | 1.197 | 1.053 | 1.411 |

# Table 31

## Varying uniaxial compressive displacement
## Unloaded sides stress-free

Side ratio $a/b = 2{\cdot}0$

| W | P | U | W | $\bar{\sigma}_{MX}$ | | $\bar{\sigma}_{MY}$ | | $\bar{M}_X$ | $\bar{M}_Y$ | $\bar{M}_{XY}$ |
|---|---|---|---|---|---|---|---|---|---|---|
| | | | | 4 | 2 | 4 | 2,3 | 4 | 4 | 1 |
| .10 | .25 | 1.005 | .120 | .536 | 1.007 | -.002 | 0.000 | .023 | .024 | .017 |
| | .50 | 1.012 | .209 | .522 | 1.021 | -.007 | 0.000 | .127 | .130 | .079 |
| | .75 | 1.024 | .360 | .498 | 1.047 | -.015 | 0.000 | .303 | .309 | .192 |
| | 1.00 | 1.070 | .659 | .428 | 1.137 | -.041 | 0.000 | .634 | .657 | .435 |
| | 1.25 | 1.135 | 1.014 | .335 | 1.261 | -.074 | 0.000 | 1.000 | 1.054 | .773 |
| | 1.50 | 1.193 | 1.282 | .264 | 1.373 | -.093 | 0.000 | 1.224 | 1.323 | 1.140 |
| .25 | .25 | 1.029 | .323 | .498 | 1.054 | -.016 | 0.000 | .086 | .088 | .044 |
| | .50 | 1.050 | .459 | .465 | 1.095 | -.028 | 0.000 | .240 | .248 | .163 |
| | .75 | 1.079 | .659 | .414 | 1.158 | -.048 | 0.000 | .465 | .482 | .322 |
| | 1.00 | 1.130 | .912 | .346 | 1.252 | -.072 | 0.000 | .728 | .763 | .560 |
| | 1.25 | 1.183 | 1.172 | .278 | 1.350 | -.093 | 0.000 | .977 | 1.043 | .851 |
| | 1.50 | 1.214 | 1.394 | .227 | 1.419 | -.104 | 0.000 | 1.158 | 1.271 | 1.163 |
| .50 | .25 | 1.092 | .625 | .393 | 1.179 | -.054 | 0.000 | .142 | .145 | .097 |
| | .50 | 1.122 | .785 | .350 | 1.238 | -.070 | 0.000 | .316 | .328 | .236 |
| | .75 | 1.157 | .979 | .306 | 1.310 | -.088 | 0.000 | .519 | .548 | .419 |
| | 1.00 | 1.194 | 1.184 | .256 | 1.378 | -.103 | 0.000 | .717 | .768 | .644 |
| | 1.25 | 1.234 | 1.389 | .210 | 1.445 | -.113 | 0.000 | .897 | .977 | .916 |
| | 1.50 | 1.273 | 1.589 | .174 | 1.518 | -.120 | 0.000 | 1.055 | 1.172 | 1.240 |
| 1.00 | .25 | 1.217 | 1.150 | .236 | 1.422 | -.116 | 0.000 | .158 | .168 | .141 |
| | .50 | 1.244 | 1.307 | .207 | 1.473 | -.125 | 0.000 | .313 | .340 | .312 |
| | .75 | 1.269 | 1.470 | .177 | 1.520 | -.131 | 0.000 | .464 | .513 | .511 |
| | 1.00 | 1.300 | 1.637 | .155 | 1.568 | -.135 | 0.000 | .606 | .688 | .744 |
| | 1.25 | 1.348 | 1.810 | .136 | 1.640 | -.138 | 0.000 | .741 | .875 | 1.056 |
| | 1.50 | 1.411 | 1.990 | .121 | 1.733 | -.142 | 0.000 | .869 | 1.073 | 1.439 |

# Table 32

## Bending displacement
## Unloaded sides restrained

Side ratio $a/b = 0{\cdot}667$

| W | P | U | W | $\bar{\sigma}$ MX | | $\bar{\sigma}$ MY | | $\bar{M}$ X | $\bar{M}$ Y | $\bar{M}$ XY |
|---|---|---|---|---|---|---|---|---|---|---|
| | | | | 4 | 4 | 2 | 3 | 4 | 4 | 1 |
| .10 | .25 | 1.000 | .129 | .403 | -.007 | -.009 | -.004 | .077 | .076 | .051 |
| | .50 | 1.000 | .199 | .397 | -.014 | -.018 | -.006 | .258 | .262 | .174 |
| | .75 | 1.000 | .309 | .385 | -.028 | -.034 | -.009 | .548 | .560 | .372 |
| | 1.00 | 1.000 | .464 | .364 | -.051 | -.062 | -.015 | .947 | .990 | .668 |
| | 1.25 | 1.000 | .636 | .339 | -.078 | -.095 | -.023 | 1.374 | 1.468 | 1.021 |
| | 1.50 | 1.000 | .777 | .318 | -.098 | -.121 | -.028 | 1.690 | 1.854 | 1.357 |
| .25 | .25 | 1.000 | .318 | .376 | -.038 | -.046 | -.012 | .179 | .183 | .122 |
| | .50 | 1.000 | .410 | .363 | -.052 | -.063 | -.016 | .417 | .433 | .294 |
| | .75 | 1.000 | .526 | .344 | -.071 | -.086 | -.021 | .717 | .751 | .518 |
| | 1.00 | 1.000 | .649 | .328 | -.089 | -.108 | -.026 | 1.018 | 1.092 | .770 |
| | 1.25 | 1.000 | .769 | .313 | -.106 | -.129 | -.030 | 1.300 | 1.435 | 1.045 |
| | 1.50 | 1.000 | .880 | .301 | -.120 | -.148 | -.034 | 1.542 | 1.751 | 1.338 |
| .50 | .25 | 1.000 | .587 | .336 | -.094 | -.114 | -.027 | .227 | .244 | .168 |
| | .50 | 1.000 | .680 | .311 | -.107 | -.130 | -.030 | .460 | .507 | .357 |
| | .75 | 1.000 | .778 | .306 | -.121 | -.146 | -.033 | .701 | .790 | .570 |
| | 1.00 | 1.000 | .873 | .297 | -.131 | -.160 | -.036 | .924 | 1.062 | .793 |
| | 1.25 | 1.000 | .965 | .286 | -.140 | -.172 | -.039 | 1.124 | 1.322 | 1.030 |
| | 1.50 | 1.000 | 1.056 | .282 | -.148 | -.183 | -.042 | 1.299 | 1.573 | 1.291 |
| 1.00 | .25 | 1.000 | 1.076 | .298 | -.168 | -.201 | -.045 | .190 | .239 | .175 |
| | .50 | 1.000 | 1.150 | .293 | -.173 | -.207 | -.046 | .366 | .469 | .358 |
| | .75 | 1.000 | 1.223 | .287 | -.178 | -.213 | -.048 | .528 | .689 | .549 |
| | 1.00 | 1.000 | 1.294 | .282 | -.180 | -.219 | -.049 | .671 | .893 | .748 |
| | 1.25 | 1.000 | 1.363 | .277 | -.182 | -.224 | -.050 | .790 | 1.085 | .958 |
| | 1.50 | 1.000 | 1.428 | .274 | -.183 | -.229 | -.051 | .881 | 1.271 | 1.189 |

# Table 33

## Bending displacement
## Unloaded sides restrained

Side ratio $a/b = 1{\cdot}0$

| W | P | U | W | $\bar{\sigma}$ MX | | $\bar{\sigma}$ MY | | $\bar{M}$ X | $\bar{M}$ Y | $\bar{M}$ XY |
|---|---|---|---|---|---|---|---|---|---|---|
| | | | | 4 | 4 | 2 | 3 | 4 | 4 | 1 |
| .10 | .25 | 1.000 | .130 | .454 | -.007 | -.007 | -.003 | .128 | .104 | .080 |
| | .50 | 1.000 | .209 | .440 | -.017 | -.018 | -.006 | .459 | .377 | .301 |
| | .75 | 1.000 | .338 | .414 | -.036 | -.037 | -.012 | 1.003 | .825 | .665 |
| | 1.00 | 1.000 | .505 | .378 | -.066 | -.066 | -.021 | 1.702 | 1.421 | 1.191 |
| | 1.25 | 1.000 | .683 | .338 | -.099 | -.100 | -.032 | 2.431 | 2.061 | 1.809 |
| | 1.50 | 1.000 | .828 | .307 | -.124 | -.125 | -.040 | 2.993 | 2.584 | 2.401 |
| .25 | .25 | 1.000 | .323 | .404 | -.043 | -.044 | -.014 | .310 | .254 | .205 |
| | .50 | 1.000 | .423 | .382 | -.061 | -.062 | -.020 | .727 | .604 | .504 |
| | .75 | 1.000 | .549 | .352 | -.085 | -.086 | -.028 | 1.255 | 1.053 | .902 |
| | 1.00 | 1.000 | .681 | .324 | -.108 | -.109 | -.035 | 1.789 | 1.522 | 1.361 |
| | 1.25 | 1.000 | .809 | .301 | -.129 | -.130 | -.042 | 2.296 | 1.982 | 1.865 |
| | 1.50 | 1.000 | .926 | .282 | -.146 | -.147 | -.048 | 2.741 | 2.403 | 2.393 |
| .50 | .25 | 1.000 | .593 | .326 | -.059 | -.109 | -.035 | .392 | .333 | .291 |
| | .50 | 1.000 | .693 | .308 | -.145 | -.125 | -.041 | .802 | .690 | .630 |
| | .75 | 1.000 | .798 | .291 | -.142 | -.141 | -.046 | 1.231 | 1.073 | 1.017 |
| | 1.00 | 1.000 | .902 | .275 | -.150 | -.155 | -.050 | 1.637 | 1.446 | 1.438 |
| | 1.25 | 1.000 | 1.001 | .263 | -.170 | -.167 | -.055 | 2.012 | 1.807 | 1.895 |
| | 1.50 | 1.000 | 1.096 | .254 | -.183 | -.178 | -.058 | 2.352 | 2.158 | 2.394 |
| 1.00 | .25 | 1.000 | 1.084 | .261 | -.192 | -.188 | -.062 | .335 | .319 | .330 |
| | .50 | 1.000 | 1.167 | .255 | -.198 | -.195 | -.064 | .654 | .629 | .686 |
| | .75 | 1.000 | 1.248 | .247 | -.204 | -.201 | -.066 | .955 | .931 | 1.067 |
| | 1.00 | 1.000 | 1.326 | .240 | -.209 | -.207 | -.068 | 1.233 | 1.218 | 1.476 |
| | 1.25 | 1.000 | 1.401 | .235 | -.213 | -.212 | -.070 | 1.484 | 1.490 | 1.922 |
| | 1.50 | 1.000 | 1.475 | .230 | -.216 | -.218 | -.071 | 1.706 | 1.753 | 2.417 |

## Table 34

### Bending displacement
### Unloaded sides restrained

Side ratio $a/b = 1{\cdot}5$

| W | P | U | W | $\bar{\sigma}_{MX}$ 4 | $\bar{\sigma}_{MX}$ 4 | $\bar{\sigma}_{MY}$ 2 | $\bar{\sigma}_{MY}$ 3 | $\bar{M}_X$ 4 | $\bar{M}_Y$ 4 | $\bar{M}_{XY}$ 1 |
|---|---|---|---|---|---|---|---|---|---|---|
| .10 | .25 | 1.000 | .130 | .383 | -.006 | -.006 | -.003 | .063 | .069 | .047 |
| | .50 | 1.000 | .206 | .379 | -.013 | -.014 | -.006 | .221 | .250 | .167 |
| | .75 | 1.000 | .328 | .367 | -.026 | -.029 | -.012 | .478 | .544 | .363 |
| | 1.00 | 1.000 | .495 | .350 | -.048 | -.053 | -.022 | .816 | .956 | .654 |
| | 1.25 | 1.000 | .678 | .331 | -.074 | -.081 | -.034 | 1.146 | 1.405 | 1.024 |
| | 1.50 | 1.000 | .824 | .311 | -.091 | -.104 | -.044 | 1.324 | 1.752 | 1.442 |
| .25 | .25 | 1.000 | .323 | .361 | -.034 | -.037 | -.015 | .154 | .174 | .118 |
| | .50 | 1.000 | .422 | .351 | -.047 | -.052 | -.022 | .355 | .414 | .288 |
| | .75 | 1.000 | .547 | .336 | -.065 | -.071 | -.030 | .606 | .723 | .513 |
| | 1.00 | 1.000 | .678 | .322 | -.082 | -.091 | -.038 | .849 | 1.049 | .781 |
| | 1.25 | 1.000 | .806 | .310 | -.097 | -.109 | -.046 | 1.044 | 1.351 | 1.107 |
| | 1.50 | 1.000 | .917 | .296 | -.107 | -.124 | -.052 | 1.140 | 1.618 | 1.516 |
| .50 | .25 | 1.000 | .594 | .324 | -.085 | -.092 | -.038 | .193 | .234 | .167 |
| | .50 | 1.000 | .693 | .316 | -.097 | -.106 | -.044 | .385 | .484 | .364 |
| | .75 | 1.000 | .798 | .307 | -.109 | -.120 | -.050 | .577 | .748 | .590 |
| | 1.00 | 1.000 | .901 | .298 | -.118 | -.132 | -.055 | .738 | 1.004 | .852 |
| | 1.25 | 1.000 | .998 | .290 | -.125 | -.142 | -.060 | .849 | 1.238 | 1.170 |
| | 1.50 | 1.000 | 1.085 | .281 | -.129 | -.151 | -.064 | .885 | 1.432 | 1.580 |
| 1.00 | .25 | 1.000 | 1.082 | .308 | -.151 | -.163 | -.067 | .152 | .229 | .189 |
| | .50 | 1.000 | 1.162 | .303 | -.155 | -.170 | -.070 | .283 | .444 | .398 |
| | .75 | 1.000 | 1.239 | .297 | -.157 | -.174 | -.072 | .393 | .646 | .629 |
| | 1.00 | 1.000 | 1.313 | .291 | -.158 | -.179 | -.075 | .470 | .833 | .899 |
| | 1.25 | 1.000 | 1.382 | .286 | -.158 | -.183 | -.077 | .506 | .998 | 1.227 |
| | 1.50 | 1.000 | 1.446 | .281 | -.155 | -.187 | -.079 | .487 | 1.134 | 1.644 |

## Table 35

### Bending displacement
### Unloaded sides stress-free

Side ratio $a/b = 0{\cdot}667$

| W | P | U | W | $\bar{\sigma}_{MX}$ 4 | $\bar{\sigma}_{MX}$ 2 | $\bar{\sigma}_{MY}$ 4 | $\bar{\sigma}_{MY}$ 2,3 | $\bar{M}_X$ 4 | $\bar{M}_Y$ 4 | $\bar{M}_{XY}$ 1 |
|---|---|---|---|---|---|---|---|---|---|---|
| .10 | .25 | 1.002 | .127 | .405 | 1.006 | -.002 | 0.000 | .071 | .070 | .046 |
| | .50 | 1.008 | .210 | .400 | 1.017 | -.005 | 0.000 | .287 | .291 | .194 |
| | .75 | 1.017 | .349 | .388 | 1.036 | -.011 | 0.000 | .651 | .667 | .447 |
| | 1.00 | 1.051 | .636 | .350 | 1.101 | -.028 | 0.000 | 1.382 | 1.464 | 1.028 |
| | 1.25 | 1.099 | .968 | .300 | 1.192 | -.050 | 0.000 | 2.183 | 2.416 | 1.840 |
| | 1.50 | 1.148 | 1.246 | .256 | 1.288 | -.067 | 0.000 | 2.771 | 3.260 | 2.776 |
| .25 | .25 | 1.020 | .325 | .380 | 1.043 | -.013 | 0.000 | .196 | .197 | .131 |
| | .50 | 1.035 | .455 | .365 | 1.071 | -.021 | 0.000 | .533 | .554 | .383 |
| | .75 | 1.059 | .642 | .341 | 1.116 | -.033 | 0.000 | 1.012 | 1.076 | .761 |
| | 1.00 | 1.093 | .874 | .307 | 1.182 | -.049 | 0.000 | 1.584 | 1.751 | 1.312 |
| | 1.25 | 1.119 | 1.123 | .271 | 1.257 | -.064 | 0.000 | 2.147 | 2.493 | 2.035 |
| | 1.50 | 1.126 | 1.370 | .236 | 1.333 | -.075 | 0.000 | 2.642 | 3.251 | 2.926 |
| .50 | .25 | 1.069 | .620 | .330 | 1.139 | -.039 | 0.000 | .313 | .334 | .233 |
| | .50 | 1.091 | .771 | .311 | 1.179 | -.049 | 0.000 | .700 | .775 | .565 |
| | .75 | 1.119 | .952 | .290 | 1.232 | -.061 | 0.000 | 1.153 | 1.320 | 1.001 |
| | 1.00 | 1.152 | 1.154 | .262 | 1.296 | -.072 | 0.000 | 1.610 | 1.927 | 1.567 |
| | 1.25 | 1.180 | 1.353 | .238 | 1.348 | -.080 | 0.000 | 2.023 | 2.550 | 2.260 |
| | 1.50 | 1.199 | 1.545 | .220 | 1.385 | -.083 | 0.000 | 2.384 | 3.179 | 3.080 |
| 1.00 | .25 | 1.166 | 1.140 | .267 | 1.325 | -.082 | 0.000 | .358 | .448 | .340 |
| | .50 | 1.189 | 1.291 | .257 | 1.370 | -.088 | 0.000 | .728 | .953 | .759 |
| | .75 | 1.213 | 1.452 | .247 | 1.414 | -.092 | 0.000 | 1.098 | 1.501 | 1.262 |
| | 1.00 | 1.238 | 1.617 | .226 | 1.446 | -.094 | 0.000 | 1.438 | 2.062 | 1.876 |
| | 1.25 | 1.265 | 1.784 | .221 | 1.490 | -.093 | 0.000 | 1.751 | 2.682 | 2.630 |
| | 1.50 | 1.294 | 1.953 | .225 | 1.539 | -.089 | 0.000 | 2.036 | 3.348 | 3.514 |

## Table 36

### Bending displacement
### Unloaded sides stress-free

Side ratio $a/b = 1{\cdot}0$

| W | P | U | W | $\bar{\sigma}_{MX}$ 4 | $\bar{\sigma}_{MX}$ 2 | $\bar{\sigma}_{MY}$ 4 | $\bar{\sigma}_{MY}$ 2,3 | $\bar{M}_X$ 4 | $\bar{M}_Y$ 4 | $\bar{M}_{XY}$ 1 |
|---|---|---|---|---|---|---|---|---|---|---|
| .10 | .25 | .998 | .127 | .456 | 1.004 | -.002 | 0.000 | .112 | .090 | .068 |
| | .50 | 1.004 | .219 | .442 | 1.015 | -.005 | 0.000 | .503 | .413 | .332 |
| | .75 | 1.021 | .379 | .417 | 1.041 | -.013 | 0.000 | 1.180 | .974 | .791 |
| | 1.00 | 1.049 | .635 | .363 | 1.091 | -.024 | 0.000 | 2.252 | 1.903 | 1.666 |
| | 1.25 | 1.085 | .917 | .302 | 1.156 | -.041 | 0.000 | 3.403 | 2.967 | 2.835 |
| | 1.50 | 1.126 | 1.143 | .264 | 1.223 | -.066 | 0.000 | 4.259 | 3.880 | 4.126 |
| .25 | .25 | 1.020 | .328 | .465 | 1.040 | -.002 | 0.000 | .117 | .094 | .215 |
| | .50 | 1.034 | .457 | .453 | 1.066 | -.006 | 0.000 | .534 | .439 | .617 |
| | .75 | 1.054 | .636 | .426 | 1.104 | -.014 | 0.000 | 1.266 | 1.048 | 1.212 |
| | 1.00 | 1.083 | .841 | .366 | 1.153 | -.027 | 0.000 | 2.386 | 2.023 | 2.041 |
| | 1.25 | 1.116 | 1.061 | .303 | 1.209 | -.045 | 0.000 | 3.528 | 3.090 | 3.094 |
| | 1.50 | 1.152 | 1.289 | .265 | 1.269 | -.070 | 0.000 | 4.354 | 3.987 | 4.356 |
| .50 | .25 | 1.064 | .618 | .335 | 1.121 | -.037 | 0.000 | .498 | .422 | .372 |
| | .50 | 1.082 | .759 | .307 | 1.154 | -.046 | 0.000 | 1.085 | .964 | .886 |
| | .75 | 1.101 | .924 | .277 | 1.186 | -.057 | 0.000 | 1.755 | 1.618 | 1.548 |
| | 1.00 | 1.129 | 1.106 | .253 | 1.235 | -.069 | 0.000 | 2.450 | 2.291 | 2.395 |
| | 1.25 | 1.163 | 1.291 | .226 | 1.287 | -.079 | 0.000 | 3.115 | 2.973 | 3.458 |
| | 1.50 | 1.197 | 1.469 | .190 | 1.335 | -.087 | 0.000 | 3.725 | 3.667 | 4.749 |
| 1.00 | .25 | 1.149 | 1.131 | .233 | 1.270 | -.080 | 0.000 | .534 | .519 | .536 |
| | .50 | 1.168 | 1.268 | .213 | 1.302 | -.087 | 0.000 | 1.075 | 1.064 | 1.188 |
| | .75 | 1.187 | 1.411 | .192 | 1.330 | -.094 | 0.000 | 1.619 | 1.636 | 1.968 |
| | 1.00 | 1.215 | 1.558 | .174 | 1.373 | -.101 | 0.000 | 2.151 | 2.235 | 2.931 |
| | 1.25 | 1.246 | 1.710 | .157 | 1.416 | -.107 | 0.000 | 2.652 | 2.838 | 4.091 |
| | 1.50 | 1.282 | 1.868 | .141 | 1.463 | -.114 | 0.000 | 3.125 | 3.451 | 5.446 |

## Table 37

### Bending displacement
### Unloaded sides stress-free

Side ratio $a/b = 1{\cdot}5$

| W | P | U | W | $\bar{\sigma}_{MX}$ 4 | $\bar{\sigma}_{MX}$ 2 | $\bar{\sigma}_{MY}$ 4 | $\bar{\sigma}_{MY}$ 2,3 | $\bar{M}_X$ 4 | $\bar{M}_Y$ 4 | $\bar{M}_{XY}$ 1 |
|---|---|---|---|---|---|---|---|---|---|---|
| .10 | .25 | .999 | .124 | .383 | 1.006 | -.001 | 0.000 | .051 | .056 | .037 |
| | .50 | 1.004 | .220 | .378 | 1.015 | -.004 | 0.000 | .253 | .284 | .191 |
| | .75 | 1.017 | .385 | .369 | 1.033 | -.009 | 0.000 | .601 | .680 | .456 |
| | 1.00 | 1.052 | .709 | .332 | 1.097 | -.024 | 0.000 | 1.256 | 1.495 | 1.082 |
| | 1.25 | 1.104 | 1.087 | .284 | 1.185 | -.039 | 0.000 | 1.922 | 2.461 | 2.106 |
| | 1.50 | 1.167 | 1.395 | .239 | 1.266 | -.045 | 0.000 | 2.271 | 3.265 | 3.532 |
| .25 | .25 | 1.022 | .329 | .366 | 1.042 | -.009 | 0.000 | .166 | .185 | .123 |
| | .50 | 1.037 | .474 | .352 | 1.070 | -.016 | 0.000 | .466 | .541 | .382 |
| | .75 | 1.059 | .687 | .329 | 1.111 | -.026 | 0.000 | .902 | 1.074 | .781 |
| | 1.00 | 1.095 | .955 | .293 | 1.172 | -.038 | 0.000 | 1.407 | 1.764 | 1.423 |
| | 1.25 | 1.144 | 1.243 | .256 | 1.244 | -.047 | 0.000 | 1.853 | 2.524 | 2.433 |
| | 1.50 | 1.203 | 1.525 | .228 | 1.318 | -.049 | 0.000 | 2.151 | 3.285 | 3.891 |
| .50 | .25 | 1.058 | .630 | .323 | 1.117 | -.028 | 0.000 | .271 | .321 | .226 |
| | .50 | 1.082 | .800 | .303 | 1.159 | -.036 | 0.000 | .611 | .757 | .581 |
| | .75 | 1.117 | 1.011 | .278 | 1.210 | -.046 | 0.000 | 1.013 | 1.305 | 1.074 |
| | 1.00 | 1.154 | 1.240 | .255 | 1.267 | -.053 | 0.000 | 1.400 | 1.924 | 1.791 |
| | 1.25 | 1.199 | 1.478 | .231 | 1.326 | -.056 | 0.000 | 1.716 | 2.585 | 2.872 |
| | 1.50 | 1.255 | 1.723 | .206 | 1.383 | -.053 | 0.000 | 1.944 | 3.281 | 4.360 |
| 1.00 | .25 | 1.156 | 1.157 | .267 | 1.287 | -.060 | 0.000 | .316 | .440 | .360 |
| | .50 | 1.182 | 1.329 | .251 | 1.329 | -.065 | 0.000 | .634 | .928 | .844 |
| | .75 | 1.213 | 1.515 | .236 | 1.369 | -.068 | 0.000 | .940 | 1.463 | 1.472 |
| | 1.00 | 1.252 | 1.709 | .222 | 1.412 | -.069 | 0.000 | 1.227 | 2.032 | 2.359 |
| | 1.25 | 1.303 | 1.913 | .212 | 1.478 | -.064 | 0.000 | 1.601 | 2.643 | 3.651 |
| | 1.50 | 1.365 | 2.128 | .205 | 1.562 | -.056 | 0.000 | 2.036 | 3.293 | 5.310 |

## Table 38

### Bending stress
### Unloaded sides stress-free

Side ratio $a/b = 0{\cdot}667$

| W | P | U | W | $\bar{\sigma}$ MX | | $\bar{\sigma}$ MY | | $\bar{M}$ X | $\bar{M}$ Y | $\bar{M}$ XY |
|---|---|---|---|---|---|---|---|---|---|---|
| | | | | 4 | 2 | 4 | 2,3 | 4 | 4 | 1 |
| .10 | .25 | 1.004 | .120 | .400 | 1.003 | -.001 | 0.000 | .038 | .048 | .056 |
| | .50 | 1.013 | .237 | .396 | 1.019 | -.008 | 0.000 | .313 | .343 | .298 |
| | .75 | 1.023 | .435 | .387 | 1.042 | -.019 | 0.000 | .801 | .854 | .692 |
| | 1.00 | 1.110 | .982 | .326 | 1.214 | -.083 | 0.000 | 1.926 | 2.120 | 1.996 |
| | 1.25 | 1.260 | 1.732 | .232 | 1.478 | -.168 | 0.000 | 3.158 | 3.703 | 4.633 |
| | 1.50 | 1.433 | 2.382 | .153 | 1.706 | -.211 | 0.000 | 3.520 | 4.755 | 9.122 |
| .25 | .25 | 1.021 | .326 | .386 | 1.036 | -.013 | 0.000 | .165 | .190 | .179 |
| | .50 | 1.049 | .528 | .369 | 1.088 | -.033 | 0.000 | .606 | .680 | .660 |
| | .75 | 1.097 | .857 | .334 | 1.180 | -.069 | 0.000 | 1.337 | 1.481 | 1.431 |
| | 1.00 | 1.197 | 1.348 | .271 | 1.359 | -.128 | 0.000 | 2.169 | 2.522 | 2.957 |
| | 1.25 | 1.340 | 1.938 | .195 | 1.582 | -.187 | 0.000 | 2.862 | 3.624 | 5.742 |
| | 1.50 | 1.500 | 2.517 | .131 | 1.770 | -.212 | 0.000 | 3.064 | 4.508 | 10.507 |
| .50 | .25 | 1.084 | .663 | .345 | 1.149 | -.053 | 0.000 | .346 | .399 | .395 |
| | .50 | 1.137 | .928 | .312 | 1.241 | -.085 | 0.000 | .851 | 1.005 | 1.168 |
| | .75 | 1.216 | 1.299 | .261 | 1.379 | -.131 | 0.000 | 1.532 | 1.831 | 2.311 |
| | 1.00 | 1.323 | 1.744 | .205 | 1.546 | -.174 | 0.000 | 2.097 | 2.679 | 4.220 |
| | 1.25 | 1.454 | 2.237 | .150 | 1.724 | -.205 | 0.000 | 2.467 | 3.499 | 7.420 |
| | 1.50 | 1.602 | 2.746 | .101 | 1.887 | -.212 | 0.000 | 2.575 | 4.255 | 12.689 |
| 1.00 | .25 | 1.261 | 1.253 | .239 | 1.437 | -.142 | 0.000 | .460 | .571 | .772 |
| | .50 | 1.337 | 1.560 | .201 | 1.552 | -.172 | 0.000 | .910 | 1.197 | 2.024 |
| | .75 | 1.426 | 1.921 | .160 | 1.690 | -.200 | 0.000 | 1.355 | 1.882 | 3.745 |
| | 1.00 | 1.533 | 2.317 | .121 | 1.812 | -.216 | 0.000 | 1.661 | 2.544 | 6.391 |
| | 1.25 | 1.655 | 2.742 | .086 | 1.933 | -.220 | 0.000 | 1.827 | 3.186 | 10.480 |
| | 1.50 | 1.784 | 3.194 | .053 | 2.089 | -.213 | 0.000 | 1.887 | 3.827 | 16.758 |

## Table 39

### Bending stress
### Unloaded sides stress-free

Side ratio $a/b = 1{\cdot}0$

| W | P | U | W | $\bar{\sigma}$ MX | | $\bar{\sigma}$ MY | | $\bar{M}$ X | $\bar{M}$ Y | $\bar{M}$ XY |
|---|---|---|---|---|---|---|---|---|---|---|
| | | | | 4 | 2 | 4 | 2,3 | 4 | 4 | 1 |
| .10 | .25 | 1.007 | .129 | .466 | 1.009 | -.001 | 0.000 | .119 | .096 | .099 |
| | .50 | 1.018 | .238 | .466 | 1.028 | -.009 | 0.000 | .585 | .485 | .454 |
| | .75 | 1.055 | .471 | .457 | 1.086 | -.027 | 0.000 | 1.579 | 1.299 | 1.226 |
| | 1.00 | 1.213 | 1.098 | .350 | 1.339 | -.105 | 0.000 | 3.998 | 3.218 | 4.091 |
| | 1.25 | 1.438 | 1.863 | .197 | 1.731 | -.226 | 0.000 | 6.386 | 5.279 | 10.676 |
| | 1.50 | 1.668 | 2.505 | .055 | 2.195 | -.370 | 0.000 | 7.306 | 6.521 | 22.265 |
| .25 | .25 | 1.035 | .332 | .448 | 1.053 | -.017 | 0.000 | .337 | .253 | .274 |
| | .50 | 1.079 | .535 | .422 | 1.123 | -.039 | 0.000 | 1.165 | .923 | 1.031 |
| | .75 | 1.163 | .872 | .373 | 1.260 | -.081 | 0.000 | 2.526 | 2.043 | 2.408 |
| | 1.00 | 1.334 | 1.401 | .284 | 1.543 | -.163 | 0.000 | 4.409 | 3.572 | 5.791 |
| | 1.25 | 1.537 | 2.004 | .172 | 1.922 | -.256 | 0.000 | 5.973 | 4.919 | 12.645 |
| | 1.50 | 1.716 | 2.570 | .058 | 2.338 | -.331 | 0.000 | 6.450 | 5.548 | 24.138 |
| .50 | .25 | 1.131 | .661 | .392 | 1.209 | -.065 | 0.000 | .662 | .539 | .583 |
| | .50 | 1.213 | .927 | .347 | 1.337 | -.105 | 0.000 | 1.689 | 1.372 | 1.855 |
| | .75 | 1.334 | 1.300 | .279 | 1.533 | -.167 | 0.000 | 3.047 | 2.494 | 3.918 |
| | 1.00 | 1.489 | 1.775 | .175 | 1.824 | -.259 | 0.000 | 4.461 | 3.743 | 8.098 |
| | 1.25 | 1.658 | 2.295 | .086 | 2.183 | -.338 | 0.000 | 5.998 | 4.786 | 16.626 |
| | 1.50 | 1.824 | 2.810 | .059 | 2.580 | -.361 | 0.000 | 7.754 | 5.336 | 31.377 |
| 1.00 | .25 | 1.383 | 1.243 | .258 | 1.601 | -.183 | 0.000 | .928 | .732 | 1.190 |
| | .50 | 1.485 | 1.550 | .193 | 1.790 | -.242 | 0.000 | 1.949 | 1.577 | 3.356 |
| | .75 | 1.607 | 1.918 | .107 | 2.035 | -.321 | 0.000 | 3.020 | 2.495 | 6.636 |
| | 1.00 | 1.743 | 2.319 | .026 | 2.303 | -.387 | 0.000 | 3.829 | 3.189 | 12.439 |
| | 1.25 | 1.872 | 2.739 | .021 | 2.612 | -.413 | 0.000 | 4.402 | 3.715 | 22.350 |
| | 1.50 | 1.973 | 3.171 | .153 | 2.983 | -.377 | 0.000 | 4.798 | 4.158 | 37.652 |

# Table 40

## Bending stress
## Unloaded sides stress-free

Side ratio $a/b = 1\cdot 5$

| W | P | U | W | $\bar{\sigma}_{MX}$ |  | $\bar{\sigma}_{MY}$ |  | $\bar{M}_X$ | $\bar{M}_Y$ | $\bar{M}_{XY}$ |
|---|---|---|---|---|---|---|---|---|---|---|
|  |  |  |  | 4 | 2 | 4 | 2,3 | 4 | 4 | 1 |
| .10 | .25 | 1.002 | .124 | .383 | 1.002 | -.001 | 0.000 | .048 | .055 | .056 |
|  | .50 | 1.009 | .236 | .378 | 1.016 | -.005 | 0.000 | .278 | .314 | .280 |
|  | .75 | 1.024 | .443 | .365 | 1.048 | -.014 | 0.000 | .702 | .790 | .698 |
|  | 1.00 | 1.114 | .978 | .312 | 1.197 | -.053 | 0.000 | 1.621 | 1.947 | 2.487 |
|  | 1.25 | 1.283 | 1.866 | .215 | 1.473 | -.124 | 0.000 | 3.067 | 3.834 | 5.768 |
|  | 1.50 | 1.512 | 3.026 | .086 | 1.841 | -.220 | 0.000 | 4.935 | 6.290 | 10.135 |
| .25 | .25 | 1.021 | .328 | .370 | 1.036 | -.010 | 0.000 | .158 | .179 | .163 |
|  | .50 | 1.046 | .525 | .354 | 1.083 | -.023 | 0.000 | .544 | .626 | .634 |
|  | .75 | 1.095 | .848 | .324 | 1.170 | -.047 | 0.000 | 1.162 | 1.350 | 1.434 |
|  | 1.00 | 1.199 | 1.348 | .264 | 1.329 | -.088 | 0.000 | 1.886 | 2.377 | 3.525 |
|  | 1.25 | 1.339 | 1.943 | .193 | 1.579 | -.119 | 0.000 | 2.445 | 3.495 | 8.669 |
|  | 1.50 | 1.482 | 2.495 | .143 | 1.939 | -.104 | 0.000 | 2.448 | 4.372 | 19.342 |
| .50 | .25 | 1.081 | .657 | .331 | 1.140 | -.040 | 0.000 | .312 | .357 | .349 |
|  | .50 | 1.131 | .913 | .303 | 1.224 | -.061 | 0.000 | .762 | .917 | 1.150 |
|  | .75 | 1.205 | 1.269 | .261 | 1.350 | -.092 | 0.000 | 1.344 | 1.678 | 2.424 |
|  | 1.00 | 1.305 | 1.706 | .210 | 1.519 | -.118 | 0.000 | 1.821 | 2.506 | 5.161 |
|  | 1.25 | 1.428 | 2.201 | .158 | 1.760 | -.129 | 0.000 | 2.125 | 3.356 | 11.041 |
|  | 1.50 | 1.565 | 2.719 | .118 | 2.120 | -.109 | 0.000 | 2.201 | 4.193 | 22.410 |
| 1.00 | .25 | 1.239 | 1.235 | .244 | 1.408 | -.106 | 0.000 | .415 | .513 | .707 |
|  | .50 | 1.304 | 1.522 | .213 | 1.509 | -.125 | 0.000 | .819 | 1.095 | 2.073 |
|  | .75 | 1.383 | 1.862 | .179 | 1.627 | -.142 | 0.000 | 1.208 | 1.744 | 4.129 |
|  | 1.00 | 1.476 | 2.240 | .144 | 1.820 | -.146 | 0.000 | 1.465 | 2.396 | 8.000 |
|  | 1.25 | 1.584 | 2.666 | .112 | 2.092 | -.137 | 0.000 | 1.624 | 3.069 | 15.223 |
|  | 1.50 | 1.708 | 3.155 | .086 | 2.432 | -.117 | 0.000 | 1.754 | 3.808 | 27.894 |

# Table 41

## Combined bending and shear displacement

Side ratio $a/b = 0\cdot 667$

| W | P | W | $\bar{\sigma}_{MX}$ |  | $\bar{\sigma}_{MY}$ |  | $\bar{\tau}_{MXY}$ |  | $\bar{M}_X$ | $\bar{M}_Y$ | $\bar{M}_{XY}$ |  |
|---|---|---|---|---|---|---|---|---|---|---|---|---|
|  |  |  | 4 | 4 | 2 | 3 | 1 | 4 | 4 | 4 | 1 | 4 |
| .10 | .25 | .134 | .352 | -.011 | -.020 | -.015 | 1.011 | 1.004 | .110 | .095 | .113 | .022 |
|  | .50 | .220 | .336 | -.026 | -.037 | -.019 | 1.012 | 1.007 | .405 | .352 | .436 | .088 |
|  | .75 | .362 | .308 | -.054 | -.068 | -.026 | 1.014 | 1.010 | .893 | .779 | .983 | .201 |
|  | 1.00 | .552 | .251 | -.102 | -.121 | -.039 | 1.021 | 1.017 | 1.637 | 1.419 | 1.876 | .400 |
|  | 1.25 | .760 | .177 | -.160 | -.183 | -.057 | 1.033 | 1.025 | 2.563 | 2.183 | 3.036 | .675 |
|  | 1.50 | .937 | .107 | -.211 | -.230 | -.076 | 1.044 | 1.027 | 3.545 | 2.932 | 4.405 | 1.005 |
| .25 | .25 | .326 | .297 | -.060 | -.074 | -.050 | 1.018 | 1.011 | .258 | .225 | .280 | .057 |
|  | .50 | .437 | .263 | -.088 | -.106 | -.058 | 1.019 | 1.012 | .665 | .578 | .751 | .160 |
|  | .75 | .583 | .221 | -.128 | -.151 | -.047 | 1.017 | 1.013 | 1.223 | 1.059 | 1.449 | .311 |
|  | 1.00 | .738 | .171 | -.156 | -.196 | -.058 | 1.030 | 1.023 | 1.902 | 1.630 | 2.340 | .516 |
|  | 1.25 | .889 | .117 | -.188 | -.236 | -.081 | 1.048 | 1.034 | 2.661 | 2.245 | 3.395 | .766 |
|  | 1.50 | 1.025 | .063 | -.248 | -.266 | -.089 | 1.052 | 1.029 | 3.448 | 2.847 | 4.353 | 1.046 |
| .50 | .25 | .600 | .207 | -.155 | -.179 | -.055 | 1.032 | 1.024 | .375 | .328 | .455 | .100 |
|  | .50 | .710 | .173 | -.184 | -.210 | -.065 | 1.036 | 1.028 | .829 | .717 | 1.067 | .235 |
|  | .75 | .830 | .135 | -.215 | -.242 | -.075 | 1.041 | 1.033 | 1.364 | 1.167 | 1.807 | .403 |
|  | 1.00 | .951 | .084 | -.246 | -.271 | -.086 | 1.052 | 1.037 | 1.976 | 1.671 | 2.687 | .607 |
|  | 1.25 | 1.068 | .040 | -.273 | -.296 | -.097 | 1.062 | 1.038 | 2.640 | 2.202 | 3.688 | .839 |
|  | 1.50 | 1.176 | .033 | -.293 | -.312 | -.107 | 1.063 | 1.030 | 3.318 | 2.721 | 4.779 | 1.090 |
| 1.00 | .25 | 1.084 | .127 | -.286 | -.309 | -.100 | 1.061 | 1.035 | .419 | .361 | .642 | .140 |
|  | .50 | 1.170 | .110 | -.306 | -.324 | -.108 | 1.065 | 1.034 | .878 | .746 | 1.377 | .303 |
|  | .75 | 1.256 | .073 | -.319 | -.338 | -.115 | 1.069 | 1.039 | 1.376 | 1.152 | 2.206 | .489 |
|  | 1.00 | 1.341 | .045 | -.334 | -.349 | -.123 | 1.075 | 1.039 | 1.916 | 1.579 | 3.127 | .691 |
|  | 1.25 | 1.423 | .020 | -.347 | -.358 | -.130 | 1.082 | 1.036 | 2.481 | 2.018 | 4.134 | .910 |
|  | 1.50 | 1.502 | -.018 | -.353 | -.363 | -.137 | 1.090 | 1.036 | 3.050 | 2.455 | 5.216 | 1.151 |

## Table 42

### Combined bending and shear displacement

Side ratio $a/b = 1{\cdot}0$

| W | P | W | $\bar{\sigma}_{MX}$ |  | $\bar{\sigma}_{MY}$ |  | $\bar{\tau}_{MXY}$ |  | $\bar{M}_X$ | $\bar{M}_Y$ |  | $\bar{M}_{XY}$ |
|---|---|---|---|---|---|---|---|---|---|---|---|---|
|  |  |  | 4 | 4 | 2 | 3 | 1 | 4 | 4 | 4 | 1 | 4 |
| .10 | .25 | .131 | .308 | -.008 | -.004 | .000 | 1.000 | 1.001 | .070 | .067 | .081 | .014 |
|  | .50 | .209 | .305 | -.017 | -.015 | -.006 | 1.003 | 1.002 | .255 | .239 | .294 | .050 |
|  | .75 | .337 | .295 | -.035 | -.037 | -.018 | 1.008 | 1.002 | .576 | .534 | .669 | .115 |
|  | 1.00 | .539 | .255 | -.072 | -.071 | -.034 | 1.018 | 1.001 | 1.136 | 1.029 | 1.379 | .241 |
|  | 1.25 | .746 | .203 | -.113 | -.109 | -.052 | 1.040 | 1.006 | 1.800 | 1.579 | 2.294 | .401 |
|  | 1.50 | .896 | .159 | -.143 | -.145 | -.072 | 1.083 | 1.023 | 2.432 | 2.037 | 3.272 | .566 |
| .25 | .25 | .323 | .286 | -.041 | -.044 | -.021 | 1.008 | 1.003 | .187 | .163 | .202 | .035 |
|  | .50 | .428 | .271 | -.061 | -.062 | -.029 | 1.014 | 1.003 | .494 | .413 | .542 | .094 |
|  | .75 | .565 | .246 | -.088 | -.087 | -.040 | 1.023 | 1.002 | .906 | .749 | 1.027 | .179 |
|  | 1.00 | .715 | .212 | -.118 | -.117 | -.054 | 1.031 | .997 | 1.305 | 1.148 | 1.689 | .293 |
|  | 1.25 | .860 | .175 | -.145 | -.145 | -.069 | 1.051 | .999 | 1.752 | 1.561 | 2.482 | .425 |
|  | 1.50 | .987 | .138 | -.166 | -.167 | -.082 | 1.092 | 1.020 | 2.316 | 1.947 | 3.363 | .564 |
| .50 | .25 | .595 | .244 | -.108 | -.099 | -.047 | 1.026 | 1.001 | .251 | .235 | .327 | .057 |
|  | .50 | .700 | .222 | -.128 | -.129 | -.056 | 1.034 | .999 | .557 | .511 | .751 | .130 |
|  | .75 | .815 | .198 | -.150 | -.147 | -.066 | 1.042 | .993 | .917 | .825 | 1.302 | .221 |
|  | 1.00 | .929 | .173 | -.169 | -.167 | -.076 | 1.049 | .987 | 1.309 | 1.158 | 1.953 | .327 |
|  | 1.25 | 1.039 | .148 | -.186 | -.186 | -.087 | 1.056 | .981 | 1.716 | 1.493 | 2.701 | .442 |
|  | 1.50 | 1.142 | .124 | -.199 | -.199 | -.096 | 1.065 | .976 | 2.122 | 1.819 | 3.534 | .565 |
| 1.00 | .25 | 1.082 | .226 | -.205 | -.192 | -.082 | 1.059 | .972 | .283 | .271 | .460 | .068 |
|  | .50 | 1.164 | .206 | -.215 | -.203 | -.089 | 1.064 | .970 | .580 | .546 | .993 | .145 |
|  | .75 | 1.246 | .187 | -.224 | -.214 | -.096 | 1.071 | .966 | .891 | .824 | 1.601 | .232 |
|  | 1.00 | 1.326 | .171 | -.233 | -.223 | -.100 | 1.075 | .964 | 1.211 | 1.108 | 2.291 | .326 |
|  | 1.25 | 1.404 | .156 | -.241 | -.230 | -.106 | 1.080 | .963 | 1.534 | 1.393 | 3.051 | .423 |
|  | 1.50 | 1.481 | .140 | -.247 | -.237 | -.118 | 1.088 | .961 | 1.858 | 1.674 | 3.872 | .522 |

## Table 43

### Combined bending and shear displacement

Side ratio $a/b = 1{\cdot}5$

| W | P | W | $\bar{\sigma}_{MX}$ |  | $\bar{\sigma}_{MY}$ |  | $\bar{\tau}_{MXY}$ |  | $\bar{M}_X$ | $\bar{M}_Y$ |  | $\bar{M}_{XY}$ |
|---|---|---|---|---|---|---|---|---|---|---|---|---|
|  |  |  | 4 | 4 | 2 | 3 | 1 | 4 | 4 | 4 | 1 | 4 |
| .10 | .25 | .138 | .287 | -.008 | -.011 | -.007 | 1.077 | 1.001 | .070 | .070 | .080 | .014 |
|  | .50 | .226 | .283 | -.019 | -.028 | -.013 | 1.078 | 1.003 | .235 | .239 | .278 | .047 |
|  | .75 | .377 | .272 | -.041 | -.045 | -.027 | 1.083 | 1.006 | .527 | .538 | .541 | .108 |
|  | 1.00 | .594 | .248 | -.080 | -.086 | -.052 | 1.098 | 1.010 | .977 | 1.001 | 1.260 | .210 |
|  | 1.25 | .784 | .213 | -.116 | -.125 | -.076 | 1.113 | 1.012 | 1.420 | 1.441 | 1.992 | .319 |
|  | 1.50 | .948 | .166 | -.149 | -.163 | -.099 | 1.130 | 1.014 | 1.860 | 1.864 | 2.838 | .438 |
| .25 | .25 | .332 | .264 | -.044 | -.048 | -.029 | 1.087 | 1.008 | .157 | .157 | .188 | .032 |
|  | .50 | .448 | .253 | -.065 | -.070 | -.042 | 1.093 | 1.009 | .391 | .396 | .495 | .082 |
|  | .75 | .594 | .237 | -.093 | -.099 | -.059 | 1.101 | 1.009 | .699 | .711 | .930 | .153 |
|  | 1.00 | .748 | .216 | -.121 | -.130 | -.078 | 1.114 | 1.012 | 1.049 | 1.072 | 1.498 | .240 |
|  | 1.25 | .893 | .193 | -.146 | -.158 | -.095 | 1.127 | 1.014 | 1.396 | 1.429 | 2.167 | .334 |
|  | 1.50 | 1.030 | .166 | -.166 | -.181 | -.109 | 1.140 | 1.014 | 1.741 | 1.783 | 2.939 | .435 |
| .50 | .25 | .602 | .236 | -.110 | -.115 | -.069 | 1.112 | 1.008 | .216 | .222 | .296 | .048 |
|  | .50 | .715 | .222 | -.130 | -.139 | -.081 | 1.115 | 1.007 | .463 | .479 | .680 | .108 |
|  | .75 | .835 | .202 | -.150 | -.160 | -.094 | 1.125 | 1.011 | .735 | .764 | 1.154 | .180 |
|  | 1.00 | .954 | .185 | -.167 | -.181 | -.107 | 1.136 | 1.014 | 1.017 | 1.061 | 1.724 | .261 |
|  | 1.25 | 1.067 | .177 | -.183 | -.198 | -.117 | 1.147 | 1.014 | 1.300 | 1.373 | 2.375 | .344 |
|  | 1.50 | 1.176 | .179 | -.198 | -.213 | -.126 | 1.158 | 1.010 | 1.585 | 1.700 | 3.108 | .429 |
| 1.00 | .25 | 1.086 | .237 | -.201 | -.205 | -.111 | 1.249 | .994 | .218 | .241 | .406 | .057 |
|  | .50 | 1.173 | .220 | -.211 | -.217 | -.120 | 1.068 | .998 | .435 | .495 | .877 | .121 |
|  | .75 | 1.261 | .206 | -.220 | -.229 | -.128 | 1.179 | 1.002 | .652 | .756 | 1.418 | .191 |
|  | 1.00 | 1.347 | .195 | -.228 | -.240 | -.135 | 1.165 | 1.006 | .866 | 1.008 | 2.031 | .264 |
|  | 1.25 | 1.428 | .185 | -.234 | -.248 | -.141 | 1.166 | 1.010 | 1.079 | 1.259 | 2.705 | .340 |
|  | 1.50 | 1.505 | .177 | -.238 | -.254 | -.146 | 1.190 | 1.015 | 1.290 | 1.508 | 3.440 | .419 |

## Table 44

### Uniform bearing stress

Side ratio $a/b = 0{\cdot}667$

| W | P | W | $\bar{\sigma}$ | | $\bar{\sigma}$ | $\bar{M}$ | $\bar{M}$ | $\bar{M}$ |
|---|---|---|---|---|---|---|---|---|
| | | | | MX | MY | X | Y | XY |
| | | | 3 | 2 | 3 | 3 | 3 | 1 |
| .10 | .25 | .121 | -.004 | .010 | .745 | .058 | .064 | .053 |
| | .50 | .226 | -.017 | .043 | .732 | .343 | .389 | .316 |
| | .75 | .402 | -.036 | .091 | .714 | .829 | .937 | .743 |
| | 1.00 | .940 | -.191 | .482 | .600 | 2.144 | 2.571 | 2.392 |
| | 1.25 | 1.702 | -.445 | 1.170 | .439 | 3.878 | 4.913 | 5.574 |
| | 1.50 | 2.388 | -.703 | 2.001 | .331 | 5.187 | 7.118 | 10.548 |
| .25 | .25 | .322 | -.035 | .088 | .723 | .203 | .232 | .175 |
| | .50 | .512 | -.081 | .203 | .686 | .701 | .821 | .697 |
| | .75 | .822 | -.164 | .406 | .620 | 1.508 | 1.779 | 1.555 |
| | 1.00 | 1.305 | -.331 | .854 | .511 | 2.614 | 3.222 | 3.354 |
| | 1.25 | 1.901 | -.554 | 1.507 | .390 | 3.859 | 5.058 | 6.499 |
| | 1.50 | 2.493 | -.773 | 2.261 | .315 | 4.982 | 7.100 | 11.507 |
| .50 | .25 | .652 | -.143 | .362 | .640 | .415 | .478 | .391 |
| | .50 | .904 | -.228 | .581 | .581 | 1.038 | 1.253 | 1.203 |
| | .75 | 1.259 | -.355 | .916 | .497 | 1.880 | 2.341 | 2.431 |
| | 1.00 | 1.703 | -.521 | 1.391 | .408 | 2.801 | 3.663 | 4.570 |
| | 1.25 | 2.207 | -.705 | 1.983 | .331 | 3.761 | 5.239 | 7.997 |
| | 1.50 | 2.728 | -.875 | 2.645 | .292 | 4.726 | 7.116 | 13.208 |
| 1.00 | .25 | 1.236 | -.332 | 1.086 | .458 | .573 | .713 | .761 |
| | .50 | 1.527 | -.558 | 1.392 | .401 | 1.208 | 1.588 | 2.016 |
| | .75 | 1.874 | -.651 | 1.772 | .346 | 1.911 | 2.629 | 3.765 |
| | 1.00 | 2.259 | -.766 | 2.201 | .306 | 2.602 | 3.802 | 6.389 |
| | 1.25 | 2.681 | -.904 | 2.693 | .276 | 3.427 | 5.326 | 10.229 |
| | 1.50 | 3.144 | -1.028 | 3.280 | .242 | 4.644 | 7.571 | 15.754 |

## Table 45

### Uniform bearing stress

Side ratio $a/b = 1{\cdot}0$

| W | P | W | $\bar{\sigma}$ | | $\bar{\sigma}$ | $\bar{M}$ | $\bar{M}$ | $\bar{M}$ |
|---|---|---|---|---|---|---|---|---|
| | | | | MX | MY | X | Y | XY |
| | | | 3 | 2 | 3 | 3 | 3 | 1 |
| .10 | .25 | .126 | -.005 | .013 | .655 | .033 | .042 | .037 |
| | .50 | .233 | -.018 | .046 | .641 | .166 | .207 | .180 |
| | .75 | .424 | -.044 | .111 | .614 | .402 | .497 | .430 |
| | 1.00 | 1.002 | -.210 | .532 | .510 | 1.050 | 1.369 | 1.363 |
| | 1.25 | 1.799 | -.473 | 1.250 | .373 | 1.876 | 2.574 | 3.146 |
| | 1.50 | 2.465 | -.731 | 2.087 | .296 | 2.422 | 3.588 | 5.916 |
| .25 | .25 | .329 | -.037 | .099 | .632 | .101 | .125 | .109 |
| | .50 | .529 | -.085 | .223 | .597 | .341 | .430 | .405 |
| | .75 | .853 | -.173 | .449 | .535 | .726 | .924 | .895 |
| | 1.00 | 1.377 | -.357 | .947 | .436 | 1.269 | 1.709 | 1.948 |
| | 1.25 | 2.024 | -.599 | 1.646 | .332 | 1.878 | 2.687 | 3.776 |
| | 1.50 | 2.648 | -.819 | 2.391 | .280 | 2.403 | 3.672 | 6.639 |
| .50 | .25 | .664 | -.149 | .396 | .563 | .204 | .258 | .238 |
| | .50 | .934 | -.240 | .641 | .509 | .512 | .670 | .716 |
| | .75 | 1.315 | -.378 | 1.012 | .430 | .927 | 1.240 | 1.442 |
| | 1.00 | 1.792 | -.556 | 1.528 | .353 | 1.384 | 1.946 | 2.693 |
| | 1.25 | 2.335 | -.754 | 2.155 | .291 | 1.866 | 2.780 | 4.695 |
| | 1.50 | 2.898 | -.937 | 2.829 | .259 | 2.360 | 3.735 | 7.757 |
| 1.00 | .25 | 1.257 | -.440 | 1.198 | .407 | .292 | .391 | .471 |
| | .50 | 1.577 | -.559 | 1.544 | .357 | .623 | .878 | 1.236 |
| | .75 | 1.960 | -.697 | 1.966 | .308 | .992 | 1.459 | 2.305 |
| | 1.00 | 2.391 | -.840 | 2.441 | .273 | 1.379 | 2.124 | 3.938 |
| | 1.25 | 2.867 | -.983 | 2.975 | .252 | 1.803 | 2.923 | 6.355 |
| | 1.50 | 3.388 | -1.122 | 3.580 | .243 | 2.293 | 3.941 | 9.856 |

## Table 46

### Uniform bearing stress

Side ratio $a/b = 1{\cdot}5$

| W | P | W | | $\bar{\sigma}_{MX}$ | $\bar{\sigma}_{MY}$ | $\bar{M}_{X}$ | $\bar{M}_{Y}$ | $\bar{M}_{XY}$ |
|---|---|---|---|---|---|---|---|---|
| | | | 3 | 2 | 3 | 3 | 3 | 1 |
| .10 | .25 | .127 | -.005 | .012 | .573 | .018 | .023 | .023 |
| | .50 | .230 | -.019 | .041 | .567 | .087 | .136 | .105 |
| | .75 | .436 | -.056 | .120 | .551 | .222 | .367 | .271 |
| | 1.00 | 1.135 | -.290 | .630 | .466 | .638 | 1.106 | .960 |
| | 1.25 | 2.093 | -.655 | 1.485 | .363 | 1.160 | 2.107 | 2.326 |
| | 1.50 | 2.922 | -1.032 | 2.502 | .324 | 1.519 | 2.972 | 4.496 |
| .25 | .25 | .327 | -.039 | .087 | .561 | .053 | .084 | .062 |
| | .50 | .534 | -.095 | .210 | .538 | .186 | .307 | .243 |
| | .75 | .885 | -.205 | .451 | .495 | .405 | .683 | .563 |
| | 1.00 | 1.531 | -.472 | 1.059 | .417 | .758 | 1.353 | 1.347 |
| | 1.25 | 2.334 | -.818 | 1.902 | .340 | 1.165 | 2.192 | 2.757 |
| | 1.50 | 3.086 | -1.119 | 2.760 | .316 | 1.515 | 3.006 | 4.947 |
| .50 | .25 | .661 | -.160 | .356 | .513 | .107 | .174 | .136 |
| | .50 | .956 | -.274 | .614 | .475 | .284 | .486 | .437 |
| | .75 | 1.387 | -.455 | 1.025 | .421 | .531 | .942 | .920 |
| | 1.00 | 1.971 | -.710 | 1.636 | .368 | .827 | 1.547 | 1.835 |
| | 1.25 | 2.648 | -.994 | 2.380 | .326 | 1.153 | 2.259 | 3.348 |
| | 1.50 | 3.336 | -1.247 | 3.156 | .304 | 1.489 | 3.025 | 5.641 |
| 1.00 | .25 | 1.269 | -.493 | 1.123 | .413 | .162 | .283 | .277 |
| | .50 | 1.628 | -.650 | 1.501 | .378 | .357 | .659 | .766 |
| | .75 | 2.077 | -.843 | 1.980 | .347 | .587 | 1.131 | 1.480 |
| | 1.00 | 2.600 | -1.047 | 2.539 | .351 | .842 | 1.688 | 2.630 |
| | 1.25 | 3.182 | -1.253 | 3.166 | .351 | 1.133 | 2.351 | 4.385 |
| | 1.50 | 3.805 | -1.449 | 3.849 | .292 | 1.469 | 3.150 | 6.936 |

## Table 47

### Non-uniform bearing stress

Side ratio $a/b = 0{\cdot}667$

| W | P | W | | $\bar{\sigma}_{MX}$ | $\bar{\sigma}_{MY}$ | $\bar{M}_{X}$ | $\bar{M}_{Y}$ | $\bar{M}_{XY}$ |
|---|---|---|---|---|---|---|---|---|
| | | | 3 | 2 | 3 | 3 | 3 | 1 |
| .10 | .25 | .133 | -.061 | .512 | .649 | .099 | .115 | .081 |
| | .50 | .212 | -.068 | .530 | .643 | .336 | .394 | .274 |
| | .75 | .419 | -.094 | .592 | .621 | .950 | .965 | .762 |
| | 1.00 | .920 | -.196 | .868 | .549 | 2.350 | 2.452 | 2.223 |
| | 1.25 | 1.721 | -.439 | 1.592 | .412 | 4.374 | 5.693 | 5.376 |
| | 1.50 | 2.818 | -.795 | 2.667 | .218 | 7.083 | 10.340 | 9.920 |
| .25 | .25 | .332 | -.088 | .585 | .631 | .245 | .287 | .202 |
| | .50 | .502 | -.120 | .666 | .606 | .740 | .877 | .640 |
| | .75 | .808 | -.190 | .852 | .557 | 1.609 | 1.927 | 1.480 |
| | 1.00 | 1.306 | -.329 | 1.251 | .468 | 2.937 | 3.654 | 3.199 |
| | 1.25 | 1.916 | -.521 | 1.878 | .377 | 4.369 | 5.849 | 6.124 |
| | 1.50 | 2.670 | -.773 | 2.727 | .270 | 6.048 | 8.597 | 10.114 |
| .50 | .25 | .652 | -.175 | .816 | .571 | .447 | .532 | .400 |
| | .50 | .899 | -.243 | 1.003 | .526 | 1.133 | 1.380 | 1.128 |
| | .75 | 1.247 | -.345 | 1.297 | .461 | 2.048 | 2.571 | 2.291 |
| | 1.00 | 1.692 | -.479 | 1.730 | .389 | 3.134 | 4.154 | 4.228 |
| | 1.25 | 2.204 | -.631 | 2.298 | .334 | 4.315 | 6.133 | 7.242 |
| | 1.50 | 2.795 | -.807 | 3.001 | .287 | 5.623 | 8.505 | 11.205 |
| 1.00 | .25 | 1.231 | -.399 | 1.450 | .428 | .624 | .794 | .740 |
| | .50 | 1.522 | -.486 | 1.728 | .384 | 1.354 | 1.807 | 1.866 |
| | .75 | 1.866 | -.581 | 2.068 | .344 | 2.168 | 3.051 | 3.453 |
| | 1.00 | 2.255 | -.679 | 2.478 | .313 | 3.070 | 4.656 | 5.790 |
| | 1.25 | 2.700 | -.783 | 2.994 | .298 | 4.165 | 6.877 | 9.211 |
| | 1.50 | 3.196 | -.892 | 3.601 | .296 | 5.411 | 9.607 | 13.575 |

## Table 48

### Non-uniform bearing stress

Side ratio $a/b = 1 \cdot 0$

| H | P | W | $\bar{\sigma}_{MX}$ 3 | $\bar{\sigma}_{MX}$ 2 | $\bar{\sigma}_{MY}$ 3 | $\bar{M}_{X}$ 3 | $\bar{M}_{Y}$ 3 | $\bar{M}_{XY}$ 1 |
|---|---|---|---|---|---|---|---|---|
| .10 | .25 | .126 | -.050 | .521 | .610 | .035 | .045 | .032 |
| | .50 | .231 | -.062 | .550 | .598 | .181 | .220 | .159 |
| | .75 | .421 | -.087 | .609 | .576 | .439 | .538 | .390 |
| | 1.00 | 1.008 | -.228 | .989 | .488 | 1.180 | 1.508 | 1.245 |
| | 1.25 | 1.810 | -.447 | 1.644 | .374 | 2.128 | 2.854 | 2.778 |
| | 1.50 | 2.464 | -.658 | 2.438 | .306 | 2.758 | 4.020 | 4.954 |
| .25 | .25 | .328 | -.077 | .598 | .593 | .335 | .133 | .095 |
| | .50 | .526 | -.118 | .706 | .564 | .608 | .465 | .357 |
| | .75 | .849 | -.192 | .906 | .513 | .798 | 1.003 | .795 |
| | 1.00 | 1.378 | -.349 | 1.361 | .427 | 1.376 | 1.879 | 1.727 |
| | 1.25 | 2.026 | -.551 | 2.011 | .337 | 2.164 | 3.000 | 3.252 |
| | 1.50 | 2.638 | -.727 | 2.719 | .293 | 2.769 | 4.188 | 5.458 |
| .50 | .25 | .632 | -.046 | .840 | .677 | .240 | .429 | .204 |
| | .50 | .906 | -.116 | 1.046 | .636 | .593 | .897 | .625 |
| | .75 | 1.328 | -.354 | 1.376 | .424 | 1.059 | 1.387 | 1.270 |
| | 1.00 | 1.828 | -.522 | 1.850 | .337 | 1.612 | 2.200 | 2.353 |
| | 1.25 | 2.384 | -.644 | 2.440 | .331 | 2.226 | 3.297 | 4.008 |
| | 1.50 | 2.976 | -.816 | 3.096 | .274 | 2.861 | 4.556 | 6.399 |
| 1.00 | .25 | 1.255 | -.413 | 1.581 | .403 | .326 | .428 | .402 |
| | .50 | 1.573 | -.511 | 1.898 | .359 | .705 | .972 | 1.042 |
| | .75 | 1.955 | -.626 | 2.286 | .316 | 1.134 | 1.634 | 1.929 |
| | 1.00 | 2.381 | -.738 | 2.744 | .288 | 1.595 | 2.437 | 3.236 |
| | 1.25 | 2.865 | -.874 | 3.270 | .270 | 2.143 | 3.581 | 5.117 |
| | 1.50 | 3.430 | -1.080 | 3.864 | .256 | 2.871 | 5.383 | 7.771 |

## Table 49

### Non-uniform bearing stress

Side ratio $a/b = 1 \cdot 5$

| H | P | W | $\bar{\sigma}_{MX}$ 3 | $\bar{\sigma}_{MX}$ 2 | $\bar{\sigma}_{MY}$ 3 | $\bar{M}_{X}$ 3 | $\bar{M}_{Y}$ 3 | $\bar{M}_{XY}$ 1 |
|---|---|---|---|---|---|---|---|---|
| .10 | .25 | .124 | -.019 | .600 | .571 | .013 | .028 | .015 |
| | .50 | .227 | -.033 | .628 | .561 | .106 | .147 | .083 |
| | .75 | .406 | -.058 | .678 | .548 | .280 | .353 | .199 |
| | 1.00 | 1.057 | -.246 | 1.123 | .471 | .699 | 1.083 | .707 |
| | 1.25 | 1.978 | -.543 | 1.889 | .368 | 1.238 | 2.118 | 1.647 |
| | 1.50 | 2.749 | -.824 | 2.763 | .312 | 1.651 | 2.997 | 2.999 |
| .25 | .25 | .323 | -.049 | .667 | .559 | .057 | .085 | .046 |
| | .50 | .526 | -.097 | .779 | .536 | .205 | .315 | .188 |
| | .75 | .859 | -.186 | .987 | .495 | .446 | .694 | .426 |
| | 1.00 | 1.457 | -.396 | 1.508 | .420 | .820 | 1.354 | .984 |
| | 1.25 | 2.211 | -.672 | 2.258 | .339 | 1.259 | 2.196 | 1.921 |
| | 1.50 | 2.913 | -.908 | 3.043 | .300 | 1.643 | 3.027 | 3.276 |
| .50 | .25 | .657 | -.155 | .919 | .510 | .118 | .180 | .103 |
| | .50 | .941 | -.251 | 1.150 | .471 | .314 | .498 | .331 |
| | .75 | 1.353 | -.399 | 1.517 | .416 | .588 | .959 | .686 |
| | 1.00 | 1.897 | -.599 | 2.055 | .379 | .911 | 1.559 | 1.315 |
| | 1.25 | 2.527 | -.821 | 2.724 | .262 | 1.267 | 2.288 | 2.293 |
| | 1.50 | 3.168 | -1.012 | 3.442 | -.101 | 1.637 | 3.131 | 3.709 |
| 1.00 | .25 | 1.262 | -.444 | 1.638 | .407 | .181 | .295 | .210 |
| | .50 | 1.606 | -.571 | 1.978 | .370 | .399 | .681 | .568 |
| | .75 | 2.032 | -.724 | 2.402 | .332 | .653 | 1.159 | 1.074 |
| | 1.00 | 2.522 | -.883 | 2.911 | .306 | .933 | 1.731 | 1.843 |
| | 1.25 | 3.065 | -1.038 | 3.494 | .291 | 1.247 | 2.423 | 2.954 |
| | 1.50 | 3.645 | -1.178 | 4.136 | .281 | 1.612 | 3.280 | 4.515 |

## Table 50

### Combined uniform bearing and bending stress

Side ratio $a/b = 0{\cdot}667$

| W | P | W | σ̄ | | σ̄ | M̄ | M̄ | M̄ |
|---|---|---|---|---|---|---|---|---|
| | | | | MX | MY | X | Y | XY |
| | | | 3 | 2 | 3 | 3 | 3 | 1 |
| .10 | .25 | .125 | .841 | 2.015 | .702 | .065 | .066 | .049 |
| | .50 | .213 | .824 | 2.057 | .683 | .294 | .303 | .229 |
| | .75 | .376 | .792 | 2.155 | .651 | .715 | .739 | .560 |
| | 1.00 | .894 | .567 | 2.766 | .422 | 1.917 | 2.029 | 1.827 |
| | 1.25 | 1.611 | .223 | 3.779 | .094 | 3.366 | 3.687 | 4.188 |
| | 1.50 | 2.182 | -.064 | 4.880 | -.121 | 4.000 | 4.715 | 7.693 |
| .25 | .25 | .321 | .788 | 2.146 | .651 | .188 | .193 | .139 |
| | .50 | .490 | .724 | 2.311 | .585 | .609 | .631 | .509 |
| | .75 | .770 | .606 | 2.627 | .462 | 1.293 | 1.346 | 1.145 |
| | 1.00 | 1.232 | .369 | 3.324 | .230 | 2.258 | 2.417 | 2.498 |
| | 1.25 | 1.802 | .074 | 4.282 | -.036 | 3.240 | 3.619 | 4.811 |
| | 1.50 | 2.333 | -.164 | 5.233 | -.196 | 3.792 | 4.549 | 8.368 |
| .50 | .25 | .641 | .623 | 2.569 | .477 | .363 | .373 | .293 |
| | .50 | .872 | .503 | 2.903 | .357 | .894 | .945 | .874 |
| | .75 | 1.198 | .330 | 3.430 | .187 | 1.600 | 1.723 | 1.774 |
| | 1.00 | 1.607 | .118 | 4.130 | -.009 | 2.323 | 2.599 | 3.289 |
| | 1.25 | 2.072 | -.105 | 4.939 | -.185 | 2.971 | 3.506 | 5.707 |
| | 1.50 | 2.547 | -.294 | 5.760 | -.278 | 3.437 | 4.368 | 9.429 |
| 1.00 | .25 | 1.216 | .219 | 3.706 | .074 | .485 | .521 | .547 |
| | .50 | 1.482 | .085 | 4.163 | -.046 | 1.000 | 1.122 | 1.424 |
| | .75 | 1.800 | -.064 | 4.721 | -.168 | 1.535 | 1.798 | 2.658 |
| | 1.00 | 2.156 | -.206 | 5.362 | -.263 | 2.007 | 2.511 | 4.528 |
| | 1.25 | 2.553 | -.364 | 6.035 | -.325 | 2.419 | 3.282 | 7.320 |
| | 1.50 | 2.991 | -.579 | 6.656 | -.349 | 2.796 | 4.153 | 11.431 |

## Table 51

### Combined uniform bearing and bending stress

Side ratio $a/b = 1{\cdot}0$

| W | P | W | σ̄ | | σ̄ | M̄ | M̄ | M̄ |
|---|---|---|---|---|---|---|---|---|
| | | | | MX | MY | X | Y | XY |
| | | | 3 | 2 | 3 | 3 | 3 | 1 |
| .10 | .25 | .124 | .628 | 2.017 | .654 | .032 | .042 | .029 |
| | .50 | .218 | .612 | 2.060 | .642 | .153 | .205 | .150 |
| | .75 | .388 | .572 | 2.147 | .617 | .373 | .499 | .368 |
| | 1.00 | .971 | .344 | 2.775 | .468 | 1.027 | 1.491 | 1.303 |
| | 1.25 | 1.790 | -.009 | 3.811 | .275 | 1.837 | 2.868 | 3.259 |
| | 1.50 | 2.470 | -.347 | 4.907 | .199 | 2.250 | 3.972 | 6.580 |
| .25 | .25 | .321 | .584 | 2.128 | .622 | .094 | .125 | .086 |
| | .50 | .501 | .524 | 2.288 | .580 | .316 | .435 | .337 |
| | .75 | .798 | .414 | 2.587 | .503 | .676 | .945 | .768 |
| | 1.00 | 1.320 | .167 | 3.293 | .365 | 1.199 | 1.823 | 1.816 |
| | 1.25 | 1.983 | -.158 | 4.267 | .221 | 1.764 | 2.921 | 3.807 |
| | 1.50 | 2.616 | -.440 | 5.215 | .174 | 2.159 | 3.949 | 7.168 |
| .50 | .25 | .645 | .432 | 2.512 | .517 | .184 | .248 | .185 |
| | .50 | .891 | .324 | 2.822 | .446 | .463 | .670 | .589 |
| | .75 | 1.242 | .158 | 3.303 | .347 | .842 | 1.274 | 1.227 |
| | 1.00 | 1.708 | -.073 | 4.003 | .248 | 1.246 | 2.044 | 2.449 |
| | 1.25 | 2.259 | -.332 | 4.839 | .172 | 1.649 | 2.946 | 4.583 |
| | 1.50 | 2.843 | -.556 | 5.657 | .150 | 2.031 | 3.935 | 8.087 |
| 1.00 | .25 | 1.225 | .078 | 3.513 | .292 | .252 | .383 | .360 |
| | .50 | 1.512 | -.062 | 3.938 | .229 | .531 | .873 | .996 |
| | .75 | 1.862 | -.217 | 4.480 | .175 | .833 | 1.473 | 1.923 |
| | 1.00 | 2.276 | -.393 | 5.094 | .142 | 1.132 | 2.163 | 3.472 |
| | 1.25 | 2.748 | -.573 | 5.752 | .128 | 1.445 | 2.959 | 5.962 |
| | 1.50 | 3.268 | -.725 | 6.420 | .131 | 1.811 | 3.895 | 9.838 |

## Table 52

### Combined uniform bearing and bending stress

Side ratio $a/b = 1{\cdot}5$

| W | P | W | $\bar{\sigma}_{MX}$ | | $\bar{\sigma}_{MY}$ | $\bar{M}_{X}$ | $\bar{M}_{Y}$ | $\bar{M}_{XY}$ |
|---|---|---|---|---|---|---|---|---|
| | | | 3 | 2 | 3 | 3 | 3 | 1 |
| .10 | .25 | .124 | .369 | 2.013 | .591 | .016 | .028 | .021 |
| | .50 | .222 | .356 | 2.047 | .588 | .083 | .146 | .107 |
| | .75 | .405 | .330 | 2.118 | .575 | .207 | .364 | .265 |
| | 1.00 | 1.148 | .059 | 2.783 | .490 | .638 | 1.234 | 1.105 |
| | 1.25 | 2.229 | -.374 | 3.914 | .384 | 1.225 | 2.515 | 3.077 |
| | 1.50 | 3.158 | -.775 | 5.155 | .358 | 1.657 | 3.672 | 6.773 |
| .25 | .25 | .320 | .338 | 2.095 | .580 | .048 | .083 | .059 |
| | .50 | .519 | .283 | 2.234 | .559 | .177 | .319 | .249 |
| | .75 | .859 | .174 | 2.504 | .518 | .394 | .723 | .584 |
| | 1.00 | 1.542 | -.124 | 3.262 | .445 | .760 | 1.520 | 1.556 |
| | 1.25 | 2.440 | -.523 | 4.349 | .370 | 1.216 | 2.595 | 3.595 |
| | 1.50 | 3.289 | -.848 | 5.411 | .347 | 1.648 | 3.697 | 7.289 |
| .50 | .25 | .650 | .212 | 2.398 | .534 | .100 | .177 | .129 |
| | .50 | .935 | .107 | 2.689 | .496 | .272 | .513 | .451 |
| | .75 | 1.365 | -.044 | 3.164 | .440 | .520 | 1.019 | .984 |
| | 1.00 | 1.987 | -.346 | 3.929 | .395 | .836 | 1.755 | 2.144 |
| | 1.25 | 2.746 | -.711 | 4.886 | .362 | 1.212 | 2.691 | 4.357 |
| | 1.50 | 3.543 | -.966 | 5.844 | .339 | 1.638 | 3.762 | 8.223 |
| 1.00 | .25 | 1.253 | -.119 | 3.257 | .427 | .153 | .299 | .267 |
| | .50 | 1.606 | -.281 | 3.695 | .397 | .346 | .720 | .802 |
| | .75 | 2.063 | -.484 | 4.261 | .372 | .580 | 1.267 | 1.625 |
| | 1.00 | 2.624 | -.703 | 4.959 | .359 | .863 | 1.973 | 3.104 |
| | 1.25 | 3.280 | -.928 | 5.744 | .340 | 1.197 | 2.823 | 5.726 |
| | 1.50 | 4.015 | -1.155 | 6.545 | .276 | 1.579 | 3.781 | 10.208 |

## Table 53

### Combined non-uniform bearing and bending stress

Side ratio $a/b = 0{\cdot}667$

| W | P | W | $\bar{\sigma}_{MX}$ | | $\bar{\sigma}_{MY}$ | $\bar{M}_{X}$ | $\bar{M}_{Y}$ | $\bar{M}_{XY}$ |
|---|---|---|---|---|---|---|---|---|
| | | | 3 | 2 | 3 | 3 | 3 | 1 |
| .10 | .25 | .131 | .755 | 2.519 | .528 | .082 | .083 | .047 |
| | .50 | .210 | .803 | 2.558 | .576 | .292 | .298 | .224 |
| | .75 | .373 | .751 | 2.636 | .524 | .722 | .738 | .545 |
| | 1.00 | .895 | .550 | 3.202 | .317 | 1.964 | 2.046 | 1.780 |
| | 1.25 | 1.614 | .261 | 4.138 | .040 | 3.465 | 3.743 | 4.044 |
| | 1.50 | 2.183 | .004 | 5.129 | -.150 | 4.143 | 4.856 | 7.315 |
| .25 | .25 | .321 | .752 | 2.632 | .525 | .190 | .194 | .135 |
| | .50 | .489 | .695 | 2.784 | .467 | .620 | .638 | .499 |
| | .75 | .766 | .593 | 3.070 | .360 | 1.317 | 1.360 | 1.115 |
| | 1.00 | 1.237 | .383 | 3.730 | .153 | 2.313 | 2.470 | 2.454 |
| | 1.25 | 1.814 | .120 | 4.627 | -.083 | 3.334 | 3.724 | 4.675 |
| | 1.50 | 2.332 | -.086 | 5.468 | -.218 | 3.918 | 4.681 | 7.918 |
| .50 | .25 | .646 | .516 | 3.020 | .281 | .378 | .324 | .286 |
| | .50 | .869 | .546 | 3.323 | .314 | .908 | .984 | .851 |
| | .75 | 1.197 | .351 | 3.787 | .118 | 1.641 | 1.746 | 1.720 |
| | 1.00 | 1.606 | .159 | 4.432 | -.059 | 2.385 | 2.637 | 3.184 |
| | 1.25 | 2.072 | -.026 | 5.191 | -.183 | 3.060 | 3.626 | 5.466 |
| | 1.50 | 2.562 | -.261 | 5.942 | -.263 | 3.583 | 4.659 | 8.853 |
| 1.00 | .25 | 1.205 | .389 | 4.060 | .157 | .474 | .502 | .533 |
| | .50 | 1.490 | .077 | 4.480 | -.148 | 1.049 | 1.172 | 1.374 |
| | .75 | 1.805 | .020 | 4.992 | -.182 | 1.600 | 1.885 | 2.542 |
| | 1.00 | 2.164 | -.142 | 5.556 | -.247 | 2.116 | 2.654 | 4.305 |
| | 1.25 | 2.563 | -.364 | 6.150 | -.336 | 2.567 | 3.442 | 6.902 |
| | 1.50 | 2.986 | -.471 | 6.752 | -.358 | 2.906 | 4.185 | 10.654 |

## Table 54

### Combined non-uniform bearing and bending stress

Side ratio $a/b = 1{\cdot}0$

| W | P | W | σ̄ MX | | σ̄ MY | M̄ X | M̄ Y | M̄ XY |
|---|---|---|---|---|---|---|---|---|
| | | | 3 | 2 | 3 | 3 | 3 | 1 |
| .10 | .25 | .124 | .625 | 2.525 | .601 | .034 | .045 | .027 |
| | .50 | .219 | .614 | 2.566 | .592 | .167 | .219 | .136 |
| | .75 | .392 | .586 | 2.651 | .569 | .408 | .536 | .337 |
| | 1.00 | .973 | .380 | 3.230 | .430 | 1.126 | 1.576 | 1.175 |
| | 1.25 | 1.784 | .060 | 4.178 | .247 | 2.004 | 2.989 | 2.840 |
| | 1.50 | 2.440 | -.225 | 5.175 | .166 | 2.420 | 4.042 | 5.513 |
| .25 | .25 | .321 | .588 | 2.628 | .572 | .101 | .131 | .079 |
| | .50 | .502 | .535 | 2.775 | .533 | .343 | .458 | .306 |
| | .75 | .801 | .435 | 3.048 | .464 | .737 | .999 | .696 |
| | 1.00 | 1.320 | .218 | 3.703 | .333 | 1.310 | 1.906 | 1.610 |
| | 1.25 | 1.968 | -.063 | 4.606 | .194 | 1.923 | 3.024 | 3.277 |
| | 1.50 | 2.570 | -.295 | 5.460 | .144 | 2.342 | 4.057 | 5.986 |
| .50 | .25 | .646 | .459 | 2.985 | .478 | .201 | .265 | .170 |
| | .50 | .892 | .355 | 3.273 | .411 | .508 | .708 | .530 |
| | .75 | 1.244 | .211 | 3.730 | .317 | .924 | 1.340 | 1.096 |
| | 1.00 | 1.704 | .013 | 4.377 | .221 | 1.370 | 2.144 | 2.139 |
| | 1.25 | 2.241 | -.207 | 5.142 | .144 | 1.810 | 3.072 | 3.906 |
| | 1.50 | 2.802 | -.393 | 5.897 | .117 | 2.202 | 4.050 | 6.738 |
| 1.00 | .25 | 1.226 | .148 | 3.937 | .270 | .280 | .406 | .327 |
| | .50 | 1.514 | .031 | 4.338 | .211 | .591 | .938 | .884 |
| | .75 | 1.864 | -.107 | 4.825 | .153 | .930 | 1.599 | 1.667 |
| | 1.00 | 2.270 | -.256 | 5.389 | .115 | 1.264 | 2.298 | 2.989 |
| | 1.25 | 2.727 | -.398 | 6.006 | .097 | 1.603 | 3.077 | 5.038 |
| | 1.50 | 3.230 | -.510 | 6.635 | .091 | 1.972 | 4.028 | 8.178 |

## Table 55

### Combined non-uniform bearing and bending stress

Side ratio $a/b = 1{\cdot}5$

| W | P | W | σ̄ MX | | σ̄ MY | M̄ X | M̄ Y | M̄ XY |
|---|---|---|---|---|---|---|---|---|
| | | | 3 | 2 | 3 | 3 | 3 | 1 |
| .10 | .25 | .124 | .434 | 2.594 | .588 | .019 | .030 | .015 |
| | .50 | .219 | .422 | 2.627 | .582 | .096 | .153 | .078 |
| | .75 | .395 | .398 | 2.702 | .572 | .236 | .378 | .193 |
| | 1.00 | 1.087 | .604 | 3.299 | .476 | .717 | 1.249 | .761 |
| | 1.25 | 2.091 | .584 | 4.320 | .350 | 1.354 | 2.524 | 2.033 |
| | 1.50 | 2.946 | -.481 | 5.477 | .310 | 1.774 | 3.661 | 4.322 |
| .25 | .25 | .320 | .406 | 2.679 | .575 | .056 | .089 | .044 |
| | .50 | .511 | .359 | 2.806 | .551 | .203 | .333 | .181 |
| | .75 | .837 | .266 | 3.053 | .508 | .450 | .749 | .421 |
| | 1.00 | 1.471 | .023 | 3.746 | .419 | .853 | 1.543 | 1.071 |
| | 1.25 | 2.303 | -.304 | 4.749 | .323 | 1.331 | 2.596 | 2.378 |
| | 1.50 | 3.098 | -.578 | 5.753 | .291 | 1.740 | 3.655 | 4.677 |
| .50 | .25 | .647 | .295 | 2.970 | .519 | .117 | .188 | .096 |
| | .50 | .921 | .200 | 3.249 | .477 | .312 | .534 | .324 |
| | .75 | 1.330 | .035 | 3.712 | .414 | .591 | 1.050 | .697 |
| | 1.00 | 1.908 | -.191 | 4.407 | .352 | .924 | 1.779 | 1.459 |
| | 1.25 | 2.606 | -.440 | 5.272 | .304 | 1.305 | 2.706 | 2.869 |
| | 1.50 | 3.337 | -.648 | 6.192 | .287 | 1.728 | 3.796 | 5.288 |
| 1.00 | .25 | 1.246 | .006 | 3.806 | .393 | .176 | .314 | .195 |
| | .50 | 1.582 | -.123 | 4.222 | .354 | .391 | .745 | .566 |
| | .75 | 2.012 | -.288 | 4.767 | .315 | .644 | 1.294 | 1.126 |
| | 1.00 | 2.531 | -.461 | 5.410 | .295 | .925 | 1.972 | 2.104 |
| | 1.25 | 3.142 | -.630 | 6.132 | .276 | 1.301 | 2.977 | 3.750 |
| | 1.50 | 3.853 | -.782 | 6.907 | .221 | 1.887 | 4.636 | 6.415 |

# References

Aalami, B. and Chapman, J. C. (1969) 'Large deflection behaviour of ship plates under normal pressure and in-plane loading', *Transactions of the Royal Institution of Naval Architects*, **114**, 151–81.

Aalami, B. and Williams, D. G. (1975) *Thin plate design for transverse loading*, CONSTRADO Monograph Series, Crosby Lockwood Staples, London.

Bergman, S. G. A. (1948) 'Behaviour of buckled rectangular plates under action of shear forces', Doctoral Thesis, Stockholm.

Bernoulli, J. (1789) *Nova Acta*, **5**, St. Petersburg.

Bulson, P. S. (1970) *The stability of flat plates*, Chatto and Windus, London.

Crisfield, M. A. (1973) 'Large-deflection elasto-plastic buckling analysis of plates using finite elements', Transport and Road Research Laboratory, Department of the Environment, England, T.R.R.L. Report LR593.

Coan, J. M. (1959) 'Large-deflection theory for plates with small initial curvature loaded in edge compression', *Journal of Applied Mechanics*, **18**(2), 143–51.

Duen Ho (1964) 'Buckling of perforated web plates', PhD Thesis, Imperial College, University of London.

Dwight, J. B. and Moxham, K. E. (1969) 'Welded steel plates in compression', *The Structural Engineer*, **47**(2), 49–66.

Falconer, B. H. and Chapman, J. C. (1953) 'Compressive buckling of stiffened plates', *The Engineer*, **195**, 789–91, 822–25.

Föppl, A. (1907) *Vorlesungen über Technische Mechanik*, **5**, 132.

Hu, P. C., Lindquist, E. E. and Batdorf, S. B. (1946) 'Effect of small deviation from flatness on effective width and buckling of plates in compression', *NACA Tech. Note 1124*.

Kaiser, R. (1936) 'Research and experimental study on stresses and deformation of square plates', *Zeitung für Angewandte Mathematik und Mechanik*, **16**, 73–98.

Keays, R. H. and Williams, D. G. (1973) 'Curves for the elastic design of initially deformed plates subject to combined uniaxial compression and shear', Institute of Engineering Conference on Stress and Strain in Engineering, Brisbane, Australia, 89–97.

Kirchhoff, G. R. (1877) *Vorlesungen über mathematische physik*, *Mechanik*, 450.

Levy, S. (1942) 'Bending of rectangular plates with large deflections', *NACA Report 737.*

Levy, S., Fienup, K. L. and Woolley, R. M. (1945) 'Analysis of square shear web above buckling load', *NACA Tech. Note 962.*

Lin, T. H., Lin, S. R. and Mazelsky, B. (1972) 'Elastoplastic bending of rectangular plates with large deflection', *Journal of Applied Mechanics*, 978–82.

Mallet, R. H. and Marcal, P. V. (1968) 'Finite element analysis of non-linear structures', *Journal of the Structural Division, ASCE*, **94**(ST9), Proc. Paper 6115, 2081–105.

Marguerre, K. (1938) 'Zur Theorie der gekruemmter platte grosser formaenderung', *Proceedings of the Fifth International Congress for Applied Mechanics*, Cambridge, England.

Massonnet, C. H. (1968) 'General theory of elasto-plastic membrane plates', In Heyman, J. and Leckie, F. A. (eds.), *Engineering plasticity*, Cambridge University Press, 443–71.

Merrison Committee (1973) 'Inquiry into basis of design and method of erection of steel girder bridges', Report of the Committee, HMSO, London.

Murray, D. W. and Wilson, E. L. (1969) 'Finite element large deflection analysis of plates', *Journal of the Engineering Mechanics Division, ASCE*, **95**(EM1), Proc. Paper 6398, 143–65.

Navier, L. (1820) Paper presented to French Academy, Abstract published in *Bull. Soc. Phil-math.*, Paris.

Rockey, K. C. (1967) 'Shear buckling of thin walled sections', *Thin Walled Structures*, Chatto and Windus, London, 248–70.

Rushton, K. R. (1969) 'Dynamic relaxation solutions of elastic plate problems', *Journal of Strain Analysis*, **3**(1), 23–32.

Saint Venant (1883) Discussion in *Theorie de l'élasticité des corps solides*, by Clebsch, 704.

Timoshenko, S. and Woinowsky-Krieger, S. (1959) *Theory of plates and shells*, McGraw-Hill Book Company, London.

Von Karman, T. (1910) 'Festigkeitsprobleme im Maschinenbau', *Encyklopaedie der Mathematischen Wissenschaften*, **4**, 349.

Way, S. (1938) 'Uniformly loaded clamped rectangular plates with large deflections', *Proceedings of Fifth International Congress on Applied Mechanics*, Cambridge, Mass., U.S.A., 123–8.

Williams, D. G. (1971) 'Some examples of the elastic behaviour of initially deformed bridge panels', *Civil Engineering and Public Works Review*, **66**(783), 1107–12.

Williams, D. G. (1973) 'The elastic design of initially deformed plates', Australian Institute of Steel Construction Conference on Steel Developments, Newcastle, Australia, 251–60.

Williams, D. G. and Walker, A. C. (1975) 'Explicit solutions for the design of initially deformed plates subject to compression', *Proceedings of the Institution of Civil Engineers*, London, **59** (Part 2), 763–87.

Williams, D. G. and Walker, A. C. (1977) 'Explicit solutions for plate buckling analysis', *Journal of the Engineering Mechanics Division, ASCE*, **103**(EM4), Proc. Paper 13122, 549–68.

Yamaki, N. (1959) 'Post-buckling behaviour of rectangular plates', *Journal of Applied Mechanics*, **26**, 407.

# Index